Computational Advances in Organic Chemistry: Molecular Structure and Reactivity

NATO ASI Series

Advanced Science Institutes Series

A Series presenting the results of activities sponsored by the NATO Science Committee, which aims at the dissemination of advanced scientific and technological knowledge, with a view to strengthening links between scientific communities.

The Series is published by an international board of publishers in conjunction with the NATO Scientific Affairs Division

A Life Sciences **B Physics**	Plenum Publishing Corporation London and New York
C Mathematical and Physical Sciences **D Behavioural and Social Sciences** **E Applied Sciences**	Kluwer Academic Publishers Dordrecht, Boston and London
F Computer and Systems Sciences **G Ecological Sciences** **H Cell Biology**	Springer-Verlag Berlin, Heidelberg, New York, London, Paris and Tokyo

Series C: Mathematical and Physical Sciences - Vol. 330

Computational Advances in Organic Chemistry: Molecular Structure and Reactivity

edited by

Cemil Ögretir

Department of Chemistry, Anadolu University,
Eskisehir, Turkey

and

Imre G. Csizmadia

Department of Chemistry, University of Toronto,
Toronto, Canada

with the assistance of

Emilia A. Lang

Department of Chemistry, University of Toronto,
Toronto, Canada

Kluwer Academic Publishers

Dordrecht / Boston / London

Published in cooperation with NATO Scientific Affairs Division

Proceedings of the NATO Advanced Study Institute on
Computational Advances in Organic Chemistry: Molecular Structure and Reactivity
Altinoluk, Edremit, Turkey
July 31 – August 12, 1989

Library of Congress Cataloging-in-Publication Data

NATO Advanced Study Institute on "Computational Advances in Organic Chemistry: Molecular Structure and Reactivity" (1989 : Altınoluk, Turkey)
Computational advances in organic chemistry : molecular structure and reactivity / edited by Cemil Ögretir and Imre G. Csizmadia with the assistance of Emilia A. Lang.
p. cm. -- (NATO ASI: Series C; Mathematical and physical sciences ; vol. 330)
"Proceedings of the NATO Advanced Study Institute on "Computational Advances in Organic Chemistry: Molecualr Structure and Reactivity", held in Altinoluk, Edremit, Turkey, July 31 - August 12, 1989".
Includes index.
ISBN 0-7923-1064-0 (U.S.)
1. Chemistry, Organic--Data processing--Congresses. 2. Chemistry, Organic--Mathematics--Congresses. I. Öğretir, Cemil. II. Csizmadia, Imre G. III. Lang, Emilia A. IV. Title. V. Series: NATO ASI series. Series C, Mathematical and physical sciences ; no.330.
QD255.5.E4N38 1989
547'.00285--dc20 90-22593

ISBN 0–7923–1064–0

Published by Kluwer Academic Publishers,
P.O. Box 17, 3300 AA Dordrecht, The Netherlands.

Kluwer Academic Publishers incorporates the publishing programmes of
D. Reidel, Martinus Nijhoff, Dr W. Junk and MTP Press.

Sold and distributed in the U.S.A. and Canada
by Kluwer Academic Publishers,
101 Philip Drive, Norwell, MA 02061, U.S.A.

In all other countries, sold and distributed
by Kluwer Academic Publishers Group,
P.O. Box 322, 3300 AH Dordrecht, The Netherlands.

Printed on acid-free paper

Printed in the Netherlands

TABLE OF CONTENTS

NATO
Advanced Study
Institute

PREFACE

The lecturers as well as the participants came from varied scientific backgrounds for the NATO - Advanced Study Institute (ASI) held at Altinoluk, Edremit, Turkey during the period of July 31 - August 12 1989.

The lecturers were University Professors from the USA, Canada, England, Germany, France and Spain and they covered a broad spectrum of specialities from methodology to appications. On the other hand students coming from the various NATO countries arrived with an inhomogeneous background to absorb the broad spectrum of material covered by the lecturers. However, by the end of the two week period of the ASI, that initial difference in scientific background had been reduced substantially.

The lecturers had covered subject matters from the most fundamental to the most applied aspects of theoretical and computational organic chemistry. The lectures were argumented with tutorial sessions and computational laboratory led by a small group of carefully selected tutors.

Overall, this NATO - ASI was a great success and the Editors are hopeful that the present volume will communicate the scientific success and will radiate the intellectual spirit of the meeting.

Cemil Ögretir　　　　　　　　　　Imre Csizmadia

SOME FUNDAMENTALS OF MOLECULAR ORBITAL COMPUTATIONS

I. G. Csizmadia
Department of Chemistry
University of Toronto
Toronto, ON, Canada

CONTENTS:

C. Ögretir and I. G. Csizmadia (eds.),
Computational Advances in Organic Chemistry: Molecular Structure and Reactivity, 1–165.

1.0 BASIC CONCEPTS

1.1 INTRODUCTION

Throughout the years some magic has been attached to everything that is theoretical in chemistry and by now a certain amount of mythology has evolved around Quantum Chemistry as it may be applied to study organic problems. For this reason some demythologization is in order. This is hoped to be achieved in five brief statements.

Statement no. 1. The term "orbital" is a synonym for the term "one–electron function (OEF)".

Statement no. 2. A single–centered OEF is synonymous with "Atomic Orbital (AO)". A multi–centered OEF is synonymous with "Molecular Orbital (MO)". The single and multi–centered nature of these one–electron functions are illustrated in Figure 1.

Figure 1. A schematic representation of Single (AO) and Multi (MO) centered one–electron functions.

An orbital (AO or MO) has as much to do with physical and chemical reality as perhaps the following functions do

$$e^x$$

$$\{x^n\}$$

$$\{\sin k\alpha, \cos k\alpha\}$$

Nevertheless, these functions (AO and MO) enable us to construct molecular wave functions that may be used to compute molecular electron density and molecular properties which are manisfestations of physical and chemical realities. It is axiomatic, thererfore, that these latter computed quantities must have a one–to–one correspondence to the corresponding physical properties of the molecule as determined experimentally.

Statement no. 3. There are three ways to express a mathematical function:

(*i*) Explicitly in analytic form

$$f(x) = e^x \qquad [1]$$

the hydrogen–like AO are usually expressed in this form.

(*ii*) As a table of numbers

x	f(x)
0.0	1.000
0.1	1.105
0.2	1.221
.	.
.	.
.	.
.	.

[2]

The Hartree–Fock type AO are usually expressed in this numerical form.

(*iii*) In the form of an expansion

$$f(x) \equiv e^x = \frac{f(0)}{0!}x^0 + \frac{f'(0)}{1!}x^1 + \frac{f''(0)}{2!}x^2 + \frac{f'''(0)}{3!}x^3 + =\ldots \quad [3]$$

$$= C_0x^0 + C_1x^1 + C_2x^2 + C_3x^3 + \ldots$$

which is analogous to the expression of MO in terms of a set of AO:

$$\phi = C_0\eta_0 + C_1\eta_1 + C_2\eta_2 + C_3\eta_3 + \ldots \quad [4]$$

Statement no. 4. The generation of MO (ϕ) from AO (η) is equivalent to the transformation of an N–dimensional vector space to another N–dimensional vector space where $\{\eta\}$ is the original set of non–orthogonal functions. After orthogonalization of the non–orthogonal AO basis set $\{\eta\}$ the orthogonal set $\{\chi\}$ is rotated to the an other orthogonal set $\{\phi\}$. This overall process is illustrated in Figure 2.

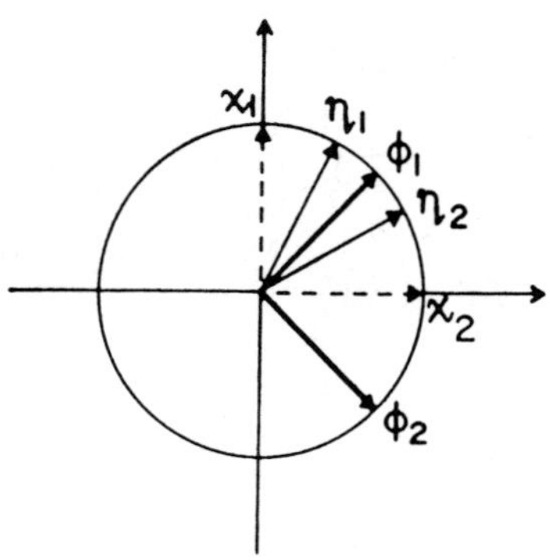

Figure 2. Two dimensional Vector Model of AO⟶MO transformation.

$$\underset{\text{AO}}{\eta} \xrightarrow{\text{orthogonalization}} \underset{\text{AO}}{\chi} \xrightarrow{\text{SCF}} \underset{\text{MO}}{\phi} \qquad [5]$$

Statement no. 5. There are certain differences between the shape of numerical Hartree–Fock atomic orbitals (HF–AO), the analytic Slater type orbitals (STO) and the analytic Gaussian type functions (GTF), however, these differences are irrelevant to the final results as the MO can be expanded in terms of any of these complete sets of functions to any desired degree of accuracy.

1. 2 ATOMIC ORBITAL BASIS SETS

The generation of MO from AO requires the generation and transformation of the Fock matrix into diagonal form. The elements of the Fock matrix are assembled from integrals in the following fashion:

$$f_{ij} = \langle i|\hat{h}|j\rangle + \sum_{k}^{N} \sum_{\ell}^{N} [2\{ij|k\ell\} - \{ik|j\ell\}] \qquad [6]$$

where the first term is a one–electron integral and the latter terms are two–electron integrals having the following forms

One–Electron Integrals

$$\langle i|\hat{h}|j\rangle \equiv \langle \eta_i|\hat{h}|\eta_j\rangle \equiv \int_1 \eta_i(1)\hat{h}_1\eta_j(1)d\tau_1 \qquad [7]$$

Two–Electron Integrals

$$\{ij|kl\} \equiv <\eta_i\eta_k|\frac{1}{r_{12}}|\eta_\ell\eta_j> \equiv$$

$$\int_1 \int_2 \eta_i(1)\eta_k(2)\frac{1}{r_{12}}\eta_\ell(2)\eta_j(1)d\tau_2 d\tau_1 =$$

$$\int_1 \eta_i(1)\left\{\int_2 \eta_k(2)\frac{1}{r_{12}}\eta'_\ell(2)d\tau_2\right\}\eta_z(1)\, d\tau_1 \qquad [8]$$

As the running indices range from 1 to N the number of one– and two–electron integrals are calculable by the following formulae

$$\text{number of 1–electron integrals: } p = \frac{N(N+1)}{2}$$
$$\text{number of 2–electron integrals: } q = \frac{p(p+1)}{2} \qquad [10]$$

The table below illustrates how rapidly the number of one– and two–electron integrals grows with the basis set size N.

Table 1. The increase of the number of one–electron (p) and two electron (q) integrals with increasing basis set size (N)

N	p	q
1	1	1
10	55	1,540
50	1,275	814,725
100	5,050	12,751,250
150	11,325	64,133,475

For the atomic orbitals $\{\eta\}$ two types of analytic functions are used in molecular computations.

1) Slater–type orbitals (STO) or Exponential type functions (ETF).

2) Gaussian–type orbitals (GTO) or Gaussian–type functions (GTF).

Some characteristics of these two sets of functions are illustrated in Figure 3. As far as STO are concerned the integral evaluation is very slow but a relatively small N gives fairly accurate results.

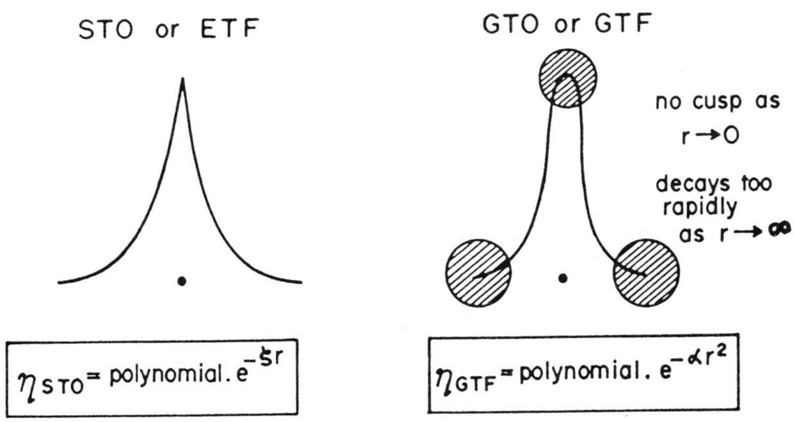

Figure 3. Some characteristics of STO and GTF.

The GTF are more popular as it is possible to compute the integrals over Gaussians very quickly, but only a relative large N gives accurate results. These AO basis sets need to be optimized for molecular calculation. This may be achieved by minimizing the electronic energy with respect to all orbital exponents.

Figure 4 illustrates the situation for a two–orbital case:

$$\eta_1 \;=\; \text{polynomial}_1.e^{-\alpha_1 r^2} \qquad [11]$$

$$\eta_2 \;=\; \text{polynomial}_2.e^{-\alpha_2 r^2} \qquad [12]$$

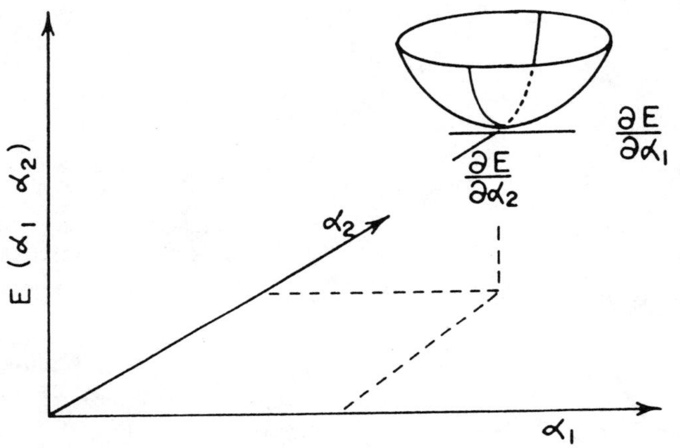

Figure 4. Electronic Energy Surface as a function of orbital exponents $\{\alpha\}$.

Very often STO are expanded in terms of GTF or, in other words, the GTF or GTO basis set is contracted to a set of STO. The contraction of a set of three Gaussian–type functions to a single Slater function is very popular (STO–3G):

$$\eta_i^{STO} \;=\; d_{i1}\eta_1^{GTO} \;+\; d_{i2}\eta_2^{GTO} \;+\; d_{i3}\eta_3^{GTO} \qquad [13]$$

This means that the total number of integrals to be stored can be reduced by contraction as illustrated below:

$$<\eta_i^{STO}|\hat{h}|\eta_j^{STO}> =$$

$$= <d_{i1}\eta_1^{GTO} + d_{i2}\eta_2^{GTO} + d_{i3}\eta_3^{GTO}|\hat{h}|d_{j1}\eta_1^{GTO} + d_{j2}\eta_2^{GTO} + d_{j3}\eta_3^{GTO}> =$$

$$= d_{i1}d_{j1}<\eta_1^{GTO}|\hat{h}|\eta_1^{GTO}>$$

$$+d_{i2}d_{j1}<\eta_2^{GTO}|\hat{h}|\eta_1^{GTO}>$$

$$+d_{i3}d_{j1}<\eta_3^{GTO}|\hat{h}|\eta_1^{GTO}>$$

$$\vdots$$

$$+d_{i3}d_{j3}<\eta_3^{GTO}|\hat{h}|\eta_3^{GTO}>=$$

$$= \sum_{k}^{3} \sum_{\ell}^{3} d_{ik}d_{j\ell}\left\langle\eta_k^{GTO}|\hat{h}|\eta_\ell^{GTO}\right\rangle \qquad [14]$$

In the above case 9 integrals have been evaluated, but only their sum total, that is, a single integral, is stored.

1.3 LOCALIZED MOLECULAR ORBITALS

CMO (ϕ). Molecular orbitals that produce a Fock matrix in canonical (diagonal) form are known as canonical molecular orbitals.

$$\mathbf{F}^{\phi} = \begin{vmatrix} <\phi_1|\hat{F}|\phi_1> & 0 & 0 & \cdots \\ 0 & <\phi_2|\hat{F}|\phi_2> & 0 & \cdots \\ 0 & 0 & <\phi_3|\hat{F}|\phi_3> & \cdots \\ \vdots & \vdots & \vdots & \end{vmatrix} \qquad [15]$$

These CMO are delocalized all over the molecule.

LMO (ψ). Localized molecular orbitals produce a non–diagonal real symmetric Fock matrix.

$$\mathbf{F}^{\psi} = \begin{vmatrix} <\psi_1|\hat{F}|\psi_1> & <\psi_1|\hat{F}|\psi_2> & <\psi_1|\hat{F}|\psi_3> & \cdots \\ \cdot & <\psi_2|\hat{F}|\psi_2> & <\psi_1|\hat{F}|\psi_3> & \cdots \\ \cdot & & <\psi_3|\hat{F}|\psi_3> & \cdots \\ \vdots & & & \ddots \end{vmatrix} \qquad [16]$$

The unitary matrix **U** that transforms ϕ to ψ:

$$\psi = \phi \, \mathbf{U} \qquad [17]$$

will change $\mathbf{F}^{\phi}$ to $\mathbf{F}^{\psi}$ via a similarity transformation:

$$\mathbf{F}^{\psi} = \mathbf{U}^{\dagger} \, \mathbf{F}^{\phi} \, \mathbf{U} \qquad [18]$$

The generation of ϕ has the unique condition that **F** must be a diagonal matrix, but the generation of ψ has no such condition. Consequently **U** could be any unitary matrix. Therefore there are literally an infinite number of possible sets of LMO. The concept of localization means the separation of electron pairs and therefore the definition of separation means the definition of **U**.

Separation has been defined by **Boys** in **stereochemical** terms and by **Edmiston–Ruedenberg** in **energetic** terms. A schematic illustration of the Boys and Edmiston–Ruedenberg localization is illustrated in Figure 5.

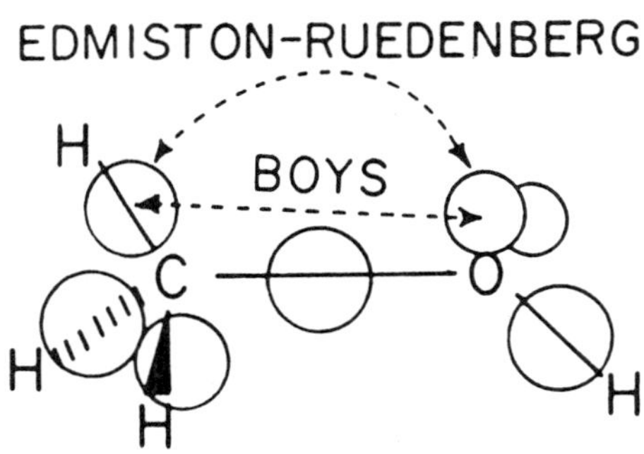

Figure 5. A schematic illustration of the Boys and Edmiston–Ruedenberg localization.

In either case the degree of degeneracy of the diagonal elements of F^{ψ} will follow the equivalency of the chemical bonds. This is illustrated for CH_4 in Figure 6.

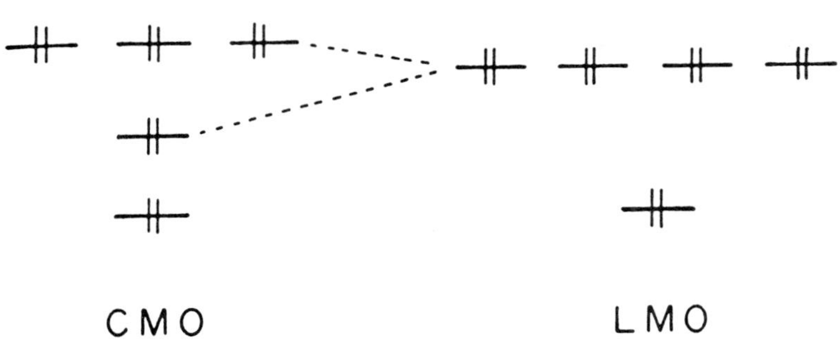

Figure 6. Schematic illustration of CMO and LMO energies for CH_4.

2. THE FORMULATION OF THE SCF THEORY

2.1 THE MOLECULAR HARTREE–FOCK PROBLEM IN TERMS OF MO

Generalization of the basic concepts of MO theory leads to the following three statements:

(*i*) The wavefuction which describes the electronic ground state of a 2M electron system has the form of a (2M) x (2M) Slater determinant

$$\Phi_0 \equiv \Phi_0 \ (1,2, \ldots\ldots 2M) =$$

$$\frac{1}{\sqrt{(2M)\,!}} \begin{Vmatrix} \phi_1(1)\,\alpha(1) & \phi_1(1)\,\beta(1) & \ldots\phi_m(1)\,\alpha(1) & \phi_m(1)\,\beta(1) \\ \phi_1(2)\,\alpha(2) & \phi_1(2)\,\beta(2) & \phi_m(1)\,\alpha(2) & \phi_m(1)\,\beta(1) \\ \vdots & \vdots & \vdots & \vdots \\ \phi_1(2M)\,\alpha(2M) & \phi_1(2M)\beta(2M)\ldots & \phi_m(2M)\,\alpha(2M) & \phi_m(2M)\beta(2M) \end{Vmatrix} \quad [19]$$

This is necessary partly because electrons are indistinguishable and therefore each electron should be assigned, on equal footing, to each of the orbitals and partly because electrons are Fermions and their wave functions must be antisymmetric with respect to the exchange of the coordinates of any pair of electrons.

(*ii*) The electronic Hamiltonian for a 2M electron system may be written as

$$\hat{H} \equiv \hat{H}\ (1,2, \ldots, 2M) = \sum_{\mu=1}^{2M} \hat{h}_\mu + \sum_{\mu>\nu}^{\frac{1}{2}M(M-1)} \hat{q}_{\mu\nu} \quad [20]$$

(*iii*) The energy expression after substitution of Φ_0 and $\hat{H}$ into the expectation value equation [21]

$$E = <\Phi_0|\hat{H}|\Phi_0>, \qquad [21]$$

and integrating out the spin variables, has the following form:

$$E = 2 \sum_{p}^{M} <\phi_p(1)|\hat{h}_1|\phi_p(1)> +$$

$$\sum_{p=1}^{M} \sum_{q=1}^{M} [2<\phi_p(1)\phi_q(2)|\hat{g}_{12}|\phi_p(1)\phi_q(2)> -$$

$$<(\phi_p(1)\phi_p(2)|\hat{g}_{12}|\phi_q(1)\phi_q(2)>] \qquad [22]$$

where the two electron integrals (the last two terms in the above equation) are the Coulomb and Exchange integrals respectively. Note that in the Coulomb integrals electron No. 1 is associated with orbital ϕ_p and electron No. 2 is associated with orbital ϕ_q. This distinction between Coulomb and Exchange terms becomes clearer in the "electron density" formalism where orbitals associated with electron No. 1 are collected to be in front of the operator while those associated with particle No. 2 are written behind the operator.

$$E = 2 \sum_{p}^{M} <\phi_p(1)|\hat{h}_1|\phi_p(1)> + \sum^{M} \sum^{M} [2\{\phi_p(1)\phi_p(1)|\phi_q(2)\phi_q(2)\} - \{\phi_p(1)\phi_q(1)|\phi_p(2)\phi_q(2)\}] \quad [23]$$

Note, that in order to distinguish the electron density formalism [23] from the traditional notation [22] the brackets were changed from < > to { } and the symbol for the electron–electron repulsion operator $\hat{g}_{\mu\nu}$ has been omitted.

The above expression: [23] can conveniently be presented in abbreviated form,

$$E = 2 \sum_{p}^{M} h^{\phi}_{pp} + \sum_{p}^{M} \sum_{p}^{M} \left[2\, J^{\phi}_{pq} - K^{\phi}_{pq}\right] \quad [24]$$

where J_{pq} and K_{pq} symbolize the Coulomb and Exchange integrals respectively. The superscript ϕ indicates that these matrix representatives are over the MO basis.

At this stage it should be pointed out that the diagonal elements of the Coulomb and Exchange integrals are identical.

$$J_{pp} = K_{pp} \quad [25]$$

It may also be noted that both J_{pq} and K_{pq} are positive quantities however K_{pq} is entered into the expression of energy with a negative sign. Therefore, this exchange term is frequently considered to be attractive. The exchange term K_{pq} has no classical analogue while the Coulombic term J_{pq} is analogous to the electrostatic Coulomb repulsion.

The following inequalities indicate the relative magnitudes of these quantitities

$$0 \leqq K_{pq} \leqq J_{pq} \leqq \frac{1}{2} (J_{pp} + J_{qq}) \qquad [26]$$

The J and K integrals are conveniently expressed as pseudo one–electron integrals by defining pseudo one–electron hermitian operators $\hat{J}_p$ and $\hat{K}_p$ such that:

$$J^{\phi}_{pq} = <\phi_p|\hat{J}_q|\phi_p> = <\phi_q|\hat{J}_p|\phi_q> \qquad [27]$$

$$K^{\phi}_{pq} = <\phi_p|\hat{K}_q|\phi_p> = <\phi_q|\hat{K}_p|\phi_q> \qquad [28]$$

The energy expression may be written as:

$$E = 2 \sum_{p}^{M} <\phi_p|\hat{h} + \sum_{q \geq p}^{M}{}' (2\hat{J}_q - \hat{K}_q)|\phi_p> \equiv$$

$$2 \sum_{p}^{M} \int \phi_p^*(\hat{h} + \sum_{p}^{M}{}' (2\hat{J}_q - \hat{K}_q)\} \phi_p d\tau \qquad [29]$$

According to the Variation Theorem the energy may be optimized by variation of ϕ. We therefore wish to minimize E with respect to ϕ_p. To do this we need δE which will be set equal to zero. To obtain δE each ϕ_p is varied by an infinitesimal amount $\delta\phi_p$. Thus the variation in energy, δE, becomes:

$$\delta E = 2 \sum_{p}^{M} \int (\delta\phi_p^*) (\hat{h} + \sum_{q}^{M}{}' (2\hat{J}_q - \hat{K}_q)\} \phi_p d\tau +$$
$$2 \sum_{p}^{M} \int (\delta\phi_p) (\hat{h}^* + \sum_{p}^{M}{}' (2\hat{J}_q^* - \hat{K}_q^*)\} \phi_p^* d\tau = 0 \qquad [30]$$

Introducing the orthonormality condition for the MO basis: $\{\phi\}$

$$S_{pq} \equiv \langle \phi_p | \phi_q \rangle \equiv \int \phi_p^* \phi_q d\tau = \delta_{pq} \qquad [31]$$

permits us to write the orthonormality restriction on $\delta\phi_p$:

$$\delta S_{pq} = \int (\delta\phi_p^*) \phi_q d\tau + \int (\delta\phi_q) \phi_p^* d\tau = 0 \qquad [32]$$

The appropiate combination of these two equations: [30] and [32], with the use of the Legrargian multipliers, ϵ, leads to the following expression:

$$\{\hat{h} + \sum_{q}^{M} (2\hat{J}_q - \hat{K}_q)\} \phi_p = \phi_p \, \epsilon_{pp} \qquad [33]$$

where the operator, involving the one–electron operator $(\hat{h})$ and the two–electron operators $(\hat{J}_q$ and $\hat{K}_q)$, is frequently called the Fock operator and abbreviated: $\hat{F}$.

$$\hat{F}\phi_p = \phi_p \epsilon_{pp} \qquad [34]$$

The elements of the diagonal matrix ϵ are, in a sense, the molecular orbital energies. This expression represents the Hartree–Fock integro–differential equation in matrix notation. To carry out the actual computation the matrix representative of the Fock operator must be generated over the MO basis. The elements of the Fock matrix over the MO basis ($\mathbf{F}^{\phi}$) may be written as

$$F^{\phi}_{st} \equiv \langle\phi_s|\hat{F}|\phi_t\rangle \qquad [35]$$

and, converting the Hartree–Fock operator equation to a matrix equation, we obtain

$$\mathbf{F}^{\phi} = \mathbf{S}^{\phi}\epsilon \qquad [36a]$$

or

$$\mathbf{F}^{\phi} = \epsilon \qquad [36b]$$

since $\qquad S^{\phi}_{st} = \langle\phi_s|\phi_t\rangle = \delta_{st}.$

2.2. THE MOLECULAR HARTREE–FOCK PROBLEM IN TERMS OF AO

Because ϕ is unkown, we substitute

$$\phi = \eta\,\mathbf{C} \qquad [37a]$$

and

$$\phi^{\dagger} = \mathbf{C}^{\dagger}\,\eta^{\dagger} \qquad [37b]$$

to equation [36a] in order to obtain the Hartree–Fock matrix equation over the AO basis

$$\mathbf{C}^{\dagger}\,\mathbf{F}^{\eta}\,\mathbf{C} \;=\; \mathbf{C}^{\dagger}\,\mathbf{S}^{\eta}\,\mathbf{C}\epsilon$$

where

$$\mathbf{F}^{\eta} \;=\; \mathbf{h}^{\eta} \;+\; 2\,\mathbf{J}^{\eta} \;-\; \mathbf{K}^{\eta} \qquad [39]$$

and

$$\mathrm{S}^{\eta}_{\;\mathrm{ij}} \;=\; <\eta_{\mathrm{i}}|\eta_{\mathrm{j}}> \qquad [40]$$

The molecular integrals, implicitely included in [39], (J and K in their pseudo one–electron form) can be written as follows:

$$\mathrm{h}^{\eta}_{\;\mathrm{ij}}(1) = <\mathrm{n}_1(1)|\hat{\mathrm{h}}|\mathrm{n}_{\mathrm{j}}(1)> \qquad [41]$$

$$\mathrm{J}^{\eta}_{\;\mathrm{ij}}(1) = \sum_{\mathrm{k}=1}^{\mathrm{N}}\;\sum_{l=1}^{\mathrm{N}}\;\{\eta_{\mathrm{i}}(1)\;\eta_{\mathrm{j}}(1)|\,\eta_{\mathrm{k}}(2)\eta_{\ell}(2)\}\;\rho_{\mathrm{k}l} \qquad [42]$$

$$\mathrm{K}^{\eta}_{\;\mathrm{ij}}(1) = \sum_{\mathrm{k}=1}\;\sum_{l=1}\;\{\;\eta_2(1)\eta_{\mathrm{k}}(1)|\,\eta_{\mathrm{j}}(2)\eta_{\ell}(2)\}\;\rho_{\mathrm{k}\ell} \qquad [43]$$

where ρ_{kl} is the k, l–th element of the density matrix:

$$\rho = \mathrm{C_M C_M}^{\dagger} \qquad [44a]$$

$$\rho \;=\; \begin{vmatrix} \mathrm{C}_{11} & \mathrm{C}_{12} & \cdots & \mathrm{C}_{1\mathrm{M}} \\ \mathrm{C}_{21} & \mathrm{C}_{22} & \cdots & \mathrm{C}_{2\mathrm{M}} \\ \vdots & \vdots & & \vdots \\ \mathrm{C}_{\mathrm{N}1} & \mathrm{C}_{\mathrm{N}2} & \cdots & \mathrm{C}_{\mathrm{NM}} \end{vmatrix} \begin{vmatrix} \mathrm{C}_{11} & \mathrm{C}_{21} & \cdots & \mathrm{C}_{\mathrm{N}1} \\ \mathrm{C}_{12} & \mathrm{C}_{22} & \cdots & \mathrm{C}_{\mathrm{N}2} \\ \vdots & \vdots & & \vdots \\ \mathrm{C}_{1\mathrm{M}} & \mathrm{C}_{2\mathrm{M}} & \cdots & \mathrm{C}_{\mathrm{NM}} \end{vmatrix} \qquad [44b]$$

N x M M x N

Thus

$$F^{\eta}_{ij} = h^{\eta}_{ij} + \sum_{k}^{N} \sum_{l}^{N} [2\{\eta_i\eta_j | \eta_k\eta_\ell\} - \{\eta_i\eta_k | \eta_j\eta_\ell\}]\rho_{k\ell} \qquad [45]$$

or briefly

$$F^{\eta}_{ij} = h^{\eta}_{ij} + 2\,J^{\eta}_{ij} - K^{\eta}_{ij} \qquad [46]$$

If we wish to write E in terms of the AO integrals we get the following expression:

$$E = 2\sum_{i=1}^{M}\sum_{j=1}^{M} \rho_{ij}h^{\eta}_{ji} + 2\sum_{i=1}^{N}\sum_{j=1}^{N} \rho_{ij}J^{\eta}_{ji} - \sum_{i=1}^{N}\sum_{j=1}^{N} \rho_{ij}K^{\eta}_{ji} \qquad [47]$$

Note that there is a factor of 2 in the first term which includes h_{ji}. One of these may be combined with the last two terms so that the Fock matrix (F_{ji}) may be explicitly included in the above expression:

$$E = \sum_{i=1}^{N}\sum_{j=1}^{N} \rho_{ij}h^{\eta}_{ji} + \sum_{i=1}^{N}\sum_{j=1}^{N} \rho_{ij}(h^{\eta}_{ji} + 2\,J^{\eta}_{ji} - K^{\eta}_{ji})$$

$$= \sum_{i=1}^{N} \sum_{j=1}^{N} \rho_{ij} h^{\eta}_{ji} + \sum_{i=1}^{N} \sum_{j=1}^{N} \rho_{ij} F^{\eta}_{ji}$$

$$= \sum_{i=1}^{N} \sum_{j=1}^{N} \rho_{ij} (h^{\eta}_{ji} + F^{\eta}_{ji}) \qquad [48]$$

If one notes that the innermost summation (over j) will eliminate the running index j and one has only running index i, then it is clear that the outermost summation (over i) is simply the summation of the diagonal elements of a **diagonal matrix** or in other words, it is the trace of the matrix.

$$E = \sum_{i=1}^{N} \left[\sum_{j=1}^{N} \rho_{ij} (h^{\eta}_{ji} + F^{\eta}_{ji}) \right]$$

$$= \text{trace} \left[\sum_{j=1}^{N} \rho_{ij} (h^{\eta}_{ji} + F^{\eta}_{ji}) \right]$$

$$= \text{trace} \, \{ \boldsymbol{\rho} \, (\mathbf{h}^{\eta} + \mathbf{F}^{\eta}) \} \qquad [49]$$

Another point of interest is to relate the orbital energies to total energy. We saw in equation [36] that the matrix representative of the Fock operator over the MO basis is simply the MO energy matrix. It should have also been clear from the earlier discussion as well as the comparison of equations [36] and [49] that when this quantity F^{ϕ}_{pp} is transformed to the AO basis one has the following equivalence:

$$\epsilon_p \equiv F^{\phi}{}_{pp} \equiv <\phi_p|\hat{F}|\phi_p> = \sum_{i}^{N} \sum_{j}^{N} C_{pi}<\eta_i|\hat{F}|\eta_j>C_{jp} \qquad [50a]$$

If one sums up all the occupied MO energies then the result is the following:

$$\sum_{p=i}^{M} \epsilon_p = \sum_{i}^{N} \sum_{j}^{N} \sum_{p}^{M} C_{ip}C_{pj}\left\langle\eta_i|\hat{F}|\eta_j\right\rangle = \sum_{i=1}^{N} \sum_{j=1}^{N} \rho_{ij}<\eta_j|\hat{F}|\eta_i> = \qquad [50b]$$

$$\sum_{i}^{N} \sum_{j}^{N} \rho_{ij}F^{\eta}{}_{ji}$$

Now we would be ready to combine equations [50b] and [48] but the latter expression also contains h^{η}_{ji} so let us express the h^{η}_{ji} element from equation [46] first:

$$h^{\eta}_{ji} = F^{\eta}_{ji} - (2\, J^{\eta}_{ji} - K^{\eta}_{ji}) \qquad [51]$$

and substitute this into [48] to get the following expression

$$E = \sum_{i=1}^{N} \sum_{j=1}^{N} \rho_{ij}(h^{\eta}{}_{ji} + F^{\eta}{}_{ji}) = \sum_{i=1}^{N} \sum_{j=1}^{N} \rho_{ij} \{2F^{\eta}_{ji} - (2J^{\eta}_{ji} - K^{\eta}{}_{ji})\} =$$

$$= 2\sum_{i=1}^{N} \sum_{j=1}^{N} \rho_{ij}F^{\eta}_{ji} - \sum_{i=1}^{N} \sum_{j=1}^{N} \rho_{ij}\,(2J^{\eta}_{ji} - K^{\eta}{}_{ji}) \qquad [52]$$

Now we may use equations [50b] and [52] to see how to relate to E to $\Sigma\epsilon_p$.

$$E = 2 \sum_{i=1}^{N} \sum_{j=1}^{N} \rho_{ij} F^{\eta}_{ji} - \sum_{i=1}^{N} \sum_{j=1}^{N} \rho_{ij} (2\, J^{\eta}_{ji} - K^{\eta}_{ji}) =$$

$$= 2 \sum_{i=1}^{M} \epsilon_i - \sum_{i=1}^{N} \sum_{j=1}^{N} \rho_{ij} (2\, J_{ji} - K_{ji}) \qquad [53]$$

This equation is quite important in the sense that it indicates that the total energy is twice the sum of the doubly occupied MO minus the electron–electron repulsion. This observation can be useful to evaluate the nature of the approximation of certain semi–empirical MO theories.

Let us return to our main line of thought that we left off at equations [45] – [47].

The basic problem is that even when all the integrals are known we cannot calculate E and $\mathbf{F}^{\eta}$ (required for the Hartree–Fock equation) because both depend on $\boldsymbol{\rho}$ which, in turn, is constructed from **C** and it is the quantity **C** which we want to obtain, as the solution, from the Hartree–Fock equation. This requires that the eigen–problem equation be solved in an iterative manner by a process referred to as the Self Consistent Field (SCF) method.

A discussion of the SCF method requires first a consideration of the orthogonality problem. The basis set $\{\eta\}$ is not orthogonal so that the coefficient matrix **C** which diagonalizes **F** cannot be found by an orthogonal transformation.

Consequently, it is necessary to transform the non–orthogonal set $\{\eta\}$ into an orthogonal set $\{\chi\}$ and, correspondingly, $\mathbf{F}^{\eta}$ into $\mathbf{F}^{\chi}$ which may be diagonalized by an orthogonal transformation (such as Jacobi's diagonalization method) and this will permit calculation of the coefficient matrix.

Since orthogonalization is a special type of transformation we separate the **C** matrix into two matrices **V** and **U**

$$\phi = \eta\, \mathbf{C} = \eta\, \mathbf{V}\, \mathbf{U} = \chi\, \mathbf{U} \qquad [54a]$$

$$\begin{array}{ccccc} & & \mathbf{C} & & \\ \underset{\text{AO}}{\eta} & \xrightarrow{\mathbf{V}} & \underset{\text{OO}}{\chi} & \xrightarrow{\mathbf{U}} & \underset{\text{MO}}{\phi} \end{array} \qquad [54b]$$

$$(\phi_1 \phi_2 \;....\; \phi_N) =$$

$$= (\eta_1 \eta_2 \;....\; \eta_N) \begin{vmatrix} V_{11} & V_{12} & \cdots & V_{1N} \\ V_{21} & V_{22} & \cdots & V_{2N} \\ \vdots & \vdots & & \vdots \\ V_{N1} & V_{N2} & \cdots & V_{NN} \end{vmatrix} \begin{vmatrix} U_{11} & U_{12} & \cdots & U_{1N} \\ U_{21} & U_{22} & \cdots & U_{2N} \\ \vdots & \vdots & & \vdots \\ U_{N1} & U_{N2} & \cdots & U_{NN} \end{vmatrix} \qquad [55]$$

also for column notation:

$$\phi^{\dagger} = \mathbf{U}^{\dagger}\, \mathbf{V}^{\dagger}\, \eta^{\dagger} = \mathbf{U}^{\dagger}\, \chi^{\dagger} \qquad [56]$$

Substituting into the Hartree–Fock equation, we have

$$\mathbf{C}^{\dagger}\, \mathbf{F}^{\eta}\, \mathbf{C} = \mathbf{C}^{\dagger}\, \mathbf{S}^{\eta}\, \mathbf{C}\, \epsilon \qquad [57]$$

$$\underbrace{U^\dagger V^\dagger F^\eta V U}_{F^\chi} = U^\dagger \underbrace{V^\dagger S^\eta V}_{S^\chi = 1} U \epsilon \qquad [58]$$

$$U^\dagger F^\chi U = \underbrace{U^\dagger U}_{1} \epsilon \qquad [59]$$

$$U^\dagger F^\chi U = \epsilon \qquad [60]$$

thus ϵ is a diagonal matrix. Consequently, F^χ is to be diagonalized by U to obtain ϵ.

2.3 THE PROBLEM OF AO ORTHOGONALIZATION

Löwdin showed that the orthogonalizing matrix may be chosen to be $S^{-1/2}$, i.e.

$$V = S^{-1/2}. \qquad [61]$$

Accepting this choice leads to the symmetric or Löwdin ortho–gonalization as illustrated by Figure 7 for the 2 equivalents orbital case (such as H_2) in accordance with equation [54].

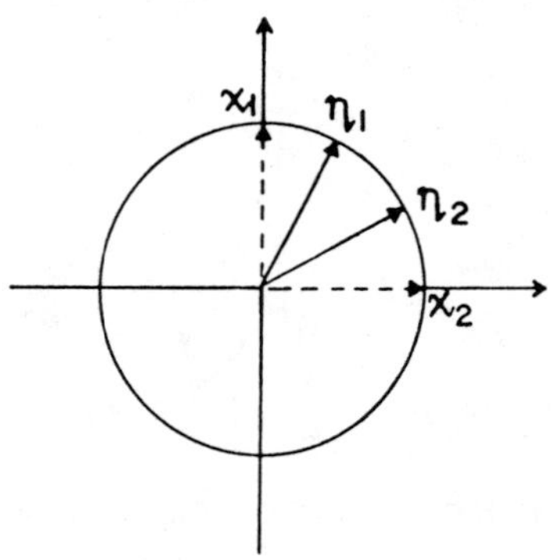

Figure 7. Symmetric or Löwdin orthogonalization of a pair of 1s orbitals associated with H_2

The process in which $S^{-1/2}$ is evaluated, is based on the special properties of diagonal matrices, according to which operation on the diagonal element is equivalent to the same operation on the whole matrix*. We therefore, convert **S** into its diagonal form **D**, take the –1/2 power of the diagonal elements, and then convert this diagonal matrix back to its non–diagonal form. Thus,

$$\mathcal{O}^{\dagger}\, \mathbf{S}\, \mathcal{O} = \mathbf{D} \qquad [62]$$

where $\mathcal{O}$ is an orthogonal matrix. The back conversion proceeds in the opposite sense

$$\mathcal{O}(\mathcal{O}^{\dagger}\, \mathbf{S}\, \mathcal{O})\, \mathcal{O}^{\dagger} = \mathcal{O}\mathbf{D}\, \mathcal{O}^{\dagger} \qquad [63]$$

$$(\mathcal{O}\mathcal{O}^{\dagger})\, \mathbf{S}\, (\mathcal{O}\mathcal{O}^{\dagger}) = \mathcal{O}\mathbf{D}\, \mathcal{O}^{\dagger} \qquad [64]$$

Since

$$\mathcal{O}\mathcal{O}^{\dagger} = \mathbf{1} \qquad [65]$$

$$\mathbf{S} = \mathcal{O}\mathbf{D}\, \mathcal{O}^{\dagger} \qquad [66]$$

The process proceeds as follows:

1st step: Diagonalization $\mathcal{O}^{\dagger}\, \mathbf{S}\, \mathcal{O} = \mathbf{D} \equiv \begin{vmatrix} d_{11} & & & 0 \\ & d_{22} & & \\ & & \ddots & \\ 0 & & & d_{nn} \end{vmatrix}$ [67]

*The following procedure is based on the general feature of matrices:

$$\mathrm{U}^{\dagger}\, \mathrm{A}\, \mathrm{U} = \mathrm{D} \quad \text{and} \quad \mathrm{A} = \mathrm{U}\, \mathrm{D}\, \mathrm{U}^{\dagger}$$

however

$$\mathrm{f}(\mathbf{A}) = \mathrm{U}[\mathrm{F}(\mathrm{D})]\mathrm{U}^{\dagger}$$

2nd step: Raising D to the −1/2 power

$$D^{-1/2} = \begin{vmatrix} 1/\sqrt{d}_{11} & & & \\ & 1/\sqrt{d}_{22} & & \\ & & \ddots & \\ & & & 1/\sqrt{d}_{nn} \end{vmatrix} \qquad [68]$$

3rd step: Back transformation

$$\mathbf{V} = \mathbf{S}^{-1/2} = \mathcal{O}\mathbf{D}^{-1/2}\,\mathcal{O}^{\dagger} \qquad [69]$$

We may now check the value of $\mathbf{V}^{\dagger}\,\mathbf{S}\,\mathbf{V}$ if $\mathbf{V} = \mathbf{S}^{-1/2}$

$$\mathbf{V}^{\dagger} = (\mathcal{O}\mathbf{D}^{-1/2}\mathcal{O}^{\dagger})^{\dagger} = (\mathcal{O}^{\dagger})^{\dagger}(\mathbf{D}^{-1/2})^{\dagger}\mathcal{O}^{\dagger} = \mathcal{O}\mathbf{D}^{-1/2}\,\mathcal{O}^{\dagger} \qquad [70]$$

since $(\mathbf{D}^{-1/2})^{\dagger} = \mathbf{D}^{-1/2}$ because it is diagonal. Furthermore,

$$\mathbf{V}^{\dagger}\,\mathbf{S}\,\mathbf{V} = (\mathcal{O}\mathbf{D}^{-1/2}\,\mathcal{O}^{\dagger})\,\mathbf{S}\,(\mathcal{O}\mathbf{D}^{-1/2}\,\mathcal{O}^{\dagger}) = \mathcal{O}\mathbf{D}^{-1/2}\,(\mathcal{O}^{\dagger}\mathbf{S}\,\mathcal{O})\,\mathbf{D}^{-1/2}\,\mathcal{O}^{\dagger} =$$

$$\mathcal{O}^{\dagger}\mathbf{D}^{-1/2}\,\mathbf{D}\,\mathbf{D}^{-1/2}\,\mathcal{O}^{\dagger} \qquad [71]$$

since diagonal matrices commute. Finally, we may write:

$$\mathbf{V}^{\dagger}\,\mathbf{S}\,\mathbf{V} = \mathcal{O}\mathbf{D}^{-1/2}\,\mathbf{D}^{-1/2}\,\mathbf{D}\,\mathcal{O}^{\dagger} = \mathcal{O}\underbrace{\mathbf{D}^{-1}\;\mathbf{D}}\,\mathcal{O}^{\dagger} = \mathcal{O}\,\mathcal{O}^{\dagger} = 1 \qquad [72]$$

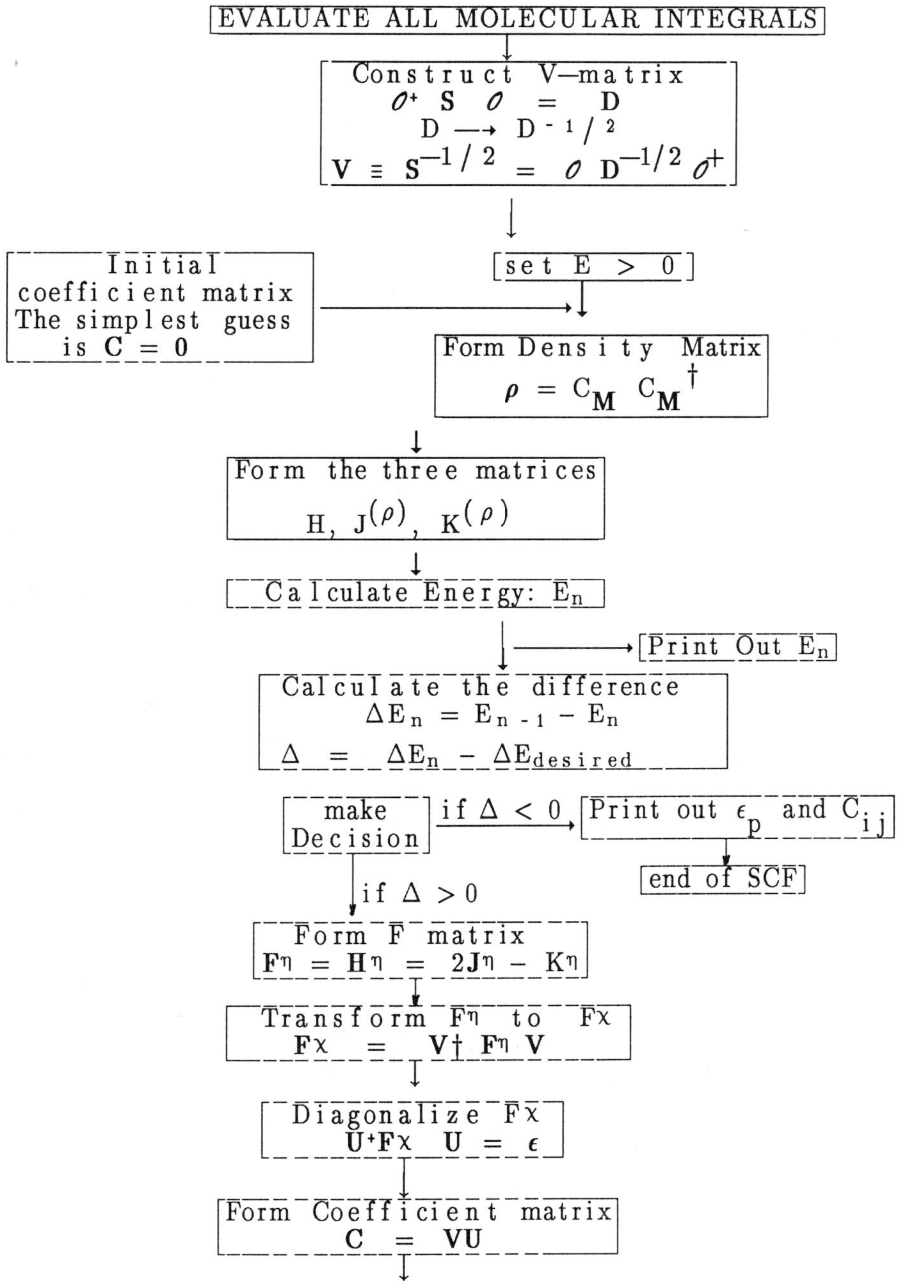

Figure 8. Flowchart for the iterative Self–Consistent–Field (SCF) method.

Now with $S^{-1/2}$ in hand

$$U^{\dagger} F^{\chi} U = \epsilon \quad [73]$$

where

$$F^{\chi} = S^{-1/2} F^{\eta} S^{-1/2} \quad [74]$$

2.4. THE ITERATIVE SCF PROCEDURE

The iterative SCF procedure can now be presented, most conveniently, in form of a self–explanatory flow–chart. (Figure 8).

In the course of the SCF procedure the total energy (E) is lowered in each iteration cycle (i.e. F_n gradually becomes a larger negative quantity as n increases). Convergence may be measured by the difference between the energy values associated with two successive iterations ($\Delta E_n = E_{n-1} - E_n$). Since we do not know the energy to be expected after the first iteration (which depends on the initial coefficient matrix) it is best to set the initial value of E greater than zero. It is also possible to judge convergence in terms of the density matrix ρ ($\Delta\rho_n = \rho_{n-1} - \rho_n$).

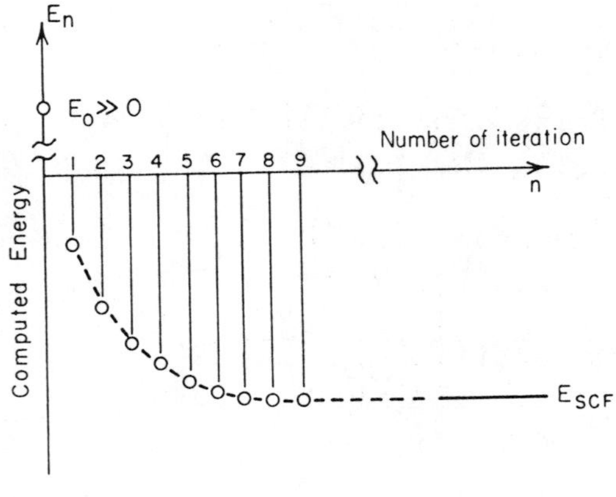

Figure 9 A schematic illustration for the convergency of an SCF calculation.

Another point of interest is the consequence of choosing the simplest possible initial coefficient matrix, i.e. C = O. In this event the density matrix becomes zero ($\rho = 0$) and terms (J, K and E) that incorporate it will vanish. However, the F–matrix will not vanish since only the last two terms in the expression F = H + 2 J – K depend on ρ. The approximation is analogous to a Hucket type calculation.

2.5 ONE AND TWO–ELECTRON FOCK MATRIX CONSTRUCTION

2.5.1 ENERGY EXPRESSION IN MO BASIS (CLOSED SHELL):

$$E = E_o + E_1 + E_2 = E_o + 2 \sum_{p=1}^{M} h^{\phi}_{pp} + \sum_{p=1}^{M} \sum_{q=1}^{M} (2 J_{pq} - K_{pq}) \quad [75]$$

where E_o is the nuclear repulsion energy

$$E_o = \sum_{A=1}^{N_{ATOMS}} \sum_{B=1}^{A-1} Z_A Z_B / R_{AB} \quad [76]$$

Z_A, Z_B: nuclear charges

R_{AB}: distance between atoms A and B

The second term in [75] is the one–electron energy (E_1) and has the following elements under the summation

$$h^{\phi}_{pp} = \langle \phi_p(1) | \hat{h} | \phi_p(1) \rangle \qquad [77]$$

where $\hat{h}$ is the "core hamiltonian":

$$\hat{h} = -\frac{1}{2} \nabla^2 - \sum_{A=1}^{N_{ATOMS}} Z_A / r_{Ap} \qquad [78]$$

In [78] the first term is the kinetic energy operator which includes the Laplace operator

$$\nabla^2 = \frac{\partial^2}{\partial x^2} + \frac{\partial^2}{\partial y^2} + \frac{\partial^2}{\partial z^2} \qquad [79]$$

and the second term in [78] is that portion of the potential energy operator that measures the unclear–electron attraction. In this latter term

r_{Ap} is the distance of atom A from the electron MO p

The last term in [75] is the two–electron energy (E_2) which is that portion of the potential energy that measures the electron–electron repulsion. This term is made up of the Coulomb (J) and exchange (K) integrals:

$$J^{\phi}_{pq} = <\phi_p(1)\phi_q(2)|\frac{1}{r_{12}}|\phi_p(1)\phi_q(2)> \equiv \{pp|qq\} \qquad [80]$$

$$K^{\phi}_{pq} = <\phi_p(1)\phi_q(2)|\frac{1}{r_{12}}|\phi_q(1)\phi_q(2)> \equiv \{pq|pq\} \qquad [81]$$

2.5.2 ENERGY EXPRESSION IN AO BASIS (CLOSED SHELL)

Let us expand the MO basis in terms of the AO basis in the following way:

$$\phi_p = \sum_{i=1}^{n} \eta_i C_{ip} \qquad [82]$$

Let us define a density matrix in the following fashion:

$$\mathbf{D} = \mathbf{CC}^{\dagger} \text{ (occupied MO only)} \qquad [83]$$

$$D_{ij} = \sum_{p=1}^{m} C_{ip} C^{\dagger}_{pj} \qquad [84a]$$

$$D_{ij} = \sum_{p=1}^{m} C_{ip} C_{jp} \qquad [84b]$$

Density matrix D: is often labeled as "ρ" in quantum meachanics books. Note the density matrix is real symmetric

$$D_{ij} = D_{ji} \qquad [85]$$

One-electron term (E_1)

$$h^{\phi}_{pp} = <\phi_p|\hat{h}|\phi_p> \qquad [86]$$

substitute for ϕ_p using [82]:

$$h^{\phi}_{pp} = \sum_{i=1}^{N} \sum_{j=1}^{N} C^{\dagger}_{pi} <\eta_i|\hat{h}|\eta_j> C_{jp} \qquad [87a]$$

$$= \sum_{i=1}^{N} \sum_{j=1}^{N} C_{ip}\, C_{jp}\, h^{\eta}_{ij} \qquad [87b]$$

Consequently, the one-electron energy can be obtained by summation over the index p and that leads to the consequence of the density matrix D.

$$E_1 = \sum_{p=1}^{M} h^{\phi}_{pp} = \sum_{p=1}^{M} \sum_{i=1}^{N} \sum_{j=1}^{N} C_{ip}\, C_{pj}\, h^{\eta}_{ij} = \sum_{i=1}^{N} \sum_{j=1}^{N} \left[\sum_{p=1}^{M} C_{ip}\, C_{pj} \right] h^{\eta}_{ij} =$$

$$= \sum_{i=1}^{N} \sum_{j=1}^{N} D_{ij}\, h^{\eta}_{ij} \qquad [88a]$$

Alternatively, we may write the above relationship simply as

$$E_1 = \mathrm{tr}\,(\mathbf{Dh}^{\eta}) \qquad [88b]$$

Note the above is **identiacal to**

$$\mathbf{h}^{\phi} = \mathbf{C}^{\dagger}\ \mathbf{h}^{\eta}\ \mathbf{C} \qquad [88c]$$

Two–electron terms (E_2):

It is necessary to introduce the following notation:

$$\{\underline{ab}\,|\,\underline{cd}\} = <\phi_a(1)\phi_c(2)|\frac{1}{r_{12}}|\phi_b(1)\phi_d(2)> \qquad [89]$$

electron 1 ↑ ↑ electron 2

Coulomb term (J):

$$J^{\phi}{}_{pq} = \{pp\,|\,qq\} = <\phi_p(1)\phi_q(2)|\frac{1}{r_{12}}|\phi_p(1)\phi_q(2)> \qquad [90]$$

i j k l

Using [82] the following substitution is possible:

$$J^{\phi}{}_{pq} = \sum_{i=1}^{N}\sum_{j=1}^{N}\sum_{k=1}^{N}\sum_{l=1}^{N} C^{\dagger}{}_{pi}\, C_{jp}\, C^{\dagger}{}_{qk}\, C_{lq}\, \{ij\,|\,kl\} \qquad [91]$$

For the double sum of J^{ϕ}_{pq} we can sum the coefficient matrix elements with indices p and q in such a way that the density matrix **D** is emerging:

$$\sum_{p=1}^{M} \sum_{q=1}^{M} J^{\phi}_{pq} = \sum_{p=1}^{M} \sum_{q=1}^{M} \sum_{i=1}^{N} \sum_{j=1}^{N} \sum_{k=1}^{N} \sum_{l=1}^{N} C_{ip} C_{jp} C_{kq} C_{lq} \{ij|kl\} =$$

$$= \sum_{i=1}^{N} \sum_{j=1}^{N} \sum_{k=1}^{N} \sum_{l=1}^{N} \{ij|kl\} \sum_{p=1}^{M} C_{ip} C_{jp} \sum_{q=1}^{M} C_{kq} C_{lq}$$

$$= \sum_{i=1}^{N} \sum_{j=1}^{N} \sum_{k=1}^{N} \sum_{l=1}^{N} D_{ij} D_{kl} \{ij|kl\} \qquad [91a]$$

$$= \sum_{i=1}^{N} \sum_{j=1}^{N} D_{ij} \sum_{k=1}^{N} \sum_{l=1}^{N} D_{kl} \{ij|kl\}$$

$$= \sum_{i=1}^{N} \sum_{j=1}^{N} D_{ij} J^{\eta}_{ij}$$

$$= \mathrm{tr}\,(\mathbf{D}\,\mathbf{J}^{\eta}) \qquad [91b]$$

where

$$J^{\eta}_{ij} = \sum_{k=1}^{N} \sum_{l=1}^{N} D_{kl} \{ij|kl\} \qquad [92]$$

Exchange term (K):

$$K^{\phi}_{pq} = \{pq|pq\} = <\phi_p(1)\phi_p(2)|\frac{1}{r_{12}}|\phi_q(1)\phi_q(2)> \qquad [93]$$

$$\quad i \quad j \; k \quad l$$

using [82] the following substitution is possible

$$K^{\phi}_{pq} = \sum_{i=1}^{N}\sum_{j=1}^{N}\sum_{k=1}^{N}\sum_{l=1}^{N} C_{pi}\, C_{pj}\, C_{kq}\, C_{lq}\, \{ik|jl\} \qquad [94]$$

For the double sum of K_{pq} we can sum the coefficient matrix elements with indices p and q in such a way that the density matrix **D** is emerges:

$$\sum_{p=1}^{M}\sum_{q=1}^{M} K^{\phi}_{pq} = \sum_{p=1}^{M}\sum_{q=1}^{M}\sum_{i=1}^{N}\sum_{j=1}^{N}\sum_{k=1}^{N}\sum_{l=1}^{N} C_{ip}\, C_{jp}\, C_{kq}\, C_{lq}\, \{ik|jl\} =$$

$$= \sum_{i=1}^{N}\sum_{j=1}^{N}\sum_{k=1}^{N}\sum_{l=1}^{N} \{ik|jl\} \sum_{p=1}^{M} C_{ip}^{\dagger}\, C_{jp}^{\dagger} \sum_{q=1}^{M} C_{kq}\, C_{lq} \qquad [95a]$$

$$= \sum_{i=1}^{N}\sum_{j=1}^{N} D_{ij} \sum_{k=1}^{N}\sum_{l=1}^{N} D_{kl}\, \{ik|jl\}$$

$$= \sum_{i=1}^{N}\sum_{j=1}^{N} D_{ij}\, K^{\eta}_{ij}$$

$$= \mathrm{tr}\,(\mathbf{D}\,\mathbf{K}^{\eta}) \qquad [95b]$$

where

$$K^{\eta}_{ij} = \sum_{k=1}^{n} \sum_{l=1}^{n} D_{kl} \{ik|jl\} \qquad [96]$$

The two electron energy E_2 may be now constructed from [91b] and [95b]

$$E_2 = 2tr(\mathbf{DJ}^{\eta}) - tr(\mathbf{DK}^{\eta}) \qquad [97]$$

Substituting into (75a) with [88b] and [97] we obtain the following:

$$E = E_o + 2\ tr\ (\mathbf{Dh}^{\eta}) + \{2tr\ (\mathbf{DJ}^{\eta}) - tr\ (\mathbf{DK}^{\eta})\} \qquad [98a]$$

$$E = E_o + tr\ [\mathbf{D}(2\mathbf{h}^{\eta} + 2\mathbf{J}^{\eta} - \mathbf{K}^{\eta})] \qquad [98b]$$

$$E = E_o + tr\ [\mathbf{D}(\mathbf{h}^{\eta} + \mathbf{h}^{\eta} + 2\mathbf{J}^{\eta} - \mathbf{K}^{\eta})]$$

$$E = E_o + tr\ [\mathbf{D}(\mathbf{h}^{\eta} + \mathbf{F}^{\eta})] \qquad [98c]$$

where $$\mathbf{F}^{\eta} = \mathbf{h}^{\eta} + 2\mathbf{J}^{\eta} - \mathbf{K}^{\eta} \qquad [99a]$$

or $$F^{\eta}_{ij} = h^{\eta}_{ij} + 2J^{\eta}_{ij} - K^{\eta}_{ij} \qquad [99b]$$

2.6 INHERENT SYMMETRY OF TWO–ELECTRON INTEGRALS

In dealing with the two–electron integrals the 4 dimensional matrix $\{\eta_i\eta_j|\eta_k\eta_l\}$ or $\{ij|kl\}$ is converted to a pseudo 2–dimensional matrix by defining the following compound indices:

$$(ij) = \frac{i(i-1)}{2} + j \quad \text{and} \quad (kl) = \frac{k(k-1)}{2} + l \qquad [100]$$

Take for example a 3–orbital case: η_1, η_2 and η_3. It should be noted that although for the 3–orbital case there are $3^2 \times 3^2 = 81$ two electron integrals as shown in Table 2, only 21 elements are distinctly different because of the following identities:

$$\{ij|kl\} = \{ji|kl\} = \{ij|lk\} = \{ji|lk\}$$

$$= \{kl|ij\} = \{lk|ij\} = \{kl|ji\} = \{lk|ji\} \qquad [101]$$

Table 2. Pseudo two–dimensional two–electron integral matrix for the 3–orbital case.

	(kl)	1	2	3	4	5	6	7	8	9
(ij)	i,j k,l	1,1	1,2	1,3	2,1	2,2	2,3	3,1	3,2	3,3
1	1,1	{11|11}	{11|12}	{11|13}	{11|21}	{11|22}	{11|23}	{11|31}	{11|32}	{11|33}
2	1,2		{12|12}	{12|13}	{12|21}	{12|22}	{12|23}	{12|31}	{12|32}	{12|33}
3	1,3			{13|13}	{13|21}	{13|22}	{13|23}	{13|31}	{13|32}	{13|33}
4	2,1				{21|21}	{21|22}	{21|23}	{21|31}	{21|32}	{21|33}
5	2,2					{22|22}	{22|23}	{22|31}	{22|32}	{22|33}
6	2,3						{23|23}	{23|31}	{22|32}	{23|33}
7	3,1							{31|31}	{31|32}	{31|33}
8	3,2								{32|32}	{32|33}
9	3,3									{33|33}

Note that the above formulae in [100] are only valid for $i \geq j$ and $k \geq l$. However, with recognition of the identities in [101] these formulae are sufficient to handle all distinctly different integrals.

Table 3 Reduced pseudo two dimensional two electron integral matrix for the 3–orbital case.

(ij) \ (kl)	1	2	3	4	5	6
1	{11|11}	{11|21}	{11|22}	{11|31}	{11|32}	{11|33}
2		{21|21}	{21|22}	{21|31}	{21|32}	{21 33}
3			{22|22}	{22|31}	{22|32}	{22|33}
4				{31|31}	{31|32}	{31|33}
5					{32|32}	{32|33}
6						{33|33}

For a 2–orbital case (e.g.: η_1 and η_2) such as it might be the case for the smallest MO computation of hydrogen molecule (H_2) or protonated helium ($He - H^+$) there are only six unique two–electron integrals

$$\begin{array}{ccc} \{11|11\} & \{21|21\} & \{21|22 \\ & \{21|21\} & \{21|22\} \\ & & \{22|22\} \end{array} \qquad [102]$$

To save time while evaluating the 2–electron integrals, and in the SCF while adding them to the appropriate F^η matrix elements, only 1 permutation of the labels will appear, and the SCF subroutine is responsible for adding that one permutation to all the appropriate F^{η} matrix elements.

2.7 THE GENERATION OF THE FOCK MATRIX

The Fock matrix [103] is a real symmetric matrix and therefore only an upper (or lower) triangular matrix needs to be generated.

$$\mathbf{F} = \begin{bmatrix} F_{11} & F_{12} & F_{13} & \cdots \\ & F_{22} & F_{23} & \cdots \\ & & F_{33} & \cdots \\ & & \vdots & \end{bmatrix} \qquad [103]$$

However, it is clear from Table 1 that, for example, in a 50–orbital problem the **F** matrix in [103] is of 50 x 50 dimension having only 50(50+1)/2 = 1275 matrix elements in its upper triangular. Nevertheless these 1275 matrix elements must be assembled from 1275 one–electron and 814,725 unique two–electron integrals. First we include the 1275 one electron (core) integrals into **F** and then we must add the two–electron integrals. The necessary 814,725 unique two–electon integrals are stored either on a magnetic tape or a magnetic disk and therefore one cannot access them randomly. For this reason the Fock matrix is not generated sequentially from element to element such as F_{11}, F_{12}, F_{13}, etc but rather the integral list is read sequentially and the two–electron integrals are added, one after the other, to the various F_{ij} elements. The following cases indicate in how many F_{ij} elements a given two–electron integral are to be included.

Case 1. {ab|cd}

This case is a general one with a,b,c,d all being different. The permutation for this general case may be written as follows:

$$\{ab|cd\} = \{ba|cd\} = \{ab|dc\} = \{ba|dc\}$$

$$= \{cd|ab\} = \{dc|ab\} = \{cd|ba\} = \{dc|ba\} \qquad [101a]$$

As a **Coulomb integral**, {ab|cd} contributes to F^{η}_{ab} as D_{cd} {ab|cd} and D_{dc} {ab|cd}, as the double sum over k,ℓ in [92] and [96] would pick up both permutations. Since J^{η}_{ij} enters into F^{η}_{ab} with a pre–factor of 2 (equation [99b]), the {ab|cd} must therefore, be added to F^{η}_{ab} 4 times:

$$F^{\eta}_{ab} \longleftarrow F^{\eta}_{ab} + 4\ Dcd\ \{ab|cd\} \qquad [104]$$

↑ this is the updated F_{ab}

↑ this is the old F_{ab} including only the one–electron (core) integrals

Note that **D** is also a symmetric matrix, therefore:

$$D_{cd} = D_{dc} \qquad [105]$$

Switching {ab|cd} to {cd|ab} gives

$$F^{\eta}_{cd} \longleftarrow F^{\eta}_{cd} + 4\ D_{ab}\ \{ab|cd\} \qquad [106]$$

As an **exchange integral**, {ab|cd} contributes to F^{η}_{ac} as $-D_{db}$ {ab|cd}, to F^{η}_{bd} as $-D_{ac}$ {ba|dc} to F^{η}_{ad} as $-D_{bc}$ {ab|cd}, and to F^{η}_{bc} as $-D_{ad}$ {ba|cd}. All the other 4 permutations contribute only to the other half of F^{η} (lower triangular, for example).

$$F^{\eta}_{ac} \longleftarrow F^{\eta}_{ac} - D_{bd} \{ab|cd\} \qquad [107a]$$

$$F^{\eta}_{bd} \longleftarrow F^{\eta}_{bd} - D_{ac} \{ab|cd\} \qquad [107b]$$

$$F^{\eta}_{ad} \longleftarrow F^{\eta}_{ad} - D_{bc} \{ab|cd\} \qquad [107c]$$

$$F^{\eta}_{bc} \longleftarrow F^{\eta}_{bc} - D_{ad} \{ab|cd\} \qquad [107d]$$

Case 2.1 {ab|ac}:

The label permutations are:

$$\{ab|cd\} = \{ab|cd\} = \{ab|cd\} = \{ab|cd\} = \qquad [101b]$$
$$\{ab|cd\} = \{ab|cd\} = \{ab|cd\} = \{ab|cd\}$$

Coulomb contributions (2J):

$$F^{\eta}_{ab} \longleftarrow F^{\eta}_{ab} - 4D_{ac} \{ab|ac\} \qquad [108a]$$

$$F^{\eta}_{ac} \longleftarrow F^{\eta}_{ac} - 4D_{ab} \{ab|ac\} \qquad [109a]$$

In [108a] the second term originates from

$$4\,\{ab|ac\} = 2\,\{ab|ac\} + 2\{ab|ca\} \qquad [108b]$$

and in [109a] the second term is due to:

$$4\,\{ab|ac\} = 2\,\{ac|ab\} + 2\{ac|ba\} \qquad [109b]$$

Exchange contributions (–K):

$$F^{\eta}_{aa} \longleftarrow F^{\eta}_{aa} - 2D_{bc}\,\{ab|ac\} \qquad [110a]$$

because $\quad 2\,\{ab|ac\} = \{ab|ac\} + \{ac|ab\} \qquad [110b]$

and $\quad F^{\eta}_{bc} \longleftarrow F^{\eta}_{bc} - D_{aa}\,\{ab|ac\} \qquad [111a]$

because $\quad \{ab|ac\} = \{ba|ca\} \qquad [111b]$

similarly $\quad F^{\eta}_{ac} \longleftarrow F^{\eta}_{ac} - D_{ab}\,\{ab|ac\} \qquad [112a]$

where $\quad \{ab|ac\} = \{ab|ca\} \qquad [112b]$

and $\quad F^{\eta}_{ab} \longleftarrow F^{\eta}_{ab} - D_{ac}\,\{ab|ac\} \qquad [113a]$

where $\quad \{ab|ac\} = \{ac|ba\} \qquad [113b]$

Combining the Coulombic ([108a] and [109b]) and Exchange (110a) – [113a]) terms one obtains [114a] – [114d]:

$$F^{\eta}_{ab} \longleftarrow F^{\eta}_{ab} + 3\, D_{ac}\ \{ab|ac\} \qquad [114a]$$

$$F^{\eta}_{ac} \longleftarrow F^{\eta}_{ab} + 3\, D_{ab}\ \{ab|ac\} \qquad [114b]$$

$$F^{\eta}_{aa} \longleftarrow F^{\eta}_{aa} - 2\, D_{bc}\ \{ab|ac\} \qquad [114c]$$

$$F^{\eta}_{bc} \longleftarrow F^{\eta}_{bc} - D_{aa}\ \{ab|ac\} \qquad [114d]$$

Case 2.2 {ab|bc}:

The Permutations for this case are written as follows:

$$\{ab|cd\} = \{ab|cd\} = \{ab|cd\} = \{ab|cd\} \qquad [101b]$$
$$\{ab|cd\} = \{ab|cd\} = \{ab|cd\} = \{ab|cd\}$$

By analogy to case 2.1 it is easy to generate the corresponding elements of the F matrix:

$$F^{\eta}_{ab} \longleftarrow F^{\eta}_{ab} + 3\, D_{bc}\ \{ab|bc\} \qquad [115a]$$

$$F^{\eta}_{bc} \longleftarrow F^{\eta}_{bc} + 3\, D_{ab}\ \{ab|bc\} \qquad [115b]$$

$$F^{\eta}_{bb} \longleftarrow F^{\eta}_{bb} - 2\, D_{ac}\ \{ab|bc\} \qquad [115c]$$

$$F^{\eta}_{ac} \longleftarrow F^{\eta}_{ac} - D_{bb}\ \{ab|bc\} \qquad [114d]$$

Case 2.3 {ac|bc}:

By analogy to case 2.1 we can update the corresponding F matrix elements as follows:

$$F^{\eta}_{ac} \longleftarrow F^{\eta}_{ac} + 3\ D_{bc}\ \{ac|bc\} \qquad [116a]$$

$$F^{\eta}_{bc} \longleftarrow F^{\eta}_{bc} + 3\ D_{ac}\ \{ac|bc\} \qquad [116b]$$

$$F^{\eta}_{cc} \longleftarrow F^{\eta}_{cc} + 2\ D_{ab}\ \{ac|bc\} \qquad [116c]$$

$$F^{\eta}_{ab} \longleftarrow F^{\eta}_{ab} - D_{cc}\ \{ac|bc\} \qquad [117]$$

Case 2.4 : {ab|ab}

The following unique permutations are possible:

$$\{ab|ab\} = \{ba|ab\} = \{ab|ba\} = \{ba|ba\} \qquad [101c]$$

Unique Coulomb contributions (2J):

$$F^{\eta}_{ab} \longleftarrow F^{\eta}_{ab} + 4\ D_{ab}\ \{ab|ab\} \qquad [118]$$

Note that $$4\ \{ab|ab\} = 2\ \{ab|ab\} + 2\ \{ab|ba\} \qquad [119]$$

Unique Exchange contributions (–K):

$$F^{\eta}_{aa} \longleftarrow F^{\eta}_{aa} - D_{bb}\ \{ab|ab\} \qquad [120a]$$

$$F^{\eta}_{bb} \longleftarrow F^{\eta}_{bb} - D_{aa}\ \{ab|ab\} \qquad [120b]$$

because $\{ab|ab\} = \{ba|ba\}$ [121a]

$$F^{\eta}_{ab} \longleftarrow F^{\eta}_{ab} - D_{ab}\ \{ab|ab\} \qquad [120c]$$

because $\{ab|ab\} = \{ab|ba\}$ [121b]

Combining [118] and [120c] results a new F_{ab} :

$$F^{\eta}_{ab} \longleftarrow F^{\eta}_{ab} + 3\ D_{ab}\ \{ab|ab\} \qquad [122a]$$

while F_{aa} and F_{bb} remain as before, in [120a] and [120b], respectively:

$$F^{\eta}_{aa} \longleftarrow F^{\eta}_{aa} - D_{bb}\ \{ab|ab\} \qquad [122b]$$

$$F^{\eta}_{bb} \longleftarrow F^{\eta}_{bb} - D_{aa}\ \{ab|ab\} \qquad [122c]$$

Case 2.5: {aa|bc}

The unique permutations here are as follow:

$$\{aa|bc\} = \{aa|cb\} = \{bc|aa\} = \{cb|aa\} \qquad [101d]$$

Unique Coulomb contributions (2J):

$$F^{\eta}_{aa} \longleftarrow F^{\eta}_{aa} + 4\, D_{bc}\ \{aa|bc\} \qquad [123a]$$

because

$$4\,\{aa|bc\} = 2\,\{aa|bc\} + 2\,\{aa|cb\} \qquad [124]$$

$$F^{\eta}_{bc} \longleftarrow F^{\eta}_{bc} + 2\, D_{aa}\ \{aa|bc\} \qquad [123b]$$

Unique exchange contributions (–K):

$$F^{\eta}_{ab} \longleftarrow F^{\eta}_{ab} - D_{ac}\ \{aa|bc\} \qquad [125a]$$

$$F^{\eta}_{ac} \longleftarrow F^{\eta}_{ac} - D_{ab}\ \{aa|bc\} \qquad [125b]$$

Case 2.6: {ab|cc}

By analogy to case 2.5:

$$F^{\eta}_{ab} \longleftarrow F^{\eta}_{ab} + 2\, D_{cc}\ \{ab|cc\} \qquad [126a]$$

$$F^{\eta}_{cc} \longleftarrow F^{\eta}_{cc} + 4\, D_{ab}\ \{ab|cc\} \qquad [126b]$$

$$F^{\eta}_{ac} \longleftarrow F^{\eta}_{ac} - D_{bc}\ \{ab|cc\} \qquad [126c]$$

$$F^{\eta}_{bc} \longleftarrow F^{\eta}_{bc} - D_{ac}\ \{ab|cc\} \qquad [126d]$$

Case 2.7.1: {ab|bb}

Here the following unique permutation are possible:

$$\begin{aligned} \{ab|bb\} &= \{ba|bb\} = \\ \{bb|ab\} &= \{bb|ba\} \end{aligned} \qquad [101e]$$

Unique Coulomb contributions (2J):

$$F^{\eta}_{ab} \longleftarrow F^{\eta}_{ab} + 2\ D_{bb}\ \{ab|bb\} \qquad [127a]$$

$$F^{\eta}_{bb} \longleftarrow F^{\eta}_{bb} + 4\ D_{ab}\ \{ab|bb\} \qquad [127b]$$

where $$4\ \{ab|bb\} = 2\ \{bb|ab\} + 2\ \{bb|ba\} \qquad [128]$$

Unique exchange contributions (–K):

$$F^{\eta}_{ab} \longleftarrow F^{\eta}_{ab} - D_{bb}\ \{ab|bb\} \qquad [129a]$$

$$F^{\eta}_{bb} \longleftarrow F^{\eta}_{bb} - 2\ D_{ab}\ \{ab|bb\} \qquad [129b]$$

where $$2\ \{ab|bb\} = \{ba|bb\} + \{bb|ba\} \qquad [130]$$

Combining the Coulomb and Exchange terms we obtain:

$$F^{\eta}_{ab} \longleftarrow F^{\eta}_{ab} + D_{bb} \{ab|bb\} \qquad [131a]$$

$$F^{\eta}_{bb} \longleftarrow F^{\eta}_{bb} + 2\, D_{ab} \{ab|bb\} \qquad [131b]$$

Case 2.7.2: {aa|bb}

The unique permutation here is very simple:

$$\{aa|bb\} = \{bb|aa\} \qquad [101f]$$

Unique Coulomb contributions (2J):

$$F^{\eta}_{aa} \longleftarrow F^{\eta}_{aa} + 2\, D_{bb} \{aa|bb\} \qquad [132a]$$

$$F^{\eta}_{bb} \longleftarrow F^{\eta}_{bb} + 2\, D_{aa} \{aa|bb\} \qquad [132b]$$

Note that $\{aa|bb\} = \{bb|aa\}$ [133]

Unique exchange contributions (–K):

$$F^{\eta}_{ab} \longleftarrow F^{\eta}_{ab} - D_{ab} \{aa|bb\} \qquad [134]$$

Case 2.7.3: {aa|ab}

By analogy to case 2.7.1 we may write:

$$F^{\eta}_{aa} \longleftarrow F^{\eta}_{aa} + 2\, D_{ab}\ \{aa|ab\} \qquad [135a]$$

$$F^{\eta}_{ab} \longleftarrow F^{\eta}_{ab} + D_{aa}\ \{aa|ab\} \qquad [135b]$$

Case 2.8: {aa|aa}

Permutations in {aa|aa} will recover {aa|aa}

Unique Coulomb contributions (2J):

$$F^{\eta}_{aa} \longleftarrow F^{\eta}_{aa} + 2\, D_{aa}\ \{aa|aa\} \qquad [136]$$

Unique exchange contributions (–K):

$$F^{\eta}_{aa} \longleftarrow F^{\eta}_{aa} - D_{aa}\ \{aa|aa\} \qquad [137]$$

Combining Coulomb and Exchange terms we obtain

$$F^{\eta}_{aa} \longleftarrow F^{\eta}_{aa} + D_{aa}\ \{aa|aa\} \qquad [138]$$

Since the diagonal elements of the Coulomb and Exchange matrices are numerically the same

$$J^{\eta}_{aa} = K^{\eta}_{aa} \qquad [139]$$

Consequently:

$$2J^{\eta}_{aa} - K^{\eta}_{aa} = D_{aa}\,[2\{aa|aa\} - \{aa|aa\}] = D_{aa}\,\{aa|aa\} \qquad [140]$$

2.8 Jacobi's Method of Matrix Diagonalization

After assembling the F^{η} matrix, as outlined in Section 2.7, from necessary one–electron and two–electron integrals we have to subject it to two consecutive similarity transformations (see Section 2.2) as specified by equations [58] – [60]. First the **V** matrix [61] is used in the similarity transformation where **V** is the inverse square root of the overlap matrix **S** [61]

$$\mathbf{V}^{\dagger}\ \mathbf{F}^{\eta}\ \mathbf{V} = \mathbf{F}^{\chi} \qquad [141]$$

Consequently $\mathbf{F}^{\chi}$ is to be diagonalized [60] in order to obtain the energies (ϵ) of the molecular orbitals

$$\mathbf{U}^{\dagger}\ \mathbf{F}^{\chi}\ \mathbf{U} = \epsilon \qquad [60]$$

Although there are a number of methods to diagonalize a real symmetric matrix, Jacobi's method is a classical algorithm which is still used in computer programs, written to carry out MO calculations. It is therefore reviewed here.

The method is based on the successive plane (i.e. 2 x 2) rotations. In fact an infinite number of rotations are required to achieve a "perfect" diagonlization but, for working accuracy, a finite number of steps is satisfactory. Jacobi's procedure may be illustrated in the following sequence where the original real symmetric matrix is $\mathbf{F}_0$, the matrix obtained after rotation in the first plane is $\mathbf{F}_1$, etc., and the final, diagonal, matrix is ϵ.

$$F_0 \longrightarrow F_1 \longrightarrow .... \longrightarrow F_{k-1} \longrightarrow F_k \longrightarrow .... \longrightarrow \epsilon \qquad [142]$$

The matrix obtained in the kth step has the following relationship to the previous matrix

$$\mathbf{F}_k = \mathbf{U}_k{}^{\dagger}\, \mathbf{F}_{k-1}\, \mathbf{U}_k \qquad [143]$$

where $\mathbf{U}_k$ is an orthogonal matrix (and $\mathbf{U}_k{}^{\dagger}$ is its transpose) associated with the rotation in the kth plane.

If the kth step involves rotation in the r,s plane then the orthogonal matrix ($\mathbf{U}_k$) has the form given in [144]. The 2x2 rotation implies the usage of the sin and cos of the rotational angle α.

$$
U_k = \left|\begin{array}{cccccccccc}
1 & 0 & \cdots & 0 & \cdots\cdots & 0 & \cdots & 0 & 0 & \\
0 & 1 & \cdots & 0 & \cdots\cdots & 0 & \cdots & 0 & 0 & \\
\vdots & \vdots & & \vdots & & \vdots & & \vdots & \vdots & \\
0 & 1 & & ..\cos\alpha & \cdots & -\sin\alpha & .. & 0 & 0 & \\
\vdots & \vdots & \vdots & \vdots & & \vdots & & \vdots & \vdots & \\
0 & 1 & & ..\sin\alpha & \cdots & \cos\alpha & .. & 0 & 0 & \\
\vdots & \vdots & \vdots & \vdots & & \vdots & & \vdots & \vdots & \\
0 & 0 & \cdots & 0 & \cdots\cdots & 0 & \cdots & 0 & 1 &
\end{array}\right|
\begin{array}{l} \\ \\ \\ \ldots.r \\ \\ \ldots.s \\ \\ \\ \end{array}
\qquad [144]
$$

$$
\begin{array}{cccccc}
 & & & r & & s
\end{array}
$$

$$
\left.\begin{array}{lllll}
U_{rr} & = & U_{ss} & = & \cos\alpha \\
U_{rs} & = & -U_{sr} & = & -\sin\alpha \\
U_{ii} & = & U_{jj} & = & 1 \quad (i \text{ or } j \neq r \text{ or } s) \\
U_{ij} & = & U_{ji} & = & 0 \quad (i \text{ or } j \neq r \text{ or } s)
\end{array}\right\} \qquad [145]
$$

Because the purpose of the plane rotation is to eliminate the off–diagonal matrix elements, the angle of rotation (α) must be chosen in such a way that the r,sth element of the matrix $\mathbf{F}_k$, i.e. f^k_{rs} , will be equal to zero.

By substituting $\mathbf{U}_k$ and $\mathbf{U}_k{}^\dagger$ into the quation

$$\mathbf{F}_k = \mathbf{U}_k^\dagger \, \mathbf{F}_{k-1} \, \mathbf{U}_k \qquad [146]$$

one obtains the following diagonal and off–diagonal elements of $\mathbf{F}_k$ which are associated with the r,s plane.

$$f_{rr}^{(k)} = f_{rr}^{(k-1)} \cos^2\alpha + 2\, f_{rs}^{(k-1)} \cos\alpha \sin\alpha + f_{ss}^{(k-1)} \sin^2\alpha \qquad [147]$$

$$f_{ss}^{(k)} = f_{rr}^{(k-1)} \sin^2\alpha - 2 f_{rs}^{(k-1)} \cos\alpha \sin\alpha + f_{ss}^{(k-1)} \cos^2\alpha \qquad [148]$$

$$f_{rs}^{(k)} = [f_{ss}^{(k-1)} - f_{rr}^{(k-1)}] \cos\alpha \sin\alpha + f_{rs}^{(k-1)}] [\cos^2\alpha - \sin^2\alpha] \qquad [149]$$

where

$$f_{sr}^{(k)} = f_{rs}^{(k)} \qquad [149a]$$

As noted above the purpose of diagonlization is to eliminate the off–diagonal elements so that the last equation must be set equal to zero

$$[f_{ss}^{(k-1)} - f_{rr}^{(k-1)}] \cos\alpha \sin\alpha + f_{rs}^{(k-1)} [\cos^2\alpha - \sin^2\alpha] = 0 \qquad [150]$$

By noting the following trigonometric relationships:

$$\cos\alpha \sin\alpha = (1/2) \sin 2\alpha$$
$$\cos^2\alpha \sin^2\alpha = \cos 2\alpha \qquad [151]$$

we obtain the following simplified expression

$$-(1/2)\,[f_{rr}^{(k-1)} - f_{ss}^{(k-1)}] \sin 2\alpha + f_{rs}^{(k-1)} \,.\, \cos 2\alpha = 0 \qquad [152]$$

or combining the appropriate terms,

$$\tan 2\alpha = \frac{2f_{rs}^{(k-1)}}{f_{rr}^{(k-1)} - f_{ss}^{(k-1)}} \qquad [153]$$

It is customary to choose α between the range of $\pm$ 45° since in the limiting cases: $2\alpha = \pm 90°$. The 45° limit is therefore appropriate whenever the denominator becomes zero $\left[\text{i.e. } f_{rr}^{(k-1)} = f_{ss}^{(k-1)}. \right]$ The sign of the angle is always the same as the sign of the numerator. The numerator will not be zero unless the diagonalization is complete.

The final orthogonal matrix **U** must incorporate all the rotational matrices $\mathbf{U}_k$ used in the process of diagonalization.

Since:

$$
\begin{aligned}
F_1 &= U_1^{\dagger} F_o U_1 \\
F_2 &= U_2^{\dagger} F_1 U_2 &&= U_2^{\dagger} U_1^{\dagger} F_o U_1 U_2 \\
&\vdots &&\vdots \\
F_n &= U_n^{\dagger} F_{n-1} U_n &&= U_n^{\dagger} \ldots\ldots U_2^{\dagger} U_1^{\dagger} F_o U_1 U_2 \ldots U_n
\end{aligned} \quad [154]
$$

It is evident that the final **U** is the matrix product of all U_k.

$$U = U_1 U_2 \ \ldots \ U_n \quad [155]$$

In practice each U_k matrix is multiplied by the previous products of U_k so that one obtains U_1, (U_1U_2), $(U_1\ U_2\ U_3)$ and so on, successively.

2.9 MULLIKEN'S POPULATION ANALYSIS OR MOLECULAR CHARGE DISTRIBUTION

Molecular orbitals are always double occupied in a closed shell system but atomic orbitals may have fractional occupancy. For this reason it is more appropriate to refer to the "population" rather than the occupancy of a set of orbitals.

Starting with a two–electron homonuclear diatomic example (such as H_2) in which only a pair of AO are used, the normalized MO may be written as a linear combination of unnormalized AOs n_1 and n_2

$$(\phi_1 \phi_2) = (n_1 n_2) \begin{bmatrix} C_{11} & C_{12} \\ C_{21} & C_{22} \end{bmatrix} \quad [156]$$

The population of the i–th MO will be N_i. Here for a single electron pair such as that may be the case for $He - H^+$, $N_1 = 2$ and $N_2 = 0$.

$$\phi_1 = \eta_1 C_{11} + \eta_2 C_{21} \qquad [157]$$

Squaring and integrating over all space gives, with a normalization to the total number of electrons N_1 rather than unity:

$$N_1 = N_1 \int \phi_1^2 d\tau = N_1 \int (\eta_1 C_{11} + \eta_2 C_{21})^2 d\tau$$

$$= N_1 C_{11}^2 \int \eta_1^2 d\tau + 2N_1 C_{11} C_{21} \int \eta_1 \eta_2 d\tau + N_1 C_{21}^2 \int \eta_2^2 d\tau$$

$$= N_1 C_{11}^2 + 2N_1 C_{11} C_{21} S_{12} + N_1 C_{21}^2 \qquad [158]$$

where S_{12} is the overlap integral $\int \eta_1 \eta_2 d\tau = \langle \eta_1 | \eta_2 \rangle$

Rewriting [158] by making use of the definition of the density matrix (eq. [44]) gives

$$N_1 = N_1 \rho_{11} + 2N_1 \rho_{12} S_{12} + N_1 \rho_{22} \qquad [159]$$

The populations $N_1\rho_{11}$ and $N_1\rho_{22}$ are called the atomic populations, and $2N_1\rho_{12}S_{12}$ is called the overlap population. Thus population matrix **P** may then be defined as

$$\mathbf{P} = N_1 \begin{bmatrix} \rho_{11} S_{11} & \rho_{12} S_{12} \\ \rho_{21} S_{21} & \rho_{22} S_{22} \end{bmatrix} \qquad [160]$$

Note that in the case of H_2: $S_{11} = S_{22} = 1$, $\rho_{11} = \rho_{22}$ as well as $\rho_{12} = \rho_{21}$ and $S_{12} = S_{21}$ and hence **P** is symmetric.

The general matrix element of the **P** matrix may be defined as

$$P = N_1 \begin{bmatrix} S_{11}\ S_{11} & S_{12}\ S_{12} \\ S_{12}\ S_{12} & S_{11}\ S_{11} \end{bmatrix} \qquad [160a]$$

$$P_{ij} = N_1 \rho_{ij} S_{ij} \qquad [161]$$

and from [159]

$$\sum_i \sum_j P_{ij} = N_1 \qquad [162]$$

In general, the MOs are expressed as a linear combination of many AOs:

$$\phi_i = \sum_j \eta_j C_{ji} \qquad [163]$$

$$\phi_i^2 = (\sum \eta_j C_{ji})^2 = (\sum_j \eta_j C_{ji})(\sum_k \eta_k C_{ki}) = \sum_j \sum_k \eta_j \eta_k C_{ji} C_{ki} \qquad [164]$$

Integrating over all space, the orbital population N_i is

$$N_i = N_i \int \phi_i^2 \, d\tau = N_i \sum_j \sum_k C_{ji} C_{ki} \int \eta_j \eta_k d\tau \qquad [165]$$

$$\sum_j \sum_k N_i C_{ji} C_{ki} S_{jk}$$

where S_{jk} is the overlap integral: $\int \eta_j \eta_k d\tau$. Summing overall occupied MOs gives

$$N = \sum_i N_i = \sum_i N_i \sum_j \sum_k C_{ji} C_{ki} S_{jk} \qquad [166]$$

$$= \sum_j \sum_k S_{jk} \quad (\sum_i N_i C_{ji} C_{ki})$$

where N is the total number of electrons. In the general case, the orbital occupancy N_i would be incorporated into the density matrix, but in a closed shell system N_i = 2 for all the occupied MOs, so a factor of 2 may be brought outside the summation.

$$\rho_{jk} = \sum^{\text{Occ MO}} C_{ji} C_{ik}^{\dagger} = \sum^{\text{Occ MO}} C_{ji} C_{ki}^{*} = \sum C_{ji} C_{ki} = \rho_{jk} \qquad [167]$$

($C_{ki}^{*} = C_{ki}$ when MO are real). Then equation [166] becomes

$$N = 2 \sum_j \sum_k \rho_{jk} S_{jk} \qquad [168]$$

A population matrix **P** may be defined by

$$P_{ij} = 2\rho_{ij} S_{ij} \qquad [169]$$

$$\mathbf{P} = 2\begin{bmatrix} \rho_{11}S_{11} & \rho_{12}S_{12} & \cdots & \rho_{1n}S_{1n} \\ \rho_{21}S_{21} & \rho_{22}S_{22} & \cdots & \rho_{2n}S_{2n} \\ \vdots & \vdots & \vdots\vdots\vdots & \vdots \\ \rho_{21}S_{n1} & \rho_{n2}S_{n2} & \cdots & \rho_{nn}S_{nn} \end{bmatrix} \qquad [170]$$

The diagonal elements P_{ii} ($=2\rho_{ii}$, since $S_{ii} = 1$) represent the "atomic charge" (in units of number of electrons) and the off diagonal elements P_{ij} called overlap populations are vaguely related to the simple idea of "bond–order". Note that $P_{ij} = P_{ji}$ since both ρ and $\mathbf{S}$ are symmetric, and that from [168] and [169]

$$\sum_{i}\sum_{j} P_{ij} = N \qquad [171]$$

Thus the final result of the Mulliken Population Analysis is

$$P_{ij} = 2\rho_{ij}S_{ij} \qquad [172]$$

The $\mathbf{P}$ matrix provides an orbital population analysis. The summation of all P_{ij} elements associated with atomic orbital i gives the gross orbital charge GOC_i of AO_i,:

$$GOC_i = P_{ii} + (1/2)\sum_{j\neq i} P_{ij} + (1/2)\sum_{j\neq i} P_{ij}$$

$$= P_{ii} + (1/2)\sum_{j\neq i} (P_{ij} + P_{ji}) \qquad [173]$$

$$= P_{ii} + \sum_{j\neq i} P_{ij} = \sum_{j} P_{ij}$$

since $P_{ij} = P_{ji}$. The factor 1/2 is introduced because only half the overlap density P_{ij} is counted with orbital i. The halving of the overlap population is somewhat arbitrary. Of course, the summation of all the P_{ij} elements associated with a given pair of atoms A and B reduces the "orbital by orbital" population matrix **P** to an "atom by atom" population matrix **R**

$$R_{AB} = \sum^{AOonA} \sum^{AOonB} P_{ij} \qquad [174]$$

The total number of electrons associated with atom A, N_A, is given by

$$N_A = R_{AA} + (1/2) \sum_{\substack{B \\ B \neq A}}^{\substack{all \\ atoms}} R_{BA}$$

$$= R_{AA} + (1/2) \sum_{\substack{B \\ B \neq A}}^{\substack{all \\ atoms}} (R_{AB} + R_{BA}) \qquad [175]$$

$$= R_{AA} + \sum_{\substack{B \\ B \neq A}}^{\substack{all \\ atoms}} R_{AB} = \sum_{B}^{\substack{all \\ atoms}} R_{AB}$$

since $R_{AB} = R_{BA}$. Once again a factor of 1/2 is needed since the electrons shared between A and B (measured by R_{AB}) and only a half is counted with A. The net charge Q_A associated with A is given by:

$$Q_A = Z_A - N_A \qquad [176]$$

where Z_A is the nuclear charge of atom A.

Note that this definition of the populations P_{ij}, R_{AB} and R_A has the property

$$N = \sum_{i,j} P_{ij} = \sum_{A,B} R_{AB} = \sum_{A} R_A \qquad [177]$$

Although population analysis is very popular since it preserves many chemical concepts, it should be noted that it is an arbitrary procedure. Any non–arbitrary method of dividing a molecule into atomic fragments is much more involved.

Mulliken first proposed his anlysis in 1955,[R. S. Mulliken, J. Chem. Phys. 23, 1833].

3. MOLECULAR INTEGRALS IN TERMS OF GAUSSIANS BASIS SETS

The Gaussian function has the explicit analytic form with x being the independent variable and α being a variable parameter that we shall call the orbital exponent:

$$f(x) = e^{-\alpha x^2} \qquad [178]$$

When we are constructing atomic basis sets for molecular calculations the independent variable becomes r, the distance measured from the atomic nucleus:

$$f(r) = e^{-\alpha r^2} \qquad [179]$$

and by the three dimensional Pythagorean relationship

$$r^2 = x^2 + y^2 + z^2 \tag{180}$$

if the atomic nucleus is located at point **A** some distance (r_A) away from the origin then the independent variable must also be shifted appropriately:

$$r_A{}^2 = (r = r_A)^2 = (x - x_A)^2 + (y - y_A)^2 + (z - z_A)^2 \tag{181}$$

The Cartesian Gaussians, that are the usual Gaussians, used as AO basis sets are defined as

$$\eta_A = (x - x_A)^k\,(y - y_A)^l\,(z - z_A)^m\, e^{\alpha r_A^2} \tag{182a}$$

or in its normalized form:

$$\eta_A = N\,(x - x_A)^k\,(y - y_A)^l\,(z - z_A)^m\, e^{-\alpha r_A^2} \tag{182b}$$

Some graphical illustration has already been given in Figure 3 for $k = l = m = 0$. The traditional s,p,d,f, etc functions can be generated from equation [182] as follows:

S–type: $k + l + m = 0$

$$s:\quad \eta = e^{-r_A{}^2} \qquad (k{=}l{=}m{=}0) \tag{183}$$

P–type: k + l + m = 1

$$p_x:\ \eta = (x = x_A)e^{-\alpha r_A^2} \quad k=1,\ l=m=0 \tag{184a}$$

$$p_y:\ \eta = (y - y_A)e^{-\alpha r_A^2} \quad l=1,\ k=m=0 \tag{184b}$$

$$p_z:\ \eta = (z - z_A)e^{-\alpha r_A^2} \quad m=1,\ k=l=0 \tag{184c}$$

d–type: k + l + m = 2

$$d_{xy}:\ \eta = (x - x_A)(y - y_A)e^{-\alpha r_A^2} \quad k=l=1,\ m=0 \tag{185a}$$

$$d_{yz}:\ \eta = (x - x_A)(z - z_A)e^{-\alpha r_A^2} \quad k=m=1,\ l=0 \tag{185b}$$

$$d_{yz}:\ \eta = (y - y_A)(z - z_A)e^{-\alpha r_A^2} \quad l=m=1,\ k=0 \tag{185c}$$

The following three cartesian type functions may look like a set of somewhat unusual d–AO:

$$d_{x}^{2}:\ \eta = (x - x_A)^2 e^{-\alpha r_A^2} \tag{186a}$$

$$d_{y}^{2}:\ \eta = (y - y_A)^2 e^{-\alpha r_A^2} \tag{186b}$$

$$d_{z}^{2}: \ \eta = (z - z_A)^2 e^{-\alpha r_A^2} \quad [186c]$$

but these may lead to 3 linear combinations two of which are conventional d–type AO:

$$d_{x^2-y^2} = d_{x^2} - d_{y^2} \quad [187a]$$

$$d_{3z^2-x^2-y^2} = 3d_{z}^{2} - d_{x^2} - d_{y^2} \quad [187b]$$

and the third one is of an s–type (i.e. 3s) Gaussian

$$d_{3s} = d_{x^2} + d_{y^2} + d_{z^2} \quad [187c]$$

Similar relationships can be given for f–type Gaussians.

3.1 Basic Theorems

Theorem. The definte integral, from $-\infty$ to $+\infty$, of the product of a Gaussian and any odd polynomial will vanish. If the odd polynomial is x^{2n+1} then we may write:

$$\int_{-\infty}^{+\infty} x^{2n+1} \ e^{-ax^2} \ dx = 0 \quad [188]$$

This is illustrated schematicaly for n=0 and n=1 in Figure 10.

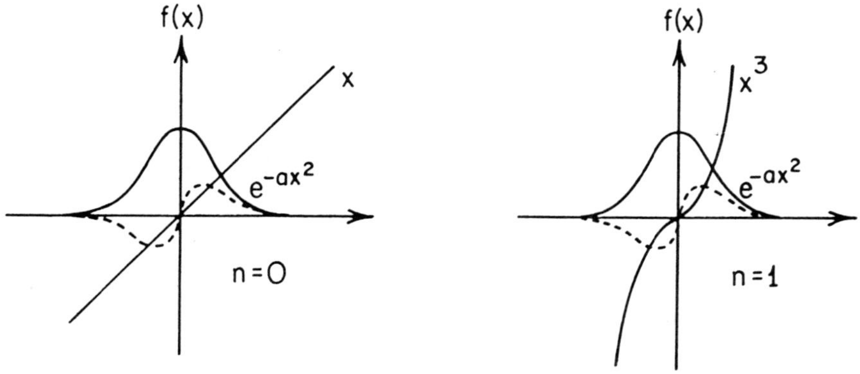

Figure 10 Functional behaviour of $f(x) = x^{2n+1}\ e^{-ax^2}$ (broken line) for n=o and n=1.

The definite integral, from $-\infty$ to $+\infty$, of the product of a Gaussian and any even polynomial is a finite value. If the even polynomial is x^{2n} as shown in Figure 11,

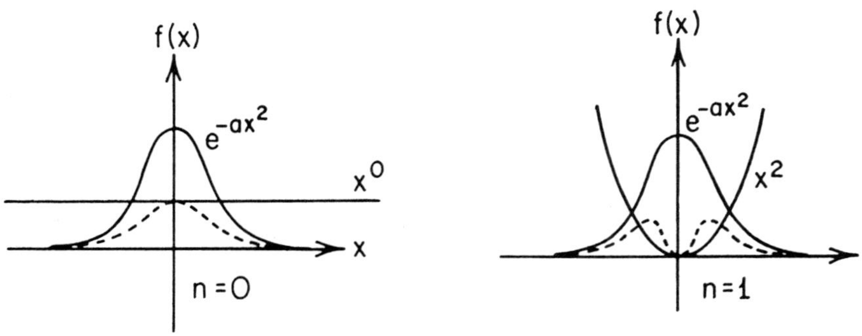

Figure 11 Functional behaviour of $f(x) = x^{2n}e^{-ax^2}$ (broken line) for n=o and n=1.

then the integral is of the following form

$$\int_{-\infty}^{\infty} x^{2n} e^{-ax^2}\, dx = \frac{1.3.5 \; . \; . \; . \; (2n-1)}{(2a)^n} \sqrt{\frac{\pi}{a}} \qquad [189]$$

For n=0 the above equation simplifies to:

$$\int_{-\infty}^{\infty} e^{-ax^2}\, dx = \sqrt{\frac{\pi}{a}} \qquad [190]$$

Proof The following proof is offered for the validity of equation [190]

let

$$I = \int_{-\infty}^{\infty} e^{-ax^2}\, dx \qquad [191]$$

$$I^2 = \int_{-\infty}^{\infty} e^{-ax^2}\, dx \int_{-\infty}^{\infty} e^{-ay^2}\, dy$$

$$= \int_{-\infty}^{\infty}\int_{-\infty}^{\infty} e^{-a(x^2 + y^2)}\, dxdy \qquad [192]$$

Substitute:

$$\left.\begin{aligned} x &= r\ \cos\,\vartheta \\ y &= r\ \sin\,\vartheta \end{aligned}\right\} \qquad [193]$$

which gives new integration limits:

$$\begin{aligned} r&: \quad 0 \longrightarrow \infty \\ \vartheta&: \quad 0 \longrightarrow 2\pi \end{aligned} \qquad [194]$$

Note that the above substitution leads to the following eqivalence:

$$x^2 + y^2 = r^2 \cos^2\vartheta + r^2 \sin^2\vartheta = r^2(\cos^2\vartheta + \sin^2\vartheta) = r^2 \qquad [195]$$

where r is the distance from the origin in a plane.

The Jacobian of the transformation is

$$J \equiv J^{xy}_{r\theta} = \frac{\partial\,(x,y)}{\partial\,(r,\theta)} = \begin{vmatrix} \frac{\partial x}{\partial r} & \frac{\partial x}{\partial \theta} \\ \frac{\partial y}{\partial r} & \frac{\partial y}{\partial \theta} \end{vmatrix} = \begin{vmatrix} \cos\theta & -r\sin\theta \\ \sin\theta & r\cos\theta \end{vmatrix}$$

$$= r\cos^2\theta - (-r\sin^2\theta) = r(\cos^2\theta + \sin^2\theta) = r \qquad [196]$$

$$I^2 = \int_0^\infty\int_0^{2\pi} e^{-ar^2}\, J^{xy}_{r\theta}\, d\theta dr = \int_0^\infty\int_0^{2\pi} e^{-ar^2} .r.d\theta dr =$$

$$\int_0^\infty d\theta \int_0^\infty e^{-ar^2}\, rdr. = \theta\Big|_0^{2\pi} \int_0^\infty e^{-ar^2}\, rdr = 2\pi \int_0^\infty e^{-ar^2} rdr \qquad [197]$$

Now let us substitute

$$s = r^2 \qquad [198]$$

$$ds = 2rdr \qquad [199a]$$

or

$$rdr = \tfrac{1}{2}\, ds \qquad [199b]$$

$$I^2 = 2\pi \int_0^\infty (\tfrac{1}{2})e^{-as}\, ds = \pi \int_0^\infty e^{-as}\, ds =$$

$$\pi\left[\frac{-e^{-as}}{a}\right]\Bigg|_{0}^{\infty} = -\frac{\pi}{a}\left[-e^{-as}\right]\Bigg|_{0}^{\infty} = -\frac{\pi}{a}(0-1) \qquad [200]$$

thus

$$I^2 = \frac{\pi}{a} \qquad [201]$$

therefore

$$I = \sqrt{\frac{\pi}{a}} \qquad [202]$$

which proves the theorem!

Theorem: The product of two Gaussians is also a Gaussian.

$$\eta_P = \eta_A \ \eta_B \qquad [203]$$

The product Gaussian (η_P) is expected to be located somewhere between A and B. Figure 12..

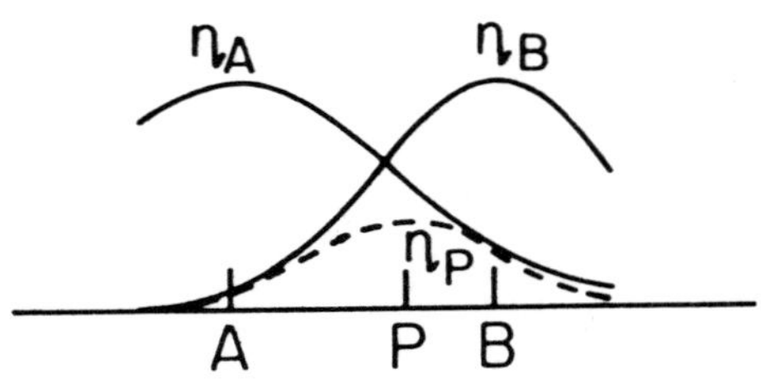

Figure 12. An illustration of the theorem that the product of two Gaussians is also a Gaussian.

Proof:

Let
$$\eta_A = e^{-\alpha_A(r - r_A)^2} \quad [204]$$

$$\eta_B = e^{-\alpha_B(r - r_B)^2} \quad [205]$$

be two Gaussians centered at r_A and r_B respectively.

Let
$$\eta_P = N_P e^{-\alpha_p(r - r_P)^2} \quad [206]$$

be the resultant product Gaussian. We desire:

$$\eta_P = \eta_A \eta_B \quad [203]$$

therefore

$$N_P e^{-\alpha_P(r - r_P)^2} = e^{-\alpha_A(r - r_A)^2} e^{-\alpha_B(r - r_B)^2} \quad [207]$$

Take the natural log of both sides:

$$\ln N_P - \alpha_P(r - r_P)^2 = -\alpha_A(r - r_A)^2 - \alpha_B(r - r_B)^2 \quad [208]$$

Expanding the above equation we obtain:

$$\ln N_P - \alpha_P(r^2 - 2rr_P + r_P^{\,2}) =$$

$$= - \alpha_A(r^2 - 2rr_A + r_A^{\,2}) - \alpha_B(r^2 - 2rr_B + r_B^{\,2}) \qquad [209]$$

Comparing terms in r^2:

$$-\alpha_p = -\alpha_A - \alpha_B \qquad [210a]$$

therefore

$$\boxed{\alpha_p = \alpha_A + \alpha_B} \qquad [210b]$$

Comparing terms in r:

$$2\alpha_P r_P = 2\alpha_A r_A + 2\alpha_B r_B \qquad [211]$$

or expressing the location on of the product Gaussian

$$r_P = \frac{\alpha_A r_A + \alpha_B r_B}{\alpha_P} \qquad [212a]$$

or in after substituting the value of α_P [210b]:

$$\boxed{r_P = \frac{\alpha_A r_A + \alpha_B r_B}{\alpha_A + \alpha_B}} \qquad [212b]$$

Since all r are vectors, therefore r_P has the following components

$$X_p = \frac{\alpha_A X_A + \alpha_B X_B}{\alpha_A + \alpha_B} \qquad [213a]$$

$$Y_p = \frac{\alpha_A Y_A + \alpha_B Y_B}{\alpha_A + \alpha_B} \qquad [214b]$$

$$Z_p = \frac{\alpha_A Z_A + \alpha_B Z_B}{\alpha_A + \alpha_B} \qquad [214c]$$

Componing constant terms:

$$\ln N_p - \alpha_p r_p^2 = -\alpha_A r_A^2 - \alpha_B r_B^2 \qquad [215]$$

$$\ln N_p = \alpha_p r_p^2 - \alpha_A r_A^2 - \alpha_B r_B^2 \qquad [216]$$

Substituting equations [210b] and [212b] we obtain:

$$\ln N_p = (\alpha_A + \alpha_B) \frac{\alpha_A^2 r_A^2 + 2\alpha_A \alpha_B r_A r_B + \alpha_B^2 r_B^2}{(\alpha_A + \alpha_B)^2} - \alpha_A r_A^2 - \alpha_B r_B^2$$

$$\frac{\alpha_A^2 r_A^2 + 2\alpha_A \alpha_B r_A r_B + \alpha_B^2 r_B^2 - \alpha_A^2 r_A^2 - \alpha_A \alpha_B r_A^2 - \alpha_A \alpha_B r_B^2 - \alpha_B^2 r_B^2}{(\alpha_A + \alpha_B)} \qquad [217]$$

$$\ln N_p = \frac{-\alpha_A \alpha_B (r_A^2 - 2 r_A r_B + r_B^2)}{\alpha_B + \alpha_B} = \frac{\alpha_A \alpha_B (r_A - r_B)^2}{\alpha_A + \alpha_B} \qquad [218]$$

$$N_P = e^{\frac{\alpha_A \alpha_B \; r^2_{AB}}{\alpha_B + \alpha_B}} \qquad [119]$$

where $r^2_{AB} = (r_A - r_B)^2 = (X_A - X_B)^2 + (Y_A - Y_B)^2 + (Z_A - Z_B)^2$ [220]

This completes the proof of the theorem.

3.2 Normalization of the basis set

Normalization helps to reduce numerical error by ensuring that all the integrals have similar magnitudes ("scaling"). The various types of Gaussians (s,p,d,etc) have different expressions for their norms: $< >$.

Overlap Integral of the type <s|s>:

$$S_{ss} = \int_0^\infty \eta_1^* \eta_1 dr = \int_0^\infty \left[N\,e^{-\alpha(r - r_A)^2}\right]\left[N\,e^{-\alpha(r - r_A)^2}\right] dr$$

$$= N^2 \left[\int_0^\infty e^{-2\alpha(x - x_A)^2} dx\right]\left[\int_0^\infty e^{-2\alpha(y - y_A)^2} dy\right]\left[\int_0^\infty e^{-2\alpha(z - z_A)^2} dz\right]$$

$$= N^2 \sqrt{\frac{\pi}{2\alpha}} \sqrt{\frac{\pi}{2\alpha}} \sqrt{\frac{\pi}{2\alpha}}$$

$$= N^2 \left[\frac{\pi}{2\alpha}\right]^{3/2} \qquad [221]$$

To normalize η_1, we want $S = 1$, so

$$N^2 \left[\frac{\pi}{2\alpha}\right]^{3/2} = 1 \qquad [222a]$$

and therefore,

$$N^2 = \left[\frac{2\alpha}{\pi}\right]^{3/2} \qquad [222b]$$

leading to

$$N = \left[\frac{2\alpha}{\pi}\right]^{3/4} \qquad [223]$$

$$\eta_s = \left[\frac{2\alpha}{\pi}\right]^{3/4} e^{-\alpha(r - r_A)^2} \qquad [224]$$

Overlap Integrals of the type <p|p>, <d|d>, <f|f> (diagonal components)

$$\eta = N\,(x - x_A)^k\, e^{-\alpha(r - r_A)^2} \qquad [225]$$

where

$$k = 1 \text{ for } p_x$$

$$k = 2 \text{ for } d_{xx}$$

$$k = 2 \text{ for } f_{xxx} \qquad [226]$$

$$S = \int_{-\infty}^{\infty} \eta^* \eta \, dr =$$

$$= N^2 \left[\int_{-\infty}^{\infty} (x - x_A)^{2k} e^{-2\alpha(x - x_A)^2} dx \right] \left[\int_{0}^{\infty} e^{-2\alpha(y - y_A)^2} dy \right] \left[\int_{0}^{\infty} e^{-2\alpha(z - z_A)^2} dz \right]$$

$$= N^2 \left[\frac{1.3.5 \; \ldots\ldots \; (2k-1)}{(2.2\alpha)^k} \sqrt{\frac{\pi}{2\alpha}} \right] \left[\sqrt{\frac{\pi}{2\alpha}} \right] \left[\sqrt{\frac{\pi}{2\alpha}} \right]$$

$$= N^2 \; \frac{1.3.5 \; \ldots\ldots \; (2k-1)}{(4\alpha)^k} \left[\frac{\pi}{2\alpha} \right]^{3/2} \qquad [227]$$

To normalize, set S = 1, leading to

$$N = \left[\frac{(4\alpha)^k \left[\frac{2\alpha}{\pi} \right]^{3/2}}{1.3.5. \; \ldots \; . \; (2k-1)} \right]^{1/2} \qquad [228a]$$

or

$$N = \frac{2^k \; \alpha^{k/2}}{\sqrt{1.3.5. \; \ldots \; .(2k-1)}} \left[\frac{2\alpha}{\pi} \right]^{3/4} \qquad [228b]$$

For $<p_x|p_x>$ where $k = 1$

$$N_{p_x} = 2 \; \alpha^{1/2} \left[\frac{2\alpha}{\pi} \right]^{3/4} \qquad [229]$$

For $<d_{xx}|d_{xx}>$ where $k = 2$

$$N_{d_{xx}} = \frac{4\alpha}{\sqrt{3}} \left[\frac{2\alpha}{\pi}\right]^{3/4} \tag{230}$$

For $<f_{xxx}|f_{xxx}>$ where $k = 3$

$$N_{f_{xxx}} = \frac{8\alpha^{3/2}}{\sqrt{15}} \left[\frac{2\alpha}{\pi}\right]^{3/4} \tag{231}$$

The same N are obtained for $<p_y|p_y>$, $<p_z|p_z>$, $<d_{yy}|d_{yy}>$, $d_{zz}|d_{zz}>$, $<f_{yyy}|f_{yyy}>$ and $<f_{zzz}|f_{zzz}>$.

Overlap Integral of the type <d|d>, <f|f> (off–diagonal components)

$$\eta = N\,(x - x_A)^k\,(y - y_A)^l e^{-\alpha(r - r_A)^2} \qquad [232]$$

$$S = \int_{-\infty}^{\infty} \eta^* \eta \, dr =$$

$$= N^2 \left[\int_{-\infty}^{\infty} (x - x_A)^{2k} e^{-2\alpha(r - r_A)^2} dx\right] \left[\int_{-\infty}^{\infty} (y - y_A)^{2l} e^{-2\alpha(r - r_A)^2} dy\right]$$

$$\left[\int_{0}^{\infty} e^{-2\alpha(r - r_A)^2} dz\right] = 2\left[\frac{1.3.5\ \ldots\ldots\ (2k-1)}{(2.2\alpha)^k}\sqrt{\frac{\pi}{2\alpha}}\right]$$

$$= \left[\frac{1.3.5\ \ldots\ldots\ (2l-1)}{(2.2\alpha)^l}\sqrt{\frac{\pi}{2\alpha}}\right]\left[\sqrt{\frac{\pi}{2\alpha}}\right] \qquad [233]$$

$$= N^2\,\frac{[1.3.5.\ \ldots\ .\ (2k-1)]\ [1.2.5.\ \ldots\ .\ (2l-1)]}{(4\alpha)^{k+l}}\left[\frac{\pi}{2\alpha}\right]^{3/2} \qquad [234]$$

Therefore to to normalize set S = 1:

$$N = \frac{2^{(k+l)}\quad \alpha^{(k+l)/2}}{\sqrt{[1.\,3.5.\ \ldots\ .\ (2k-1)]\ [1.3.5.\ \ldots\ .\ (2l-1)]}}\left[\frac{2\alpha}{\pi}\right]^{3/4} \qquad [235]$$

For $<d_{xy}|d_{xy}>$ where $k = 1, l = 1$

$$N_{d_{xy}} = 4\alpha \left[\frac{2\alpha}{\pi}\right]^{3/4} \quad [236]$$

$$= \sqrt{3}\, N_{d_{xx}} \quad [237]$$

For $<f_{xxy}|f_{xxy}>$ where $k = 2, l = 1$

$$N_{f_{xxy}} = \frac{8\alpha^{3/2}}{\sqrt{3}} \left[\frac{2\alpha}{\pi}\right]^{3/4} \quad [238]$$

$$= \sqrt{5}\ N_{f_{xxx}} \quad [239]$$

Overlap Integral of the type $<f_{xyz}|f_{xyz}>$

$$\eta = N\,(x = x_A)^k\,(y - y_A)^l\,(z - z_A)^m\, e^{-\alpha(r - r_A)^2} \quad [240]$$

$$S = \int_{-\infty}^{\infty} \eta^* \eta\, dr =$$

$$N^2 \left[\int_{-\infty}^{\infty} (x - x_A)^{2k} e^{-2\alpha(r - r_A)^2} dx\right] \left[\int_{0}^{\infty} (y - y_A)^{2l} e^{-2\alpha(y - y_A)^2} dy\right]$$

$$\left[\int_{0}^{\infty} (z - z_A)^{2m} e^{-2\alpha(z - z_A)^2} dz\right]$$

$$= N^2 \frac{[1.3.5. \ldots .(2k-1)]\,[1.3.5. \ldots .(2l-1)]\,[1.3.5. \ldots .(2m-1)]}{(2.2\alpha)^{k+l+m}} \left[\frac{\pi}{2\alpha}\right]^{3/4} \quad [241]$$

To normalize set S = 1:

$$N = \frac{2^{(k+l+m)}\ \alpha^{(k+l+m)/2}}{\sqrt{[1.3.5. \ldots .(2k-1)][1.3.5. \ldots .(2l-1)][1.3.5. \ldots (2m-1)]}} \left[\frac{2\alpha}{\pi}\right]^{3/4} \quad [242]$$

For $<f_{xyz}|f_{xyz}>$ where $k = 1, l = 1, m = 1$

$$N_{f_{xyz}} = 8\alpha^{3/2} \left[\frac{2\alpha}{\pi}\right]^{3/4} = \sqrt{15}\ N_{f_{xxx}} \quad [243]$$

Note: different components of a "shell" (d,f,.. may have different normalization constants these must be taken into account in all integral and integral derivative packages! All is OK for s and p shells though, because there are <u>no</u> "cross–terms" like d_{xy}, d_{xz}. f_{xxy} and f_{xyz}.

3.3 DIFFERENTIAL OVERLAPS THE $\langle \eta_{s_A} | \eta_{s_B} \rangle$ CASE:

Here η_1 and η_2 are defined as follows:

$$\eta_1 = N_1 \, e^{-\alpha_1 (r - r_A)^2} \qquad [244]$$

$$\eta_2 = N_2 \, e^{-\alpha_2 (r - r_B)^2} \qquad [245]$$

$$S_{12} = \int_{-\infty}^{\infty} \eta_1^{*} \eta_2 \, dr =$$

$$= N_1 N_2 \int_{-\infty}^{\infty} e^{-\alpha_1 (r - r_A)^2} \, e^{-\alpha_2 (r - r_B)^2} \, dr$$

$$= N_1 N_2 N_P \int_{-\infty}^{\infty} e^{-\alpha_P (r - r_P)^2} \, dr \qquad [246]$$

Substituting the results of equation [202] we obtain

$$S_{12} = N_1 N_2 N_P \left[\frac{\pi}{\alpha_P} \right]^{3/2} \qquad [247]$$

From [223] we substitute for N_1 and N_2 and from [210b] and [219] we obtain:

$$S_{12} = \left[\frac{2\alpha 1}{\pi}\right]^{3/4} \left[\frac{2\alpha 2}{\pi}\right]^{3/4} e^{-\frac{\alpha_1 \alpha_2 r_{AB}^2}{\alpha_1+\alpha_2}} \left[\frac{\pi}{\alpha_1+\alpha_2}\right]^{3/2} \qquad [248]$$

or

$$S_{12} = \frac{\left[4\alpha_1\alpha_2\right]^{3/2}}{(\alpha_1 + \alpha_2)^{3/2}} e^{-\frac{\alpha_1 \alpha_2 r_{AB}^2}{\alpha_1+\alpha_2}} \qquad [249]$$

The $<\eta_{p_{x_B}} | \eta_{s_B}>$ case:

$$\eta_1 = N_1(x - x_A)\, e^{-\alpha_1(r - r_A)^2} \qquad [250]$$

$$\eta_2 = N_2 e^{-\alpha_2(r - r_B)^2} \qquad [251]$$

$$S_{12} = \int_{-\infty}^{\infty} \eta_1^{*} \eta_2 \, dr =$$

$$= N_1 N_2 \int_{-\infty}^{\infty} (X - X_A)\, e^{-\alpha_1(r - r_A)^2} e^{-\alpha_2(r - r_B)^2} dr \qquad [252]$$

Replace product of 2 Gaussians by the product–Gaussian using [206], and substitute [210b], [212b] and [219] as well as replacing $(x - x_A)$ by $[(x - x_P) + (x_P - x_A)]$:

we obtain

$$S_{12} = N_1 N_2 N_P \int_{-\infty}^{\infty} [(x - x_P) + (x_P - x_A)]\, e^{-\alpha_p (x - x_p)^2}\, dx$$

$$\int_{-\infty}^{\infty} e^{-\alpha_P (y - y_A)^2}\, dy \int_{-\infty}^{\infty} e^{-\alpha_P (z - z_A)^2}\, dz =$$

$$N_1 N_2 N_P \left[\int_{-\infty}^{\infty} (x - x_P)\, e^{-\alpha_P (x - x_P)^2}\, dx + (x_P - x_A) \int_{-\infty}^{\infty} e^{-\alpha_P (x - x_P)^2}\, dx \right]$$

$$\left[\frac{\pi}{\alpha_P}\right]^{1/2} \left[\frac{\pi}{\alpha_P}\right]^{1/2} \qquad [253]$$

but the first $\int dx$ is 0 because $(x - x_P)$ is an odd function of x, leaving

$$S_{12} = N_1 N_2 N_p\, (x_p - x_A) \left[\frac{\pi}{\alpha_P}\right]^{1/2} \left[\frac{\pi}{\alpha_P}\right]^{1/2} \left[\frac{\pi}{\alpha_P}\right]^{1/2} \qquad [254]$$

Substitute for N_1, N_2, N_p, and write

$$x_P - x_A = \frac{\alpha_1 x_A + \alpha_2 x_B}{\alpha_1 + \alpha_2} - x_A$$

$$\frac{\alpha_1 x_A + \alpha_2 x_B - \alpha_1 x_A + \alpha_2 x_A}{\alpha_1 + \alpha_2} = \frac{\alpha_2 (x_B - x_A)}{\alpha_1 + \alpha_2} \quad [255]$$

then [254] becomes:

$$S_{12} = \left[\frac{2\alpha_1}{\pi}\right]^{3/4} \left[\frac{2\alpha_2}{\pi}\right]^{3/4} \left[e^{-\frac{\alpha_1 \alpha_2 r_{AB}^2}{\alpha_1+\alpha_2}}\right] \frac{\alpha_2 (x_B - x_A)}{\alpha_1 + \alpha_2} \left[\frac{\pi}{\alpha_1+\alpha_2}\right]^{3/2}$$

$$S_{12} = \frac{\left[4\alpha_1\alpha_2\right]^{3/2}}{(\alpha_1 + \alpha_2)^{5/2}} \alpha_2 \, e^{-\frac{\alpha_1 \alpha_2 r^2_{AB}}{\alpha_1+\alpha_2}} [x_B - x_A] \quad [256]$$

Note in [256], if r_{AB} goes to 0, then $S_{12} = 0$ (as expected from atomic orbital overlap). Alternatively, if r_{AB} goes to ∞, then S_{12} goes to 0 because $e^{-r^2_{AB}}$ goes to 0 faster than $(x_B - x_A)$ goes to ∞.

Similar expressions may be found for $<\eta_{p_y}|\eta_s>$ and $<\eta_{p_z}|\eta_s>$, replacing $(x_B - x_A)$ by $(y_B - y_A)$ and $(z_B - z_A)$ respectively. The same general procedure can be followed to drive any desired overlap integral.

More general formulae may be found in J. C. Browne and R. D. Poshusta, J. Chem. Phys., 36, 1933–1937 (1962).

3.4 OVERLAP INTEGRALS OVER CONTRACTED FUNCTIONS

Normally, basis sets are constructed from contracted functions:

$$\eta_1 = (x - x_A)^{k_1} (y - y_A)^{l_1} (z - z_A)^{m_1} \sum_{i=1}^{n_1} d_{1i} N_{1i} e^{-\alpha_{1i}(r - r_A)^2}$$

$$\eta_2 = (x - x_B)^{k_2} (y - y_B)^{l_2} (z - z_B)^{m_2} \sum_{i=1}^{n_2} d_{2j} N_{2j} e^{-\alpha_{2j}(r - r_B)^2}$$

where d_{1i} and d_{2j} are the contraction coefficients.

$$S_{12} = \int_{-\infty}^{\infty} \eta_1^* \eta_2 \, dr =$$

$$= \sum_{i=1}^{\eta_1} \sum_{j=1}^{\eta_2} d_{1i} d_{2j} N_{1i} N_{2j} \int_{-\infty}^{\infty} (x - x_A)^{k_1} (x - x_B)^{k_2}$$

$$(y - y_A)^{\ell_1} (y - y_B)^{\ell_2} (z - z_A)^{m_1} (z - z_B)^{m_2}$$

$$e^{-\alpha_{1i}(r - r_A)^2 - \alpha_{2j}(r - r_B)^2} \, dr$$

$$= \sum_{i=1}^{\eta_1} \sum_{j=1}^{\eta_2} d_{1i} d_{2j} S_{1i,2j} \qquad [257]$$

where $S_{1i,2j}$ is given by [249] for $<S_A|S_B>$ overlap.

3.5 Electron Kinetic Energy Integrals

$$\hat{T} = -1/2 \left[\frac{\partial^2}{\partial x^2} + \frac{\partial^2}{\partial y^2} + \frac{\partial^2}{\partial z^2}\right] = -1/2\nabla^2 \qquad [258]$$

Apply $\hat{T}$ to the s basis function

$$\eta_2 = N_2 e^{-\alpha_2(r - r_B)^2} \qquad [259]$$

$$\hat{T}\,\eta_2 = -1/2 \left[\frac{\partial^2}{\partial x^2} + \frac{\partial^2}{\partial y^2} + \frac{\partial^2}{\partial z^2}\right] N_2 e^{-\alpha_2(x - x_B)^2} e^{-\alpha_2(y - y_A)^2} e^{-\alpha_2(z - z_B)^2}$$

$$= -1/2\, N_2 \left[\frac{\partial^2}{\partial x^2} e^{-\alpha_2(x - x_B)^2} e^{-\alpha_2(y - y_B)^2} e^{-\alpha_2(z - z_B)^2} + \right.$$

$$\left. + \text{ similar term in y} + \text{similar term in z}\right] \qquad [260]$$

$$\hat{T}\,\eta_2 = \frac{N_2}{2} \left[-2\alpha_2 \frac{\partial}{\partial_x} e^{-\alpha_2(y - y_B)^2} e^{-\alpha_2(z - z_B)^2} (x - x_B) e^{-\alpha_2(x - x_B)^2} + \right.$$

$$\left. + \text{ term in y} + \text{term in z}\right] \qquad [261]$$

$$= - N_2\alpha_2 \left\{ e^{-\alpha_2(y - y_B)^2} e^{-\alpha_2(z - z_B)} \left[- 2_{\alpha_2}(x - x_B)^2 e^{-\alpha_2(x - x_B)^2} + \right.\right.$$

$$\left.\left. e^{-\alpha_2(x - x_B)^2} \right] + \text{terms in y} + \text{terms in z} \right\} +$$

$$= N_2\alpha_2 \left[3\, e^{-\alpha_2(r - r_B)^2} - 2\alpha_2(x - x_B)^2\, e^{-\alpha_2(r - r_B)^2} + \right.$$

$$\left. - \text{terms in y} - \text{terms in z} \right] \quad [262]$$

$$\hat{T}\eta_2 = 3\alpha_2\,\eta_2 - 2\alpha_2^{\,2}\,(x - x_B)^2\,\eta_2 - 2\alpha_2^{\,2}\,(y - y_B)^2\,\eta_2$$

$$- 2\alpha_2^{\,2}\,(z - z_B)^2\eta_2 \quad [263]$$

Letting $\eta_1 = N_1\, e^{-\alpha_1(r - r_A)^2}$ [264]

Then

$$T_{12} = \int_{-\infty}^{\infty} \eta_1^{\,*}\,(-1/2\,\nabla^2)\,\eta_2 dr =$$

$$= \int_{-\infty}^{\infty} \eta_1^{\,*} \left[3\alpha_2\eta_2 - 2\alpha_2^{\,2}\,(x - x_B)^2\eta_2 - z\alpha_2^{\,2}\,(y - y_B)^2\eta_2 \right.$$

$$\left. - 2\alpha_2^{\,2}\,(z - z_B)^2\eta_2 \right] dr \quad [265]$$

$$= 3\alpha_2 \int_{-\infty}^{\infty} \eta_1^* \eta_2 dr - 2\alpha_2^2 \left[\int_{-\infty}^{\infty} (x - x_B)^2 \eta_1^* \eta_2 dr + \int_{-\infty}^{\infty} (y - y_B)^2 \eta_1^* \eta_2 dr \right] \quad [266]$$

$$= 3\alpha_2 S_{12} - 2\alpha_2^2 \left[S_{12,(x_2 + 2)} + S_{12,(y_2+2)} + S_{12,(z_2+2)} \right] \quad [267]$$

where $S_{12,(x_2+2)}$ means that the power of $(x - x_B)$ is incremented by 2 before evaluating the overlap integral, and similarly for y and z. Thus the kinetic energy integrals can be written as simple linear combinations of overlap integrals.

4. GRADIENT OPTIMIZATION OF MOLECULAR GEOMETRY

There are three types of internal motions of molecules: stretch, bend and torsion as illustrated by Figure 13. The torsion is a periodic motion even if its periodicity low so that it reapeats itself only after a 360° rotation. The bending motion usually governs a double–well potential. If the bonds that undergo bending motion are attached to N or O then the barrier to inversion is quite low (from a few to a few tens kcal/mole) but if it involves C then the inversion potential is very high as the inversion would pass through a planar carbon.

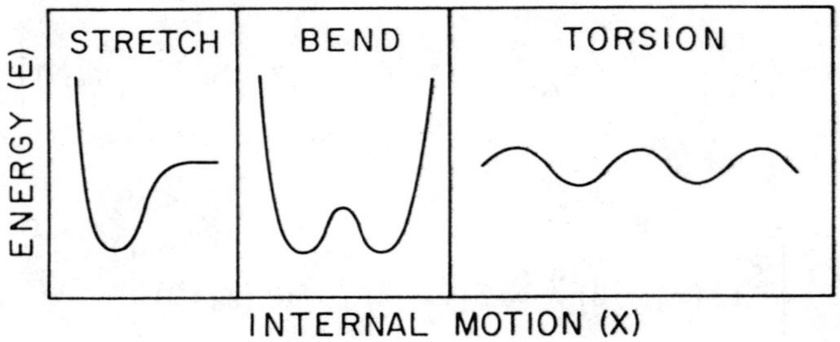

Figure 13. Three types of potentials associated with three internal modes of motion.

If we knew the potential curves that are characteristic to a given molecular system then we could determine the whereabout of the minimum energy points that in fact correspond to the equilibrium geometry. Unfortunately, these potential functions are not only unknown but they change from molecule to molecule and from bond to bond. However, we do know that they may be approximated, near any of their minima by same quadratic potential, as illustrated by Figure 14, since quadratic and true functions osculate at the minimum.

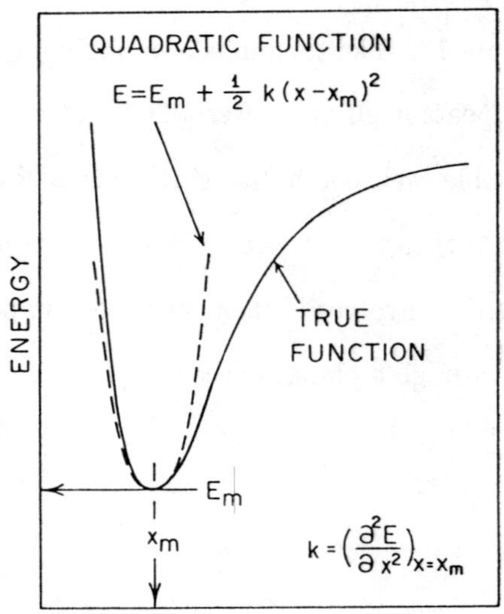

Figure 14. The osculation of the quadratic and the true functions at the minimum as illustrated for a bond stretch.

The quadratic function is the traditional Hooke's law:

$$E = E_m + \frac{1}{2} G (x - x_m)^2 \qquad [268]$$

where G, the second derivative of E with respect to x, is the force constant usually denoted by k

$$\frac{d^2E}{dx^2} = G \equiv k \qquad [269]$$

The minimum energy point is denoted by E_m and x_m in equation [268] as well as in Figure 14.

For a multi–dimensional problem the generalized Hooke's law may be written as follows:

$$E = E_m + + \frac{1}{2} (\underline{x} - \underline{x}^m)^\dagger \; G \; (\underline{x} - \underline{x}^m) \qquad [268a]$$

$$E = E_m + \frac{1}{2}([x_1 - x_1^m], [x_2 - x_2^m], \ldots.) \begin{bmatrix} G_{11} & G_{12} & \cdots \\ & G_{22} & \cdots \\ & \vdots & \ddots \end{bmatrix} \begin{bmatrix} [x_1 - x_1^m] \\ [x_2 - x_2^m] \\ \vdots \\ \vdots \end{bmatrix}$$

[268b]

where $(x - x^m)$ is the displacement vector and G, the Hessian matrix, collects all diagonal and off–diagonal or interaction force constants

$$\begin{bmatrix} \frac{\partial^2 E}{\partial x_1^2} & \frac{\partial^2 E}{\partial x_1 \partial x_2} & \cdots \\ & \frac{\partial^2 E}{\partial x_2^2} & \cdots \\ & \vdots & \cdots \\ & & \cdots \end{bmatrix} = \begin{bmatrix} G_{11} & G_{12} & \cdots \\ & G_{22} & \cdots \\ & & \cdots \\ & & \cdots \end{bmatrix} \qquad [270]$$

For a two dimensional case the situation is illustrated in terms of a schematic energy contour diagram in Figure 15. This figure also shows that the internal coordinates $\{x_i\}$ are not the same as the normal coordinates $\{y_i\}$.

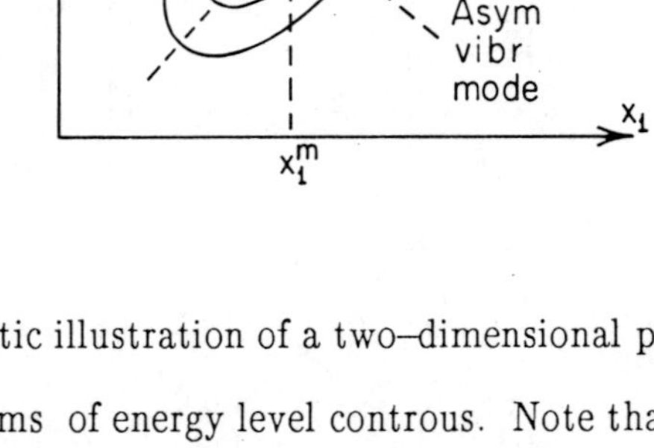

Figure 15. A schematic illustration of a two–dimensional potential energy surface: $E = E(x_1, x_2)$ in terms of energy level controus. Note that the normal coordinates $\{y_i\}$ are different from the internal coordinates $\{x_i\}$

Molecular geometry optimization involves the finding of the minimum energy (E_m) point or in otherwords locating x_1^m, x_2^m, ... This can most effectively be done by evaluating the gradient vector:

$$<g| \equiv g = \left(\frac{\partial E}{\partial x_1}, \frac{\partial E}{\partial x_2}, \ldots\ldots\right) \qquad [271]$$

and searching for the point where the gradient vector is a zero vector

$$<g| = (0, 0, \ldots\ldots) \qquad [272]$$

since at the minimum the gradient vanishes. This is illustrated for the minima and higher order critical points for one– and two–dimensions in Figures 16 and 17 respectively.

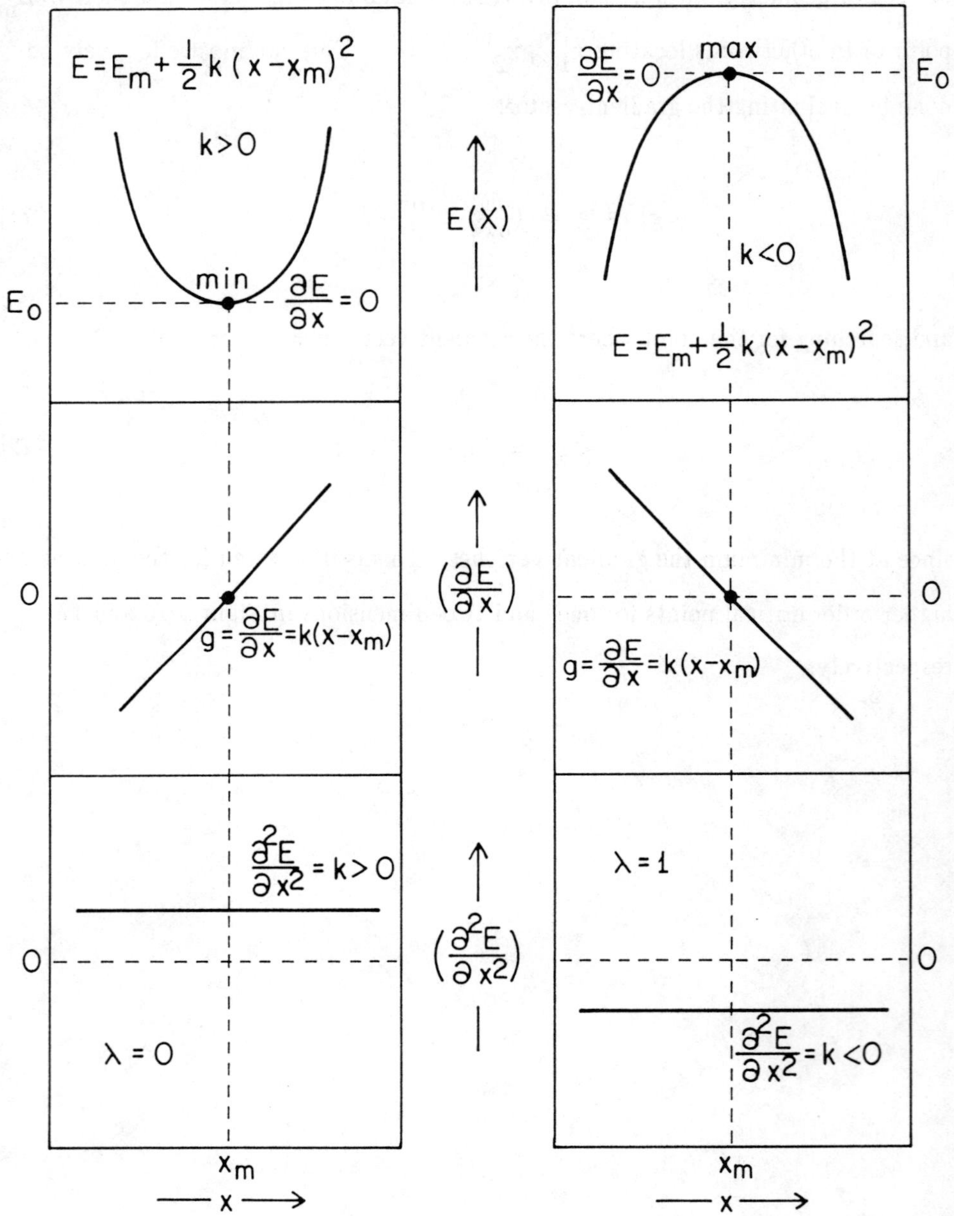

Figure 16 Characteristics of minima and maxima of a potential energy curve, [Note that g is the gradient, k is the force constant of the potential energy function and λ is the index of the critical point in question.

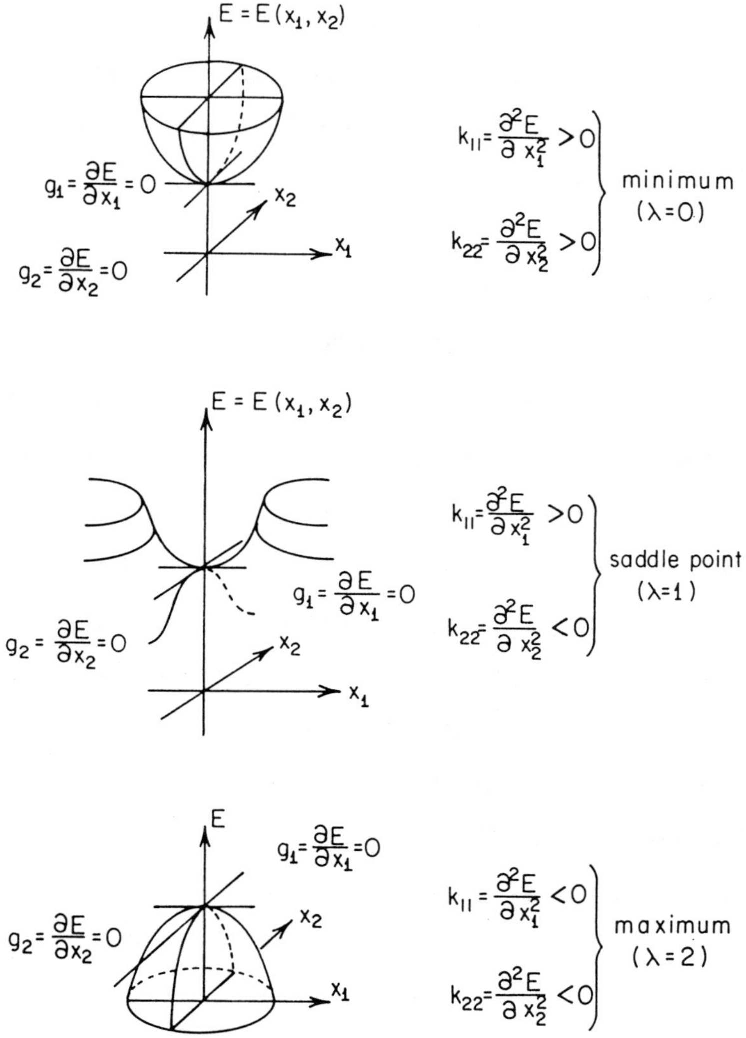

Figure 17 Three types of critical points of a potential energy surface and their characteristics in terms of second partial derivatives. [Note that the index (λ) of a critical point is the number of negative second derivatives.]

Even though the gradient vector provides a powerful tool in locating the minimum we alwyas carry out the multi–dimensional optimization as a sequence of one–dimensional optimizations. The one–dimensional optimization, however, involves a quadratic or parabolic interpretation.

4.1 PARABOLIC INTERPOLATION

4.1.1 BASIC PROCEDURE FOR ONE–DIMENSION

First of all, we must choose an initial point: x_o and a distance: d which lead to three equidistant points $(x_o + d)$, x_o, $(x_o - d)$. Secondly, we need to evaluate E(x) at the three equidistance points leading to three energy values E_+, E_o and E_-. Thirdly, we locate the minimum energy (E_m) point which is located at x_m If x_m falls within the range $x_o - d$ to $x_o + d$, and if d is sufficiently small then quit. Of course we want to avoid extrapolation. Alternatively we should get enough (3) equidistant points to bracket $x_m^{(1)}$, and redetermine $x_m^{(2)}$. This may involve using some of the original points again if $x_m^{(1)}$ was close to x_o.

4.1.2 MINIMUM OF A PARABOLA

A general equation of a parabola may be written as:

$$E(x) = a\,x^2 + b\,x + c \qquad [273]$$

and the function may be evaluated at three equidistant points $(x_o - d)$, x_o, $(x_o + d)$

$$E_+ = E(x_o + d) = a(x_o + d)^2 + b(x_o + d) + c \qquad [274]$$

$$E_o = E(x_o) = ax_o^2 + bx_o + c \qquad [275]$$

$$E_- = E(x_o - d) = a(x_o - d)^2 + b(x_o - d) + c \qquad [276]$$

Expanding [274], [275] and [276] gives:

$$E_+ = ax_o^2 + 2adx_o + ad^2 + bx_o + bd + c \qquad [277]$$

$$E_o = ax_o^2 + bx_o + c \qquad [278]$$

$$E_- = ax_o^2 - 2adx_o + ad^2 + bx_o - bd + c \qquad [279]$$

Adding [277] and [279] together we obtain:

$$E_+ + E_- = 2ax_o^2 + 2ad^2 + 2bx_o + 2c \qquad [280]$$

Subtracting from [280] twice of equation [278] leads to [281]

$$E_+ + E_- - 2E_o = 2ad^2 \qquad [281]$$

This may be solved for the unknown parameters: a

$$a = \frac{E_+ + E_- - 2E_o}{2d^2} \qquad [282]$$

We may obtain parameter b, as [284], after subtracting [279] from [277]

$$E_+ - E_- = 4adx_o + 2bd \qquad [283]$$

$$b = \left[\frac{E_+ - E_-}{2d}\right] - 2ax_o \qquad [284]$$

Finally parameter c may be obtainedfrom equation [275]:

$$c = E_o - ax_o^2 - bx_o \qquad [285]$$

The location of minimum, where $x = x_m$ may be achieved by setting the gradient equal to zero

$$g \equiv \frac{\partial E}{\partial x} = 2ax_m + b = 0 \qquad [286]$$

$$x_m = -\frac{b}{2a} \qquad [287]$$

After substituting to [287] the values of a and b from equations [282] and [284] respectively we obtain:

$$x_m = x_o - d\left[\frac{E_+ - E_-}{2(E_+ + E_- - 2E_o)}\right] \qquad [288]$$

It is convenient to introduce the following notations:

$$t = E_+ - E_- \qquad [289]$$

$$s = E_+ - E_- - 2E_o \qquad [290]$$

$$x_m = x_o - d\left(\frac{t}{2s}\right) \qquad [291]$$

This last equation resulted from [288].

The three parameters a (from [282]), b (from [284]) and c (from [285]) may be rewritten in terms of t and s.

$$a = \frac{s}{2d^2} \tag{292}$$

$$b = \frac{t}{2d} - \frac{s}{d^2} x_o \tag{293}$$

$$c = E_o - \frac{s}{2d^2} x_o^2 - \frac{t}{2d} x_o + \frac{s}{d^2} x_o^2 =$$

$$= E_o + \frac{s}{2d^2} x_o^2 - \frac{t}{2d} x_o \tag{294}$$

The energy at x_m may be written according to [273] as:

$$E(x_m) = ax_m^2 + bx_m + c \tag{295}$$

using [292], [293] and [294] the above expression may be rewritten as follows:

$$E(x_m) = \frac{s}{2d^2} x_m^2 + \frac{t}{2d} x_m - \frac{s}{d^2} x_o x_m + E_o + \frac{s}{2d^2} x_o^2 - \frac{t}{2d} x_o$$

$$= \frac{s}{2d^2} (x_m^2 - 2x_o x_m + x_o^2) + \frac{t}{2d} (x_m - x_o) + E_o$$

$$= \frac{s}{2d^2} (x_m - x_o)^2 + \frac{t}{2d} (x_m - x_o) + E_o \tag{296}$$

Finally, the deviation $(x_m - x_o)$ may be expressed from equation [291]

$$(x_m - x_o) = -\frac{dt}{2s} \qquad [297]$$

Substituting [297] into [296] we obtain:

$$E(x_m) = \frac{s}{2d^2}\left[\frac{d^2t^2}{4s^2}\right] + \frac{t}{2d}\left[\frac{-dt}{2s}\right] + E_o \qquad [298]$$

which simplifies to [299]

$$E(x_m) = E_o - \frac{t^2}{8s} \qquad [299]$$

This predicted $E(x_m)$ is the minimum energy of the quadratic function. This value can be compared to $E(x_m)$ calculated from the function. Note also that an estimate of the force constant, the parameter a, is readily obtained.

$$\frac{\partial^2 E}{\partial x^2} = a \equiv k \equiv G \qquad [300]$$

As it is clear form Figure 16

for a minimum $a>0$ [301a]

for a maximum $a<0$ [301b]

4.1.3 PROCEDURE FOR N–DIMENSIONS

First of all, from the start position, one should find the minimum along the first direction(x_1). Subsequently, one must cycle through the remaining independent variables, (x_2, x_3, ...:, x_n), fitting parabolas to sets of 3 points in each direction in turn. To reach M_1, the best point after the 1st cycle through all the parameters. An example with 2 independent variables, $E = E(x_1,x_2)$ is graphically illustrated by Figure 18. If M_i and M_{i-1} are sufficiently close together then quit. Alternatively the use M_i as the new start position, and continue. As the optimization progresses, it is desirable to cut down the step size d.

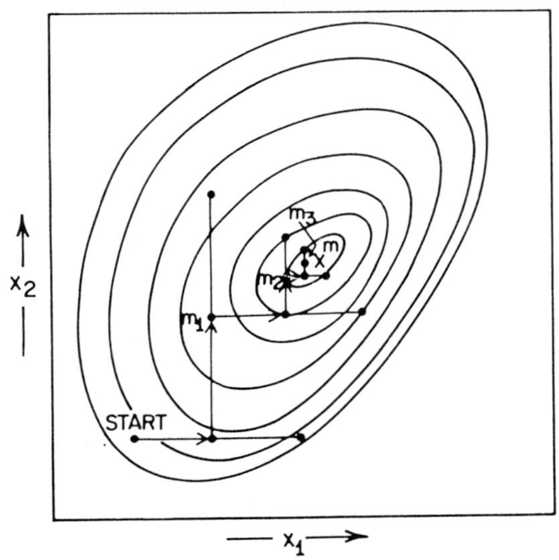

Figure 18. Typical parabolic interpolationoptimization of $E(x_1,x_2)$ (Note that S: start position, M_i: end of cycle i, M: true minimum.)

While this type of sequential optimization along, one internal coordinate before the next one, is a working method but it is rather pedastrian. Also, if the variables are coupled, as they very often are, the method is very inefficient as illustrated by Figure 19.

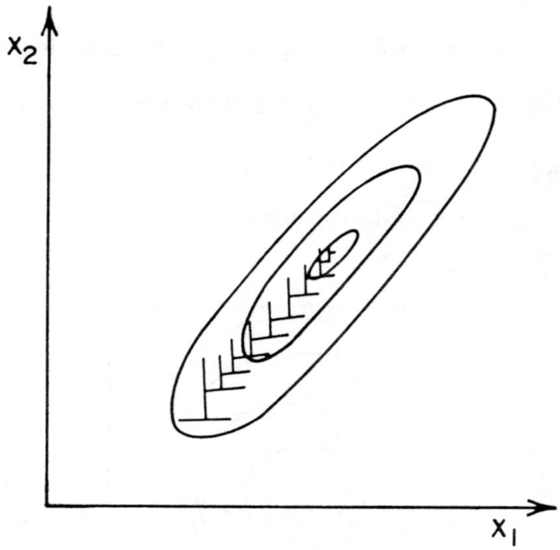

Figure 19. A schematic illustration of the inefficiency of sequential optimization method along one one of the independent coordinates at a time is shown for a strongly coupled system.

4.1.4 CHARACTERISTICS OF POTENTIAL ENERGY SURFACE CRITICAL POINTS

The force constant or the Hessian matrix is a real symmetric matrix [270]. In diagonal form the diagonal elements are the eigen values of the Hessian. Sometime all of these eigen values are positive, other times some of them are negative while some of them are positive and other times all of the eigen values are negative. Equation [302] shows a general case, for a particular critical point, where the first λ diagonal elements are negative and the rest of them are positive. Of course, λ may assume values between 0 and n.

$$G^{diag} = \begin{bmatrix} -a_1^2 & & & & & & 0 \\ & -a_2^2 & & & & & \\ & & \ddots & & & & \\ & & & -a_\lambda^2 & & & \\ & & & & +a_{\lambda+1}^2 & & \\ & & & & & \ddots & \\ 0 & & & & & & +a_n^2 \end{bmatrix} \qquad [302]$$

The parameter λ, the number of negative eigen values, is called the index of the particular critical point. For a minimum $\lambda = 0$, that is all diagonal elements of [302] are positive. For a saddle point, corresponding to a chemical transition state $\lambda = 1$. For a maximum $\lambda = n$. For a one– and a two–dimensional case this is illustrated by Figures 16 and 17, respectively. Table 4 summarizes the situation up to four independent variables.

Table 4. The Variety of Critical point types up to four independent variables.

	Number of independent variables (n)			
Index	1^a	2^b	3^c	4^c
$\lambda = 0$	min	min	min	min
$\lambda = 1$	max	1°saddle	1°saddle	1°saddle
$\lambda = 2$		max	2°saddle	2°paddle
$\lambda = 3$			max	3°saddle
$\lambda = 4$				max

a) potential energy curve
b) potential energy surface
c) potential energy hyper surface

Since the fundamental frequency is related to the square root of the force constant

$$\nu = \frac{1}{2\pi}\sqrt{\frac{k}{\mu}} \qquad [303a]$$

or, in a multidimensional case, for the m–th diagonal element of [302]

$$\nu_m \sim \frac{1}{2\pi}\sqrt{\frac{\pm a_m^2}{\mu}} \qquad [303b]$$

therefore, the first λ eigen values of [302] correspond to imaginary frequency. For $m<\lambda$ we obtain [304]

$$\nu_m \sim \frac{1}{2\pi}\sqrt{\frac{-a_m^2}{\mu}} = \frac{a_m}{2\pi\sqrt{\mu}}\sqrt{-1} = \frac{a_m}{2\pi\sqrt{\mu}}\, i \qquad [304]$$

At the end of the optimization it is advisable to check the order (λ) of the critical point. If $\lambda = 0$ then we can rest assured that the critical point in our optimization we have converged to is indeed a minimum.

4.2 GENERAL INTRODUCTION TO GRADIENT METHODS OF OPTIMIZATION

Methods for unconstrained optimization of n variables (n dimensions) are designed to produce the answer in a finite number of steps when the function[1], f(x), is quadratic. It is then hoped that the method will be efficient on more general functions, but especially those with slowly varying second derivatives.

[1]*For a general function the notation f(x) is used rather than the E(x) specified before. This is done to emphasize that the method is applicable to any routine differentiable function. We return to the E(x) notation in Section 5.

At the minimum or very close to it the functions has a quadratic form, in addition to the constant $f(x_m)$, without a linear term. This function may be written, according to equation [268a] as follows:

$$f(x) = f(x^m) + \frac{1}{2}(x - x^m)^{\dagger} G (x - x^m) \qquad [305]$$

However, some distance away from the minimum at point x^k the inclusion of a linear term is advisable (c.f. Figure 20) as specified by the next equation

$$f_k(x) = \frac{1}{2}(x - x^k)^{\dagger} G (x - x^k) + (x - x^k)^{\dagger} g_k + f(x^k) \qquad [306]$$

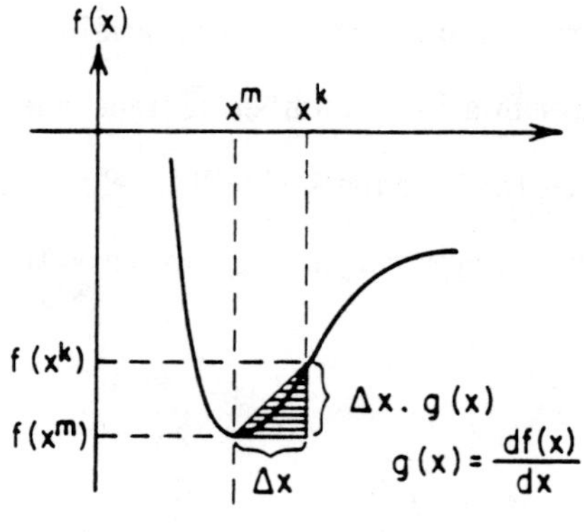

Figure 20. A linear segment that takes the minimum energy point from x^m to x^k.

The components of the gradient vector g(x) are the first partial derivatives of f(x) written in a fashion that is analogous to [271]:

$$\text{grad } f(x) = \nabla f(x) = g = <g| = \left[\frac{\partial f}{\partial x_1}, \frac{\partial f}{\partial x_2}, \ldots\ldots\right] \qquad [307]$$

In general, the objective function f(x) is

$$f(x) = f_m + \sum_{i=1}^{n} a_i x_i + \frac{1}{2} \sum_{i,j}^{n} G_{ij} x_i x_j \qquad [308a]$$

or in Dirac notation

$$f(x) = f_m + >x|a> + \frac{1}{2} <x|G|x> \qquad [308b]$$

where

$$G_{ij} = \frac{\partial^2 f}{\partial x_i \partial x_j} \qquad [309]$$

and the Hessian matrix" is constructed according to [270]

The gradient of f(x), g(x) is:

$$g_i(x) = a_i + \sum_j G_{ij} x_j \qquad [310a]$$

or

$$|g> = |a> + G|x> \qquad [310b]$$

where $g_i(x)$ is the i'th component of $|g>$.

At the the extremum of f,

$$|g\rangle = 0 \qquad [311]$$

Letting $|x_m\rangle$ be the coordinates of the extremum. From [310b] we may express $|a\rangle$.

$$|a\rangle + G|x_m\rangle = 0 \qquad [312]$$

therefore
$$|x_m\rangle = - G^{-1}|a\rangle \qquad [313]$$

From [310b], substitute for $|a\rangle = |g\rangle - G|x\rangle$:

$$|x_m\rangle = -G^{-1}[|g\rangle - G|x\rangle]$$

$$= -G^{-1}|g\rangle + G^{-1}G|x\rangle$$

$$= -G^{-1}|g\rangle + |x\rangle \qquad [314a]$$

$$|x_m> = |x> - G^{-1}g> \qquad [314b]$$

which is the Newton (or Newton–Raphson) equation. Given exact G and $|g>$, the minimum can be found in one step for a quadratic function.

Methods which do not use an exact G often use an approximate matrix H (a positive definite symmetric matrix) such that

$$\mathbf{H}\ \mathbf{G} = \mathbf{1} \qquad [315]$$

In these cases, the step size equation [314b] is usually written:

$$|x^{(i+1)}> = |x^{(i)}> - \lambda_i H|g^{(i)}> \qquad [316]$$

where λ_i is to be determined. Note that $|x^{(i+1)}>$ will not necessarily be the extremum desired for any value of λ_i, as H is <u>not</u> exact. Besides, if f is not quadratic, [314a] is not valid anyways so [316] is always used in practice. Such methods, (based on [316], are called quasi–Newton methods.

4.2.1 STEEPEST DESCENTS

A very simple optimization method called steepest descents arises from [316] by assuming **H** is the unit matrix, so that the search direction is always $-|g^{(i)}>$. A line search is carried out along the direction $-|g^{(i)}>$ to obtain λ_i. This method, however, poses some problems. Subsequent search directions tend to be linearly dependent, so only a small subspace of the total space is explored.

Directions with large components of $|g^{(i)}>$ are always favoured whereas progress can sometimes be made by searching in orthogonal directions to reach a part of the surface that would allow better progress. Also, the method converges very slowly near the minimum as g(x) is getting smaller in that vicinity.

4.2.2 THE DFP (DAVIDON–FLETCHER–POWELL)* METHOD

This method uses a variant of [316] given by equation [317].

$$|x^{(i+1)}> = |x^{(i)}> - \lambda_i H^{(i)}|g^{(i)}> \qquad [317]$$

where **H** is different for each iteration – thus this is a "variable metric" method because the "metric" matrix **G** (estimated by $\mathbf{H}^{-1}$) varies during the optimization, using information gained during the iteration.

The search direction for iteration i is

$$|s^{(i)}> = -\mathbf{H}^{(i)}|g^{(i)}> \qquad [318]$$

and the step taken along that direction is

$$|\sigma^{(i)}> = \lambda_i|s^{(i)}> \qquad [319]$$

is closed so that $|\sigma^{(i)}>$ is an eigen function of $\mathbf{H}^{(i+1)}\mathbf{G}$

[2]R. Fletcher and M.J.D. Powell, Comp. J., 6, 163–168 (1963).

$$H^{(i+1)}\,G|\sigma^{(i)}> \;=\; |\sigma^{(i)}> \qquad [320]$$

This ensures that as the optimization progresses, **H** tends to $\mathbf{G}^{-1}(x^m)$ that is the **G** matrix evaluated at the extremum $|x_m>$.

<u>Procedure</u>:

Set $|s^{(i)}>$ from [318], obtain λ_i specified in [317] such that

$$f(|x^{(i)}> \;+\; \lambda_i|s^{(i)}>) \qquad [321]$$

is a minimum with respect to λ_i. This is called the "line search".

Set $|\sigma^{(i)}>$ from [319].

Set $$|x^{(i+1)}> \;=\; |x^{(i)}> \;+\; |\tau^{(i)}> \qquad [322]$$

Evaluate $f(|x^{(i+1)}>)$ and $|g^{(i+1)}>$ which is $g(|x^{(i+1)}>)$, noting that

$$<g^{(i+1)}|\sigma^{(i)}> \;=\; 0 \qquad [323]$$

Since we have minimized f along $|\sigma^{(i)}>$, $|g^{(i+1)}>$ must be orthogonal to the search direction.

Set $|y^{(i)}> \;+\; |g^{(i+1)}> \;-\; |g^{(i)}>$ [324]

Set $\mathbf{H}^{(i+1)} \;=\; \mathbf{H}^{(i)} \;+\; \mathbf{A}^{(i)} \;+\; \mathbf{B}^{(i)}$ [325]

where:

$$\mathbf{A}^{(i)} = \frac{|\sigma^{(i)}> < \sigma^{(i)}|}{<\sigma^{(i)}|y^{(i)}>} \qquad [326]$$

$$\mathbf{B}^{(i)} = -\frac{H^{(i)}|y^{(i)}> < y^{(i)}|H^{(i)}}{<y^{(i)}|H^{(i)}|y^{(i)}>} \qquad [327]$$

Repeat the above sequence until $|s^{(i)}>$ is small enough, either for each component, or for $<s^{(i)}|s^{(i)}>^{1/2}$. In practice $|\sigma^{(i)}>$ is also tested in case no further progress can be made ($\lambda_i \longrightarrow 0$).

In the DFP method the **H** matrix is updated in every cycle of the iteration. This is in direct contrast to the "steepest descents" method in which the **H** matrix is never updated but in fact is kept constant ($\mathbf{H} = \mathbf{1}$). Other optimization methods, to be discussed later use different algorithms to update the **H** matrix.

Example. Let us consider the following problem to examplify the use of the DFP method of optimization:

$$f(x) = x_1^2 - x_1x_2 + 3x_2^2 \qquad [328]$$

where

$$x = (x_1,x_2) \qquad [329]$$

This is a simple problem so one may locate the minimum, evaluate the gradient and Hessian analytically. Consequently, we might know the answer before starting the DFP gradient optimization.

The gradient vector is:

$$g = \left[\frac{\partial f}{\partial x_1}, \frac{\partial f}{\partial x_2}\right] = [(2x_1 - x_2), (-x_1 + 6x_2)] \tag{330}$$

The Hessian matrix is:

$$\left| \begin{matrix} \dfrac{\partial^2 f}{\partial x_1^{\,2}} = 2 & \dfrac{\partial^2 f}{\partial x_2 \partial x_1} = -1 \\ \\ \dfrac{\partial^1 f}{\partial x_1 \partial x_2} = -1 & \dfrac{\partial^2 f}{\partial x_2^{\,2}} = 6 \end{matrix} \right| \tag{331}$$

The minimum is:

$$x = (x_1^{\,m}, x_2^{\,m}) = (0,0) \tag{332}$$

At the minimum:

$$f(0) = 0 \tag{333}$$

$$g(0) = ([2x_1 - x_2], [-x_1 + 6x_2]) = (0,0) \tag{334}$$

$$\mathbf{G}_o = \begin{bmatrix} 2 & -1 \\ -1 & 6 \end{bmatrix} \text{ and } \mathbf{H}_o \equiv \mathbf{G}_o^{-1} = \begin{bmatrix} 6/11 & 1/11 \\ 1/11 & 2/11 \end{bmatrix} \tag{335}$$

and therefore, the original equation may be rewritten as

$$f(x) = \frac{1}{2}(x_1 \;\; x_2) \begin{bmatrix} 2 & -1 \\ -1 & 6 \end{bmatrix} \begin{bmatrix} x_1 \\ x_2 \end{bmatrix} \tag{336}$$

So much prior knowledge is not necessary for the gradient optimization but in this case it is nice to know the final results so that we might know where we are converging to.

Let us start the optimization at

$$x^{(1)} = (1, 2) \tag{337}$$

Thus

$$g^{(1)} = (0, 11) \tag{338}$$

At this stage the unit matrix is assumed for **H**.

$$\mathbf{H}^{(1)} = \begin{bmatrix} 1 & 0 \\ 0 & 1 \end{bmatrix} \tag{340}$$

$$|s^{(1)}> = -H^{(\lambda)}|g^{(1)}> = |g^{(1)} = (0, -11) \tag{341}$$

The next step is to determine λ_1 (in general $\lambda_k > 0$). This is usually done through a numerical "line search". It should perhaps be emphasized that while in Section 4.1.3 the search has been carried out along one independent variable ($x_1 \longrightarrow x_2 \longrightarrow$... etc.) at a time (c.f. Figure 18) and in Section 4.2.1 the search was performed along the line of the gradient vector $|g>$, here the search line, $|s^{(i)}>$ is a transformed gradient vector $-\mathbf{H}^{(i)}|g^{(i)}>$ as illustrated by Figure 21

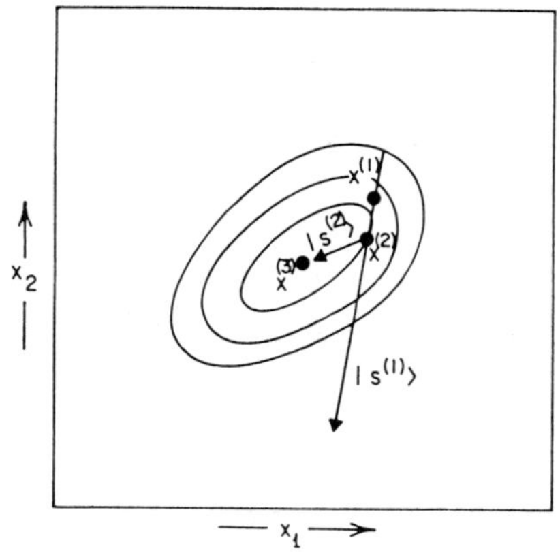

Figure 21. Search directions $|s^{(i)}>$ in the DFP gradient optimization process of $f(x) = x_1^{\;2} - x_1x_2 + 3x_2^{\;2}$ starting with $|x> = (x_1,x_2) = (1,2)$.

It is perhaps important to note that the variable parameter along the search direction is λ and the function evaluation at three different λ–values will locate the lowest functional value along the search direction according to Section 4.1.1. Thus the optimum λ value is found by optimizing

$$f(x^{(i)} + \lambda_i s^{(i)}) \qquad [342]$$

as depicted in Figure 22.

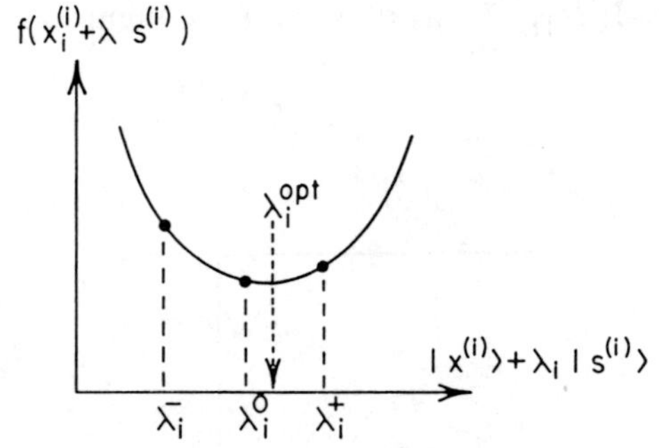

Figure 22. Schematic illustration of "line search" to find the optimum λ_i value in the DFP gradient optimization

Again this process is normally performed numerically as illustrated above but in this particular case we can do it analytically thus:

$$\lambda_1 s(1) = \lambda_1 g^{(1)}$$

$$= \lambda_1 \begin{array}{rcl} x^{(1)} & = & (1,2) \\ (0,-11) & = & (0,11\lambda) \\ \hline x^{(1)}-\lambda_1 s^{(1)} & = & [1,(2-11\lambda)] \end{array}$$

$$f(x^{(1)} - \lambda_1 g^{(1)}) = 1-(2-11\lambda_1) + 3(2-11\lambda_1)^2$$

$$= 11 - 121\lambda_1 + 363\lambda_1^{\,2}$$

$$\frac{df}{d\lambda} = 726\lambda_1 - 121 = 0 \qquad [343]$$

therefore, $$\lambda_1^{\,opt} = \frac{121}{726} = \frac{1}{2}$$

Consequently: $$\lambda_1 = 1/6$$ [344]

Therefore:

$$|\sigma^{(1)}> = \lambda_1 |ds^{(1)}> = 1/6\,(0, -11) = (0, -11/6)$$ [345]

now we can determine $|x^{(2)}>$

$$|x^{(2)}> = |x^{(1)}> + \sigma^{(1)}> = (1,2) + (0, -11/6) = [1,(\tfrac{12}{6} - \tfrac{11}{6})]$$ [346]

thus finally we have: $$|x^{(2)}> = (1,\ 1/6)$$ [347]

With this result at hand we can declare at this point the <u>end of the first cycle</u> as well as the <u>start of the second cycle</u>.

From [347]: $$|x^{(2)}> = (1,\ 1/6)$$

from [330]: $$g(x) = ([2x_1 - x_2], [-x_1 + 6x_2])$$

may evaluate $|g^{(2)}>$:

$$|g^{(2)}> = ([2 = 1/6], [-1 + 6\tfrac{1}{6}]) = (11/6,\ 0)$$ [348]

The update of the **H** matrix may be written in accordance with the [325] in the following form:

$$\mathbf{H}^{(2)} = \mathbf{H}^{(1)} + \mathbf{A}^{(1)} + \mathbf{B}^{(1)}$$ [349]

requiring the explicit form of $\mathbf{A}^{(1)}$ and $\mathbf{B}^{(1)}$ in addition to $\mathbf{H}^{(2)}$ which is given by [340].

$$\mathbf{A}^{(1)} = \begin{bmatrix} 0 & 0 \\ 0 & 1/6 \end{bmatrix} \qquad [350]$$

$$\mathbf{B}^{(1)} = \begin{bmatrix} -1/37 & 6/37 \\ 6/37 & 36/37 \end{bmatrix} \qquad [351]$$

$$\mathbf{H}^{(2)} = \begin{bmatrix} 36/37 & 6/37 \\ 6/37 & 43/222 \end{bmatrix} \qquad [352]$$

Applying the form for $|s^{(i)}>$ from [318].

$$|s^{(2)}> \; = \; -\mathbf{H}^{(2)}|g^{(2)}> \qquad [353]$$

$$|s^{(2)}> \; = \; (-66/37 \; - \; 11/37) \qquad [354]$$

Optimizing again λ in the fashion of [343] we find.

$$\lambda_2{}^{\mathrm{opt}} = 37/66$$

With this result one can calculate $|\sigma^{(2)}>$ and $|x^{(3)}>$:

$$|\sigma^{(2)}> \; = \; \lambda_2|s^{(2)}> \; = \; (-1, -1/6) \qquad [355]$$

$$|x^{(3)}> \; = \; |x^{(2)}> + |\sigma^{(2)}> \; = \; (0,0) \qquad [356]$$

This is the end of the second cycle and the start of the third cycle.[3]

---

[3]C. G. Broyden, J. Inst. Maths. Appl., 6, 76–90 222–231 (1970), R. Fletcherr, Conp. J., 13, 317–322 (1970), D. Goldfarb, Math. Comp., 24, 23–26 (1970); D. F.

From $|x^{(3)}>$ given in [349] and from the anlytic expression for the gradient vector [330] one may obtain after substitution the new gradient vector $|g^{(3)}>$:

$$|g^{(3)}> \ = \ (0,0) \qquad [358]$$

Since the gradient vector is a zero vector, as illustrated by Figure 23, the optimization is now completed.

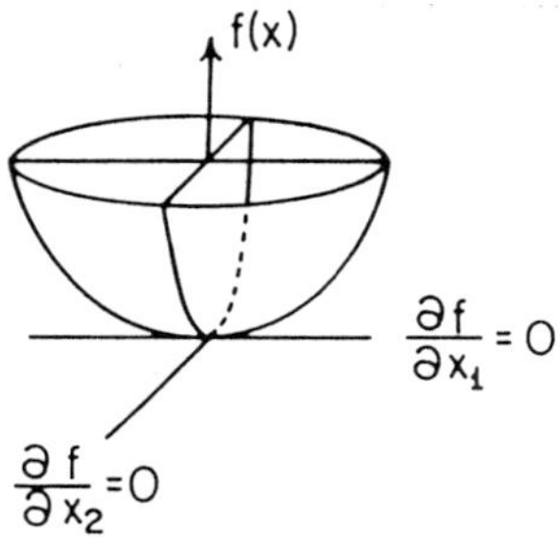

Figure 23. At the minimum of the f(x) the gradient vector is zero.

4.2.3 THE BFGS (BROYDEN–FLETCHER–GOLDFARB–SHANNO)* METHOD

This method, discovered independently by four authors (B, F, G, S), is also a variable metric method, based on [317] like DFP. The main incentive to find a new algorithm was to eliminate the requirement of an exact line search, required in the DFP method, which cost a lot of f(x) and $|g(x)>$ evaluations along one search direction, yet retain as many desirable properties as possible. It turns out the property lost is finite convergence for quadratic f, but $H^{(k)}$ update retain the positive definiteness of $H^{(k+1)}$, and $H^{(k)}$ still tends to G^{-1}, and the method is stable.

Slanno, Math. Comp., 2, 647–656 (1970).

Although finite convergence will be sacrificed, having $\mathbf{H}^{(k)}$ tend to $\mathbf{G}^{-1}$ should ensure good convergence properties (**ultimate** convergence for quadratic f can still be proved). Also, extra conditions will have to be imposed on the step size taken on each iteration to ensure that f decreases by a "sufficient" amount, and that λ_i does not become too small.

The new updating formula, following Fletcher's "derivation", is defined as follows:

$$\mathrm{H}^{(k+1)} = \left[1 - \frac{|\sigma^{(k)}\rangle \langle y^{(k)}|}{\langle y^{(k)} | \sigma^{(k)}\rangle} \right] \mathbf{H}^{(k)} \left[\mathrm{I} - \frac{|y^{(k)}\rangle \langle \sigma^{(k)}|}{\langle y^{(k)} | \sigma^{(k)}\rangle} \right] + \\ + \frac{|\sigma^{(k)}\rangle \langle \sigma^{(k)}|}{\langle y^{(k)} | \sigma^{(k)}\rangle} \qquad [359]$$

or in details:

$$\mathrm{H}^{(k+1)} = \mathrm{H}^{(k)} + \frac{1}{\langle y^{(k)}|\sigma^{(k)}\rangle} \left[1 + \frac{\langle y^{(k)}|\mathbf{H}^{(k)}|y^{(k)}\rangle}{\langle y^{(k)}|\sigma^{(k)}\rangle} |\sigma^{(k)}\rangle \langle \sigma^{(k)}| - \\ - |\sigma^{(k)}\rangle \langle y^{(k)}|\mathrm{H}^{(k)} - \mathrm{H}^{(k)}|y^{(k)}\rangle \langle|\sigma^{(k)}| \right] \qquad [360]$$

which is the **BFGS formula** [it is often also called the **complementary DFP formula** due to the way it is divided].[4]

---

[4]W.C. Davidon, Math. Programming, 9, 1–30 (1975), W.C. Davidon and L. Nazareth, Technical Memoranda #303 and #306, Argonne National Laboratories, Argonne, Illinois (1977).

4.2.4. THE OC (OPTIMALLY CONDITIONED)* METHOD

This algorithm is also a quasi–Newton variable metric method, which does no line search, yet maintains quadratic termination and the positive definiteness of the Hessian update. The updates are "optimally conditioned" in the sense that they minimize the ratio of the largest to smallest eigenvalue of $[\mathbf{H}^{(x)}]^{-1}\ \mathbf{H}^{(k+1)}$; i.e. minimize ϵ_1/ϵ_n where ϵ_1 is the largest eigenvalue of $[\mathbf{H}^{(k)}]^{-1}\ \mathbf{H}^{(k+1)}$ and ϵ_n is the smallest eigenvalue. Note that since $[\mathbf{H}^{(k)}]^{-1}\ \mathbf{H}^{(k+1)}$ should approach the unit matrix **1**, the ratio ϵ_1/ϵ_n should approach 1. In addition, the algorithm uses projections of $|\Delta x\rangle$ and $|\Delta g\rangle$

$$|\Delta x\rangle \;=\; -\mathbf{H}\,|g\rangle \;=\; |x^{(k+1)}\rangle - |x^{(k)}\rangle \qquad [361]$$

$$|\Delta g\rangle \;=\; |g(|x\rangle + |\Delta x\rangle)\rangle \;-\; |g(|x\rangle)\rangle \qquad [362]$$

or

$$|\Delta g\rangle \;=\; |g^{(k+1)}\rangle \;-\; |g^{(k)}\rangle \qquad [363]$$

such that while only projections of $|\Delta x>$ and $|\Delta g>$ are used to update $\mathbf{H}^{(k)}$, allowing quite arbitrary step sizes and directions, the updated matrix $\mathbf{H}^{(k+1)}$ still has the property

$$H^{(k+1)}|\Delta g^{(k)}> \;=\; |\Delta x^{(k)}> \qquad [366]$$

for the **full** step $|\Delta x>$. Note that DFP and BFGS have this property.

Update formula:

Let $|y>$ and $|\sigma>$ be any column vectors with $<y|\sigma> \; > 0$, and let $\mathbf{H}$ be any positive definite matrix. Define

$$a \;=\; <y|H|y> \;\; > \; 0 \qquad [367]$$

$$b \;=\; <y|\sigma> \;\; > \; 0 \qquad [368]$$

Then choosing

$$\mathbf{H}_{\phi}^{(k+1)} \;=\; \mathbf{H}^{(k)} \;-\; \frac{\mathbf{H}|y> \; <y|\mathbf{H}}{a} \;+\; \frac{|z> \; <z|}{b}$$

$$+\; \phi\, a\left[\frac{|z>}{b} \;-\; \frac{H|y>}{a}\right]\left[\frac{<z|}{b} \;-\; \frac{<y|H}{a}\right] \qquad [369]$$

for some number ϕ will also satisfy

$$\mathbf{H}_{\phi}^{(k+1)}|y\rangle = |\sigma\rangle \qquad [370]$$

$$\mathbf{H}_{\phi}^{(k+1)} = \mathbf{H}_{\phi}^{(k+1)^{+}} \qquad [371]$$

In a limiting sense

$\phi = 0$ gives DFP update and

$\phi = 1$ gives BFGS update

provided that the $|y\rangle$ and $|\sigma\rangle$ used here are the same as $|y\rangle$ and $|\sigma\rangle$ used before.

4.2.5 "SUM OF THE SQUARES OF THE GRADIENTS" METHOD[5]

The gradient vector is a zero vector not only for minima but for all critical points including transition states as illustrated in Figure 24.

$$\text{grad } E \equiv \underset{\sim}{g} = \left(\frac{\partial E}{\partial x_1}, \frac{\partial E}{\partial x_2}, \ldots\right) = (0, 0, \ldots) \qquad [372]$$

For this reason a method based on this principle could be used for the location a transition states since the previous methods always converged to minima. We may define the gradient vector length (F) which is zero at any critical point:

[5]*M.J.D. Powell in P. Rabinowits (Ed), Numerical Methods for Nonlinear Algebraic Equations, Gordon and Breach, London, 1970, p. 87.

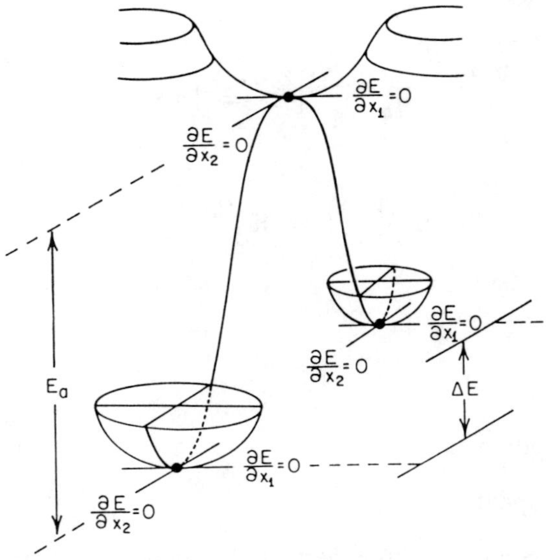

Figure 23 A schematic illustration of the principle that the gradient vector is a zero vector for the transition states as well as for reactants and products.

$$F = \sum_{i=1}^{\text{all modes}} \left[\frac{\partial E}{\partial x_i}\right]^2 = \sum_{i=1}^{\text{all modes}} g_i^{\;2} = 0 \qquad [373]$$

At some point somewhat away from the critical point the sum of the squares of the gradients (F) is greater than zero:

$$F = \sum_{i=1}^{\text{all modes}} \left[\frac{\partial E}{\partial x_i}\right]^2 = \sum_{i=1}^{\text{all modes}} g_i^2 > 0 \qquad [374]$$

and the minimization of F will yield the location of the critica point (minimum, saddle point, etc) in question. This equation forms the basis of a gradient geometry optimization algorithm since the gradients of the energy can be computed analytically at the SCF level.

This method is called the "Sum of the Squares of the Gradients" method. Although it can be used for the location of minima (reactants, products and reactive intermediates), but its real value is in the search for the location of saddle points or transition states.

4.2.6. ORDER CHECK OF CRITICAL POINTS

The second partial derivatives of energy are important for the characterization of the critical points into minima, saddle points and maxima. The number of negative second partial derivatives, called the index (λ) of the critical points.

The second derivatives of E with respect to x, that in fact corresponds to the force constants (k), may be estimated numerically as follows:

$$k = \frac{\partial^2 E}{\partial x_i}1 = \frac{\partial E}{\partial x_1}1 \approx \frac{\Delta F}{\Delta x_i} \qquad [375]$$

This procedure is quite useful if k is large but for a small force constant it is very hard to know if a $k>0$ or $k<0$ is not the result of the inaccuracy of numerical method specified in [365]. Sophisticated computer programs, nowadays, have analytic second derivatives.

4.3 ENERGY GRADIENTS FOR CLOSED SHELL SCF

Equation [49] from Section 2.2 gives the molecular SCF energy as follows

$$E = \mathrm{tr}\,\{\mathbf{D}(\mathbf{h} + \mathbf{F})\} \qquad [376]$$

where

$$\mathbf{F} = \mathbf{h} + 2\mathbf{J}[D] - \mathbf{K}[D] \qquad [377]$$

and the matrix elements are defined below:

$$h_{pq} = \langle \eta_p | \hat{h} | \eta_q \rangle \qquad [378]$$

$$\begin{aligned} J[D]_{pq} &= \sum_{k=1} D_{kl} \langle \eta_p(1)\eta_q(1) | g_{12} | \eta_k(2)\eta_l(2) \rangle \\ &= \sum_{k,l.} D_{kl} \{pq|kl\} \end{aligned} \qquad [379]$$

$$\begin{aligned} K[D]_{pq} &= \sum_{k=1} D_{kl} \langle \eta_p(1)\eta_q(2) | g_{12} | \eta_k(1)\eta_l(2) \rangle \\ &= \sum_{k,l.} D_{kl} \{pk|ql\} \end{aligned} \qquad [380]$$

where $g_{12} = 1/r_{12}$.

The density matrix **D** (sometime denoted as ρ) are given as

$$\mathbf{D} = \mathbf{C}\ \mathbf{L}^n\ \mathbf{C}^\dagger \qquad [381]$$

where

$$L_{ij}^{\ n} = \delta_{ij} f(n) \qquad [382]$$

and the fractional occupancy is:

$$f(n,i) = 1 \text{ if } i \leq n \qquad [383]$$

$$= 0 \text{ if } i > n \qquad [384]$$

MATHEMATICAL REVIEW NO. 1 SOME PROPERTIES OF tr { }

$$\mathrm{tr}\{A\} = \sum_{p}^{m} A_{pp} \qquad [MR1\text{–}1]$$

$$\mathrm{tr}\,\{A\,B\} = \sum_{p}^{m} (AB)_{pp} = \sum_{p,s}^{m} A_{ps}B_{sp} = \mathrm{tr}\,(BA) \qquad [MR1\text{–}2]$$

To generalize the above finding we may write

$$\mathrm{tr}\,\{ABCD\} = \mathrm{tr}\{DABC\} = \mathrm{tr}\{CDAB\} = \mathrm{tr}\{BCDA\} \qquad [MR1\text{–}3]$$

In other words the reationship is "cyclic".

$$\mathrm{tr}\{AB + CD\} = \mathrm{tr}\{AB\} + \mathrm{tr}\{CD\} \qquad [MR1\text{–}4]$$

$$\frac{\partial}{\partial\lambda}\,\mathrm{tr}\{A\} = \mathrm{tr}\left\{\frac{\partial A}{\partial\lambda}\right\} = \mathrm{tr}\{A'\} \qquad [MR1\text{–}5]$$

where

$$A'_{pq} = \frac{\partial A_{pq}}{\partial\lambda} \qquad [MR1\text{–}6]$$

Differentiate the energy expression, equation [366], with respect to some parameter λ and denote $\partial/\partial\lambda$ with a prime$'$ (i.e. $\partial E/\partial\lambda = E'$) then we obtain for the energy gradient:

$$\text{grad}\,E = E' = \text{tr}\{\mathbf{D}\,(\mathbf{h} + \mathbf{f}) + \mathbf{D}(\mathbf{h}' + \mathbf{F}')\} \qquad [385]$$

Differentiating the Fock matrix we obtain:

$$\mathbf{F}' = \mathbf{h}' + 2\mathbf{J}'[\mathrm{D}] + 2\mathbf{J}[\mathrm{D}'] - \mathbf{K}'[\mathrm{D}] - \mathbf{K}[\mathrm{D}'] \qquad [386]$$

with the following matrix elements:

$$h'_{pq} = \frac{\partial}{\partial\lambda} <\eta_p|\hat{h}|\eta_q> \qquad [387]$$

$$J'[D]_{pq} = \sum_{k,l} D_{kl} \frac{\partial}{\partial\lambda} \{pq|kl\} \qquad [388]$$

$$J[D']_{pq} = \sum_{k,l} \{pq|kl\} \frac{\partial D_{kl}}{\partial\lambda} \qquad [389]$$

Similarly to [377] and [378] we can derive expressions for $K'[\mathbf{D}]$ and $K[\mathbf{D}]$. Substituting [375] into [374] gives after expansion:

$$E' = \text{tr}\{\mathbf{D}'\mathbf{h} + \mathbf{D}'\mathbf{F} + \mathbf{D}\mathbf{h}' + \mathbf{D}\mathbf{h}' + 2\mathbf{D}\mathbf{J}'[\mathrm{D}] + 2\mathbf{D}\mathbf{J}[\mathrm{D}'] - \mathbf{D}\mathbf{K}'[\mathrm{D}]$$

$$- \mathbf{D}\mathbf{K}[\mathrm{D}']\} \qquad [390]$$

QUANTUM MECHANICS REVIEW NO. 1
DETAILED EXPRESSIONS FOR $tr\{DJ[D']\}$

$$tr\{DJ[D']\} = \sum_{p} \sum_{v} D_{pv} \; J[D']_{vp}$$

$$= \sum_{p,v} D_{pv} \sum_{k,l} D'_{kl} \{vp|kl\} \qquad [QM1\text{–}1]$$

Because of the symmetry of 2–electron integrals

$$\{vp|kl\} = \{kl|vp\} = \{lk|vp\} = \{lk|pv\} \qquad [QM1\text{–}2]$$

as defined by [101] in Section 2.5 we may write the following

$$tr\{DJ[D']\} = \sum_{p,v} \sum_{k,l} D_{pv} D'_{kl}\{lk|pv\} \qquad [QM1\text{–}3]$$

An alternative sequence of summation is also possible that leads to the reversal of the functional dependence D and D'.

$$tr\{DJ[D'] = \sum_{k,l} D'_{kl} \sum_{p,v} D_{pv} \{lk|pv\}$$

$$= \sum_{k,l} D'_{kl} \; J[D]_{lk} = tr\{D'J\,[D]\} \qquad [QM1.4]$$

Similarly we can write for Exchange term:

$$\mathrm{tr}\{\mathbf{D}\,\mathbf{K}\,[\mathrm{D}']\} \;=\; \mathrm{tr}\,\{\mathbf{D}'\mathbf{K}[\mathrm{D}]\} \qquad [QM1\text{–}5]$$

Using [MR2–4] and [MR2–5] we can express [379] as an alternative sum of traces:

$$\mathrm{E}' = \mathrm{tr}\{\mathbf{D}'\mathbf{F} \;+\; \mathbf{D}(2\mathbf{h}' + 2\mathbf{J}'[\mathrm{D}] - \mathbf{K}'[\mathrm{D}]) + \mathbf{D}'(\mathrm{h} + 2\mathbf{J}[\mathrm{D}] - \mathbf{K}[\mathrm{D}])\} \qquad [391]$$

or $$\mathrm{E}' = 2\,\mathrm{tr}\{\mathbf{D}'\mathbf{F}\} + \mathrm{tr}\,\{\mathbf{D}(2\mathbf{h}' + 2\mathbf{J}'[\mathrm{D}] - \mathbf{K}'[\mathrm{D}])\} \qquad [391a]$$

$$\mathrm{E}' = \mathrm{tr}\,\{\mathbf{D}\,(\mathbf{h}' + \mathrm{F}')\} \qquad [391b]$$

QUANTUM MECHANICS REVIEW NO. 2

THE INTERDEPENDENCE OF D′ AND S′

From [371]: $$\mathbf{D} \;=\; \mathbf{C}\,\mathbf{L}^{n}\,\mathbf{C}^{\dagger} \qquad [QM2\text{–}1]$$

$$\mathbf{D}' = \mathbf{C}'\,\mathbf{L}^{n}\,\mathbf{C}^{\dagger} + \mathbf{C}^{n}\,\mathbf{C}'^{\dagger} \qquad [QM2\text{–}2]$$

since $$(\mathrm{L}^{n})' \;=\; 0 \qquad [QM2\text{–}3]$$

$$\mathrm{tr}\{\mathbf{D}'\,\mathbf{F}\} \;=\; \mathrm{tr}\{(\mathbf{C}'\mathbf{L}^{n}\mathbf{C}^{\dagger} \;+\; \mathbf{C}\,\mathbf{L}^{n}\mathbf{C}'^{\dagger})\,\mathbf{F}\}$$

$$=\; \mathrm{tr}\{(\mathbf{C}'\mathbf{L}^{n}\mathbf{C}^{\dagger}\mathbf{F}\} \;+\; \mathrm{tr}\{\mathbf{C}\,\mathbf{L}^{n}\mathbf{C}'^{\dagger}\mathbf{F}\} \qquad [QM2\text{–}4]$$

or $$\mathrm{tr}\{\mathbf{D}' \ \mathbf{F}\} = \mathrm{tr}\{(\mathbf{L}^{n}\mathbf{C}^{\dagger}\mathbf{F}\,\mathbf{C}'\} + \mathrm{tr}\{\mathbf{L}^{n}\mathbf{C}'^{\dagger}\mathbf{F}\,\mathbf{C}\}) \qquad \text{[QM2–5]}$$

Using [38] from Section 2.2

$$\mathbf{FC} = \mathbf{SC}\epsilon \qquad \text{[QM2–6]}$$

We may take the adjoint of both sides

$$(\mathbf{FC})^{\dagger} = (\mathbf{SC}\epsilon)^{\dagger} \qquad \text{[QM2–7]}$$

We obtain: $$= \mathbf{C}^{\dagger}\mathbf{F}^{\dagger} = \epsilon^{\dagger}\,\mathbf{C}^{\dagger}\,\mathbf{S}^{\dagger} \qquad \text{[QM2–8]}$$

but $$\mathbf{F} = \mathbf{F}^{\dagger} \qquad \text{[QM2–9]}$$

$$\epsilon = \epsilon^{\dagger}, \qquad \text{[QM2–10]}$$

and $$\mathbf{S} = \mathbf{S}^{\dagger} \qquad \text{[QM2–11]}$$

Therefore: $$\mathbf{C}^{\dagger}\mathbf{F} = \mathbf{E}\,\mathbf{C}^{\dagger}\,\mathbf{S} \qquad \text{[QM2–12]}$$

Substituting [QM2–6] and [QM2–12] into [QM2–5]

We obtain

$$\mathrm{tr}\{\mathbf{D}' \ \mathbf{F}\} = \mathrm{tr}\{(\mathbf{L}^{n}\epsilon\,\mathbf{C}^{\dagger}\,\mathbf{S}\,\mathbf{C}'\} + \mathrm{tr}\{\mathbf{L}^{n}\mathbf{C}'^{\dagger}\mathbf{S}\,\mathbf{C}\,\mathbf{E}\}$$

$$= \mathrm{tr}\{(\mathbf{L}^{n}\,\epsilon\,\mathbf{C}^{\dagger}\,\mathbf{S}\,\mathbf{C}'\} + \epsilon\,\mathbf{L}^{n}\,\mathbf{C}'^{\dagger}\,\mathbf{S}\,\mathbf{C}\} \qquad \text{[QM2–13]}$$

But $$\mathbf{L}^{n}\,\epsilon \;=\; \epsilon\,\mathbf{L}^{n}$$ [QM2–14]

since $\mathbf{L}^{n}$ is diagonal and ϵ is also diagonal

and therefore: $$\mathrm{tr}\{\mathbf{D}'\,\mathbf{F}\} \;=\; \mathrm{tr}\{\mathbf{E}\,\mathbf{L}^{n}(\mathbf{C}^{\dagger}\mathbf{S}\,\mathbf{C}' \;+\; \mathbf{C}'^{\dagger}\,\mathbf{S}\,\mathbf{C})\}$$ [QM2–15]

From the orthogonality condition, (as shown in [57] and [59]):

$$\mathbf{C}^{\dagger}\,\mathbf{S}\,\mathbf{C} \;=\mathbf{1}$$ [QM2–16]

Differentiating [QM2–16] we obtain

$$\mathbf{C}'^{+}\,\mathbf{S}\,\mathbf{C} \;=\; \mathbf{C}^{\dagger}\,\mathbf{S}'\,\mathbf{C} \;+\; \mathbf{C}^{\dagger}\,\mathbf{S}\,\mathbf{C}' \;=\; \phi$$ [QM2–17]

or $$-\,\mathbf{C}^{\dagger}\,\mathbf{S}'\,\mathbf{C} \;=\; \mathbf{C}^{\dagger}\,\mathbf{S}\,\mathbf{C}' \;+\; \mathbf{C}'^{\dagger}\,\mathbf{S}\,\mathbf{C}$$ [QM2–18]

Substituting [QM2–18] into [QM2–15] gives,

$$\mathrm{tr}\{\mathbf{D}'\;\mathbf{F}\} \;=\; \mathrm{tr}\{\epsilon\,\mathbf{L}^{n}(-\mathbf{C}^{\dagger}\;\mathbf{S}'\;\mathbf{C})\} \;=\; -\;\mathrm{tr}\{\underbrace{\mathbf{C}\;\mathbf{E}\;\mathbf{L}^{n}\;\mathbf{C}^{\dagger}}_{\mathbf{W}}\;\mathbf{S}'\}$$ [QM2–19]

Consequently:

$$\mathrm{tr}\{\mathbf{D}'\,\mathbf{F}\} \;=\; \mathrm{tr}\;\{\mathbf{W}\,\mathbf{S}'\}$$ [QM2–20]

with $$\mathbf{W} \;=\; \mathbf{C}\,\epsilon\,\mathbf{L}^{n}\,\mathbf{C}^{\dagger}$$ [QM2–21]

which is an energy–weighed density matrix.

Substituting [QM2–20] into [381] finally gives,

$$E' = \mathrm{tr}\{\mathbf{D}(2\mathbf{h}' + 2\mathbf{J}'[\mathrm{D}] - \mathbf{K}'[\mathrm{D}]) - 2\,\mathbf{WS}'\} \qquad [392]$$

This relationship clearly indicates that the gradient of the density matrix is not required for the calculation of the energy gradient.

The total energy for the system is defined by,

$$E_{Tot} = E + V_{nn} \qquad [393]$$

with E given by equation [366] and V_{nn} the nuclear–nuclear repulsion defined by [384]:

$$V_{nn} = \sum_{i=1}^{N} \sum_{j>i}^{N} Z_i Z_j \frac{1}{|R_i - R_j|} = \frac{1}{2} \sum_{i,\,j=1}^{N} (1 - \delta_{ij}) Z_i Z_j \frac{1}{|R_i - R_j|}$$

[394]

Note that $R_i - R_j$ in the denominator is a vectorial difference as it will be discussed below in MATHEMATICAL REVIEW NO. 2.

The change in V_{nn} with the components of the kth nuclei, R_{wk} with w = x,y,z, is given by,

$$\frac{dV_{nn}}{dR_{wk}} = \frac{1}{2} \sum_{i\,,\,j}^{N} (i - \delta_{ij}) Z_i Z_j \frac{d}{dR_{wk}} \left[\frac{1}{|R_i - R_j|}\right] \qquad [395]$$

which is in fact the components of the nuclear gradient vector.

MATHEMATICAL REVIEW NO. 2

Derivative of $|R_i - R_j|^{1/2}$.

$$|R_i - R_j| = \sqrt{(R_{xi} - R_{xj})^2 + (R_{yi} - R_{yj})^2 + (R_{zi} - R_{zj})^2} = u^{1/2} \qquad \text{[MR2–1]}$$

where

$$u = (R_{xi} - R_{xj})^2 + (R_{yi} - R_{yj})^2 + (R_{zi} - R_{zj})^2 \qquad \text{[MR2–2]}$$

$$\frac{d}{dR_{wk}}\left[\frac{1}{|R_i - R_j|}\right] =$$

$$\frac{d}{dR_{wk}}\left[\frac{1}{u^{1/2}}\right] = \frac{du^{-1/2}}{dR_{wk}} = \frac{du^{-1/2}}{du}\,\frac{du}{dR_{wk}} = -\frac{1}{2}\,u^{-3/2}\,\frac{du}{dR_{wk}} \qquad \text{[MR2–3]}$$

But

$$\frac{du}{dR_{wk}} = \frac{d}{dR_{wk}}(R_{xi} - R_{xj})^2 + \frac{d}{dR_{wk}}(R_{yi} - R_{yj})^2 + \frac{d}{dR_{wk}}(R_{zi} - R_{zj})^2 \qquad \text{[MR2–4]}$$

or

$$\frac{1}{2}\frac{du}{dR_{wk}} =$$

$$= (R_{xi} - R_{xj})\left\{\frac{dR_{xi}}{dR_{wk}} - \frac{dR_{xj}}{dR_{wk}}\right\} + (R_{yi} - R_{yi})\left\{\frac{dR_{yi}}{dR_{wk}} - \frac{dR_{yj}}{dR_{wk}}\right\} +$$

$$+ (R_{zi} - R_{zj})\left\{\frac{dR_{zi}}{dR_{wk}} - \frac{dR_{zj}}{dR_{wk}}\right\} \qquad \text{[MR2–5]}$$

Let us introduce the notation:

$$\frac{dR_{v\ell}}{dR_{wk}} = \delta_{vw}\,\delta_{\ell k} \qquad [MR2\text{–}6]$$

where $$v = x,y,z \text{ and } w = x,y,z \qquad [MR2\text{–}7]$$

Consequently:

$$\frac{1}{2}\,\frac{du}{dR_{wk}}\;(R_{xi} - R_{xj})\,(\delta_{xw}\delta_{ik} - \delta_{xw}\delta_{jk}) + (R_{yi} - R_{yj})\,(\delta_{yw}\delta_{ik} - \delta_{yw}\delta_{jk})$$

$$+ (R_{zi} - R_{zj})\,(\delta_{zw}\delta_{ik} - \delta_{zw}\delta_{jk}) \qquad [MR2\text{–}8]$$

On the basis of [MR2–8] we may construct du/dR_{xk} as well as du/dR_{yk} and du/dR_{zk} as follows:

$$\frac{du}{dR_{xk}} = 2(R_{xi} - R_{xj})\,(\delta_{ik} - \delta_{jk}) \qquad [MR2\text{–}9]$$

$$\frac{du}{dR_{yk}} = 2(R_{yi} - R_{yj})\,(\delta_{ik} - \delta_{jk}) \qquad [MR2\text{–}10]$$

$$\frac{du}{dR_{zk}} = 2(R_{zi} - R_{zj})\,(\delta_{ik} - \delta_{jk}) \qquad [MR2\text{–}11]$$

Substituting [MR2–9], into [MR2–3] with $w = x$ gives,

$$\frac{d}{dR_{xk}} \left[\frac{1}{|R_i - R_j|}\right] =$$

$$\left[(R_{xi} - R_{xj})^2 + (R_{yi} - R_{yj})^2 + (R_{zi} - R_{zj})^2\right]^{-3/2} (R_{xi} - R_{xj}) (\delta_{ik} - \delta_{jk})$$

[MR2–12]

Substituting [MR2–12] into equation [285] with w = x gives,

$$\frac{dV_{nn}}{dR_{xk}} =$$

$$-\frac{1}{2} \sum_{i,j}^{N} (1 - \delta_{ij}) Z_i Z_j \left[(R_{xi} - R_{xj})^2 + (R_{yi} - R_{yj})^2 + (R_{zi} - R_{zj})^2\right]^{-3/2}$$

$$(R_{xi} - R_{xj}) \delta_{ik} +$$

$$+ \frac{1}{2} \sum_{i,j}^{N} (1 - \delta_{ij}) Z_i Z_j \left[(R_{xi} - R_{xj})^2 + (R_{yi} - R_{yj})^2 + (R_{zi} - R_{zj})^2\right]^{-3/2}$$

$$(R_{xi} - R_{xj}) \delta_{ik} \qquad [396]$$

The first term will not vanish if i = k while the second term will not vanish if j = k.

Consequently:

$$\frac{dV_{nn}}{dR_{xk}} =$$

$$-\frac{1}{2}\sum_{j=i}^{N} (1-\delta_{kj})\, Z_k\, Z_j \left[(R_{xk}-R_{xj})^2 + (R_{yk}-R_{yj})^2 + (R_{zk}-R_{zj})^2\right]^{-3/2}$$

$$(R_{xk} - R_{kj})$$

$$+\ \frac{1}{2}\sum_{i=1}^{N} (1-\delta_{ik})\, Z_i\, Z_k \left[(R_{xi}-R_{xk})^2 + (R_{yi}-R_{yk})^2 + (R_{zi}-R_{zk})^2\right]^{-3/2}$$

$$(R_{xi} - R_{ik}) \qquad [397]$$

But i and j are indices running from 1 to N so that we can replace j in the 1st term with i.

However, $$(R_{xk} - R_{xi})^2 = (R_{xi} - R_{xk})^2 \qquad [398]$$

and $$(R_{xi} - R_{xk}) = -(R_{xk} - R_{xi}) \qquad [399]$$

Taking this into consideration, the first term becomes positive and therefore, we may say that

$$\frac{dV_{nn}}{dR_{xk}} =$$

$$\sum_{\substack{i=1 \\ i \neq k}}^{N} Z_i Z_k \left[(R_{xi} - R_{xk})^2 + (R_{yi} - R_{yk})^2 + (R_{zi} - R_{zk})^2\right]^{-3/2} (R_{xi} - R_{xk})$$

[400]

The procedure may be repeated for dV_{nn}/dR_{yk} and dV_{nn}/dR_{zk} utilizing [MR2–10] and [MR2–16], giving the following general result.

$$\frac{dV_{nn}}{dR_{xk}} =$$

$$\sum_{\substack{i=1 \\ i \neq k}}^{N} Z_i Z_k (R_{wi} - R_{wk}) \left[(R_{xi} - R_{xk})^2 + (R_{yi} - R_{yk})^2 + (R_{zi} - R_{zk})^2\right]^{-3/2}$$

[401]

for $w = x,y,z$

According to equation [383], one can evaluate the change in the total energy of the system with a change in the components of the kth nuclei by:

$$\frac{dE_{tot}}{dR_{wk}} = \frac{dE}{dR_{wk}} + \frac{dV_{nn}}{dR_{wk}} \qquad [402]$$

where $\partial V_{nn}/dR_{wk}$ is given by [391] and $dE/\partial R_{wk}$ is given by equation [382], where,

$$h'_{pq} = \frac{dh_{pq}}{dR_{wk}} = \frac{d}{dR_{wk}} \left\langle \eta_p | \hat{h} | \eta_q \right\rangle \quad [403]$$

$$J'[D]_{pq} = \sum_{k,l} D_{kl} \frac{d}{dR_{wk}} \left\langle \eta_p(1)\eta_q(1) | \hat{q}_{12} | \eta_k(2)\eta_l(2) \right\rangle \quad [404]$$

$$K'[D]_{pq} = \sum_{k,l} D_{kl} \frac{d}{dR_{wk}} \left\langle \eta_p(1)\eta_k(2) | \hat{q}_{12} | \eta_q(1)\eta_l(2) \right\rangle \quad [405]$$

and

$$S'_{pq} = \frac{d}{dR_{wk}} \left\langle \eta_p | \eta_q \right\rangle \quad [406]$$

OVERLAP GRADIENT: S'_{pq}.

Applying the chain rule of differentiation to [396] will yield:

$$S'_{pq} = \frac{d}{dR_{wk}} \left\langle \eta_p | \eta_q \right\rangle = \left\langle \frac{d\eta_p}{dR_{wk}} \middle| \eta_q \right\rangle + \left\langle \eta_p \middle| \frac{d\eta_q}{dR_{wk}} \right\rangle \quad [407]$$

However $\partial/\partial R_{wk}$ is equivalent to $\partial/\partial A_{wk}$, therefore:

$$S'_{pq} = \left\langle \frac{d\eta_p}{dA_{wk}} \middle| \eta_q \right\rangle + \left\langle \eta_p \middle| \frac{d\eta_q}{dA_{wk}} \right\rangle \quad [408]$$

This result indicates that for the gradient of the overlap integral requires the gradients of the Gaussians.

Considering a simple unnormalized s–type Gaussian,

$$\eta_s = e^{-\alpha[(x - A_x)^2 + (y - A_y)^2 + (z - A_z)^2]} \qquad [409]$$

its gradient along a given coordinate, say x, gives a p_x type Gaussian

$$\frac{\partial}{\partial A_x} \eta_s = \frac{\partial}{\partial A_x} \left\{ e^{-\alpha[(x - A_x)^2 + (y - A_y)^2 + (z - A_z)^2]} \right\} =$$

$$= e^{-\alpha[(x - A_x)^2 + (y - A_y)^2 + (z - A_z)^2]} \bullet \frac{\partial}{\partial A_x} \{-\alpha(x - A_x)^2\} =$$

$$\eta_s \; \frac{\partial}{\partial A_x} \{-\alpha(x - A_x)^2\} = -2\alpha(x - A_x) \; \eta_s \frac{\partial}{\partial A_x} (x - A_x) =$$

$$= +2\alpha(x - A_x)\eta_s =$$

$$= 2\alpha\eta_{p_x} \qquad [410]$$

Returning to S'_{pq} as given in [398] we can distinguish 3 cases for <s–type|s–type> overlaps.

CASE I

If both η_p and η_q belong to the kth center, the integral gradient has the following:

$$S'_{pq} = \frac{d}{dR_{xk}} \left\langle \eta_p | \eta_q \right\rangle = \frac{d}{dR_{xk}} \text{<s–type|s–type>} =$$

$$= \text{<}p_x\text{–type|s–type>} + \text{<s–type|}p_x\text{–type>} \qquad [411]$$

However, due to s/p orthogonality on the same centre both terms will vanish (c.f. Figure 24).

CASE II

If only one of the two Gaussians, say η_p belongs to the kth centre, the integral gradient formalism is simpler,

$$S'_{pq} = \frac{d}{dR_{xk}} \left\langle \eta_p | \eta_q \right\rangle = \frac{d}{dR_{xk}} \langle s\text{–type} | s\text{–type} \rangle = \langle p_x\text{–type} | s\text{–type} \rangle \qquad [412]$$

since the second term in [401] is zero because η_q is constant with respect to the variable along which the differentiation is taking place.

CASE III.

If neither of the two Gaussians belong to the k–th centre then the integral will vanish because both η_p and η_q are constant with respect to the variable along which the differentiation is taking place.

$$S'_{pq} = \frac{d}{dR_{xk}} \left\langle \eta_p | \eta_q \right\rangle = 0 \qquad [413]$$

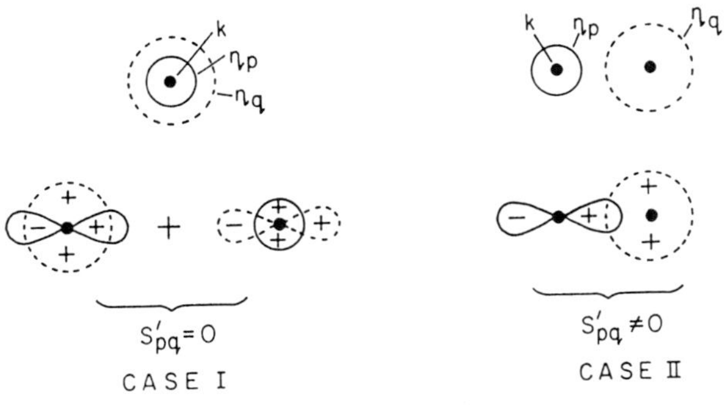

Figure 24. Two cases of overlap integral gradients.

CORE INTEGRAL GRADIENT: h'_{pq}:

Applying the chain rule to [393] will give,

$$h'_{pq} = \frac{d}{dR_{wk}} \left[\left\langle \eta_p | \hat{h} | \eta_q \right\rangle \right] =$$

$$\left\langle \frac{d\,\eta_p}{dR_{wk}} \middle| \hat{h} \middle| \eta_q \right\rangle + \left\langle \eta_p \middle| \frac{d\,\hat{h}}{dR_{wk}} \middle| \eta_q \right\rangle + \left\langle \eta_p \middle| \hat{h} \middle| \frac{\partial\,\eta_q}{dR_{wk}} \right\rangle \qquad [414]$$

Again, we may say that,

$$\frac{d\,\eta_p}{dR_{wk}} = \frac{d\,\eta_p}{dA_{wk}} \text{ and } \frac{d\,\eta_q}{dR_{wk}} = \frac{d\,\eta_q}{dA_{wk}} \qquad [415]$$

The first and last terms in [404] are regular one–electron integrals but, of course, the Gaussian derivatives are also Gaussians of higher angular type (i.e. s ⟶ p, p ⟶ d, d ⟶ f, etc.). The second term in [404], however, requires the derivation of an operator. Remembering that the Core–Hamiltonian operator consists of the kinetic energy and the nuclear–electron attraction operators:

$$\hat{h} = \frac{1}{2} \nabla^2 - \sum_{i=1}^{N} Z_i \frac{1}{|r - R_i|} \qquad [416]$$

we may write:

$$\frac{d\hat{h}}{dR_{wk}} = -\frac{1}{2} \frac{d\nabla^2}{dR_{wk}} - \sum_{i=1}^{N} Z_i \frac{d}{dR_{wk}} \left[\frac{1}{|r - R_i|} \right] \qquad [417]$$

However $d\nabla^2/dR_{wk} = 0$ and therefore:

$$\frac{d\hat{h}}{dR_{wk}} = - \sum_{i=1}^{N} Z_i \frac{d}{dR_{wk}} \left[\frac{1}{|r - R_i|}\right] \quad [418]$$

The derivative in the second term of [407] which is also given in [408] is similar to equation [MR2–3]. Considering first d/dR_{xk} gives,

$$\frac{d}{dR_{xk}} \left[\frac{1}{|r - R_i|}\right] = \quad [419]$$

$$- \left[(r_x - R_{xi})^2 + (r_y - R_{yi})^2 + (r_z - R_{zi})^2\right]^{-3/2} (r_x - R_{xi}) \left[\frac{dr_x}{dR_{xk}} - \frac{dR_{xi}}{dR_{xk}}\right]$$

But $dr_x/dR_{xk} = 0$ and $dR_{xi}/dR_{xk} = -\delta_{ik}$. Therefore:

$$\frac{d}{dR_{xk}} \left[\frac{1}{|r - R_i|}\right] =$$

$$\delta_{ik}(r_x - R_{xi}) \left[(r_x - R_{xi})^2 + (r_y - R_{yi})^2 + (r_z - R_{zi})^2\right]^{-3/2} \quad [420]$$

Substituting this into [408] with w = x will give:

$$\frac{d\hat{h}}{dR_{xk}} =$$

$$- \sum_{i=1}^{N} Z_i \delta_{ik}(r_x - R_{xi}) \left[(r_x - R_{xi})^2 + (r_y - R_{yi})^2 + (r_z - R_{zi})^2\right]^{-3/2}$$

[421]

Because of δ_{ik} only the i = k term is non–zero, generalizing this to $d\hat{h}/dR_{yk}$ and $d\hat{h}/dR_{zk}$ will give:

$$\frac{d\hat{h}}{dR_{wk}} =$$

$$Z_k \; (r_w - R_{wk}) \; \left[(r_x - R_{xk})^2 + (r_y - R_{yk})^2 + (r_z - R_{zk})^2\right]^{-3/2} \qquad [422]$$

with w = x,y,z.

We may evaluate the second term in [404] using [412] as the operator

COULOMB AND EXCHANGE GRADIENTS: $J'[D]$ AND $K'[D]$

According to [394] and [395] both $J'[D]$ and $K'[D]$ require derivatives of the form,

$$\frac{d}{dR_{wk}}\left[\left\langle \eta_p \eta_q | \hat{g}_{12} | \eta_k \eta_\ell \right\rangle\right] \qquad [423]$$

Applying the chain rule of differentiation to [313], one obtains the following expression:

$$\frac{d}{dR_{wk}}\left[\left\langle \eta_p \eta_q | \hat{g}_{12} | \eta_k \eta_\ell \right\rangle\right] =$$

$$\left\langle \frac{d\eta_p}{dA_{wk}} \eta_q | \hat{g}_{12} | \eta_k \eta_e \right\rangle + \left\langle \eta_p \frac{d\eta_q}{dA_{wk}} | \hat{q}_{j2} | \eta_k \eta_\ell \right\rangle$$

$$+ \left\langle \eta_p \eta_q \left| \frac{d\hat{g}_{12}}{dA_{wk}} \right| \eta_k \eta_\ell \right\rangle + \left\langle \eta_p \eta_q \left| \hat{g}_{12} \right| \frac{d\eta_k}{dA_{wk}} \eta_e \right\rangle + \left\langle \eta_p \eta_q \left| \hat{g}_{12} \right| \eta_k \frac{d\eta_\ell}{dA_{wk}} \right\rangle \qquad [424]$$

However $\hat{g}_{12}$ is the electron–electron repulsion operator and, therefore, does not depend on R_{wk}. Consequently:

$$\frac{d}{dR_{wk}} \left[\left\langle \eta_p \eta_q \left| \hat{g}_{12} \right| \eta_k \eta_\ell \right\rangle \right] =$$

$$\left\langle \frac{d\eta_p}{dA_{wk}} \eta_q \left| \hat{g}_{12} \right| \eta_k \eta_e \right\rangle + \left\langle \eta_p \frac{d\eta_p}{dA_{wk}} \left| \hat{g}_{12} \right| \eta_k \eta_\ell \right\rangle +$$

$$\left\langle \eta_p \eta_q \left| \hat{g}_{12} \right| \frac{d\eta_k}{dA_{wk}} \eta_\ell \right\rangle + \left\langle \eta_p \eta_q \left| \hat{g}_{12} \right| \eta_k \frac{d\eta_\ell}{dA_{wk}} \right\rangle \qquad [425]$$

Using [415] equations [394] and [395] may be evaluated.

4.4 GRADIENTS OF ATOMIC ORBITALS

It became clear in Section 4.3 that to compute the gradient of the energy, one needs the gradients of the molecular integrals. The formulation of the gradients of molecular integrals also showed that the gradients of atomic orbitals are essential. This final section deals therefore with the differentiation of atomic orbitals constructed from primitive Cartesian Gaussians.

4.4.1 ATOMIC ORBITALS

The ith AO denoted as

$$\eta_i = \eta_i[l_i, A_j, \{d_{ik}\}, \{\alpha_{ik}\}, r] \qquad [426]$$

which involves the following parameters.

$l_i = (l_{xi}, l_{yi}, l_{zi}) \equiv$ the type of A.O. (i.e. s,$p_{xi}p_{yi}p_z$...)

$A_j = (A_{xj}, A_{yj}, A_{zj}) \equiv$ the center of the A.O. (i.e. j^{th} nucleus)

$\{d_{ik}\} \equiv$ a set of contraction coefficients necessary to from η_i

$\{\alpha_{ik}\} \equiv$ a set of exponents for the primitive Gaussians to be contracted to η_i

$r = (r_x, r_y, r_z) \equiv$ distance measured from the atomic nucleus

Each atomic orbital is formed from a linear combination of primitive Gaussians:

$$\eta_i[l_i, A_j, \{d_{ik}\}, \{\alpha_{ik}\}, r] = \sum_{k=1}^{a_i} d_{ik}\, P[l_i, A_j, r] \,.\, E[A_j, \alpha_{ik}, r] \qquad [427]$$

where,

a_i ≡ the number of primitive Gaussian contracted to the ith A.O.

k ≡ an index over the primitive Gaussians included in η_i.

The polynomial $P = P[l_i, A_j, r\,]$ and the exponential $E = E[A_j, \alpha_{ik}, r\,]$ of [417] are given by,

$$P[l_i, A_j, r\,] = (r_x - A_{xj})^{l_{x_i}} (r_y - A_{yj})^{l_{y_i}} (r_z - A_{zj})^{l_{z_i}} \qquad [428]$$

$$E[A_j, \alpha_{ik}, r\,] = \exp\{-\alpha_{ik} |r - A_j|^2\} \qquad [429]$$

The exponential may also be written as the arithmetic product of three Gaussians:

$$E[A_j, \alpha_{ik}, r\,] =$$

$$= \exp\{-\alpha_{ik}(r_x - A_{x_j})^2\} \exp\{-\alpha_{ik}(r_y - A_{y_j})$$

$A_{yj})^2\}$

$$= \exp\{-\alpha_{ik}(r_z - A_{zj})^2\} \qquad [430]$$

4.4.2 DERIVATIVES OF ATOMIC ORBITALS:

According to equation [417], the derivative of η_i, with respect to some parameter λ, is given by,

$$\frac{\partial \eta_i}{\partial \lambda} = \frac{\partial}{\partial \lambda} \left[\sum_{k=1}^{a_i} d_{ik}\, P\, E \right] \qquad [431a]$$

Keeping a_i constant (i.e. the number of Gaussians in η_i does not depend on λ?

gives,

$$\frac{\partial \eta_i}{\partial \lambda} = \sum_{k=1}^{a_i} \left[\frac{dd_{ik}}{d\lambda} P.E + d_{ik} \frac{\partial P}{\partial \lambda} E + d_{ik} P \frac{\partial E}{\partial \lambda} \right] \quad [431b]$$

EVALUATION OF $\partial P/\partial \lambda$

Using [418] we may set out to differentiate P against λ as follows:

$$\frac{\partial P}{\partial \lambda} \left[\frac{e_i, A_j, r}{\partial \lambda}\right] = \frac{\partial}{\partial \lambda} \left[(r_x - A_{xj})^{lx_i} (r_y - A_{yj})^{ly_i} (r_z - A_{zj})^{lz_i} \right] \quad [432]$$

$$\frac{\partial P}{\partial \lambda} \left[\frac{e_i, A_j, r}{\partial \lambda}\right] =$$

$$\left[\frac{\partial}{\partial \lambda}(r_x - A_{xj})^{lx_i}\right] (r_y - A_{yj})^{ly_i} (r_z - A_{zj})^{lz_i}$$

$$+ (r_x - A_{xj})^{lx_i} \left[\frac{\partial}{\partial \lambda}(r_y - A_{yj})^{ly_i}\right] (r_z - A_{zj})^{lz_i}$$

$$+ (r_x - A_{xj})^{lx_i} (r_y - A_{yj})^{ly_i} \left[\frac{\partial}{\partial \lambda}(r_z - A_{zj})^{lz_i}\right] \quad [433]$$

Each of the three terms in [423] involve derivatives of the form given by [424] with

w = x,y or z.

$$\frac{\partial}{\partial \lambda} (r_w - A_{wj})^{lw_i} \quad [434]$$

Let

$$u = (r_w - A_{wj}) \quad [435]$$

and $$v = l_{wi} \quad [436]$$

so that $$u^v = (r_w - A_{wj})^{l_{w_i}} \quad [437]$$

Let $$u^v = e^y \quad [438]$$

so that $$\ln(u^v) = v \ln u = \ln(e^y) = y \quad [439]$$

Hence,

$$(r_w - A_{wj})^{l_{wi}} = u^v = e^y = e^{v \ln u} \quad [440]$$

Using [430] to evaluate [424] gives,

$$\frac{\partial(r_w - A_{wj})}{\partial\lambda} = \frac{\partial}{\partial\lambda}(r_w - A_{wj})^{lwi} = \frac{\partial e^y}{\partial\lambda} = \frac{\partial e^y}{\partial y} \cdot \frac{\partial y}{\partial\lambda} = e^y \frac{\partial}{\partial\lambda}(v \ln u)$$

$$= u^v \ln u \frac{\partial v}{\partial\lambda} + u^v V \frac{\partial l_{nu}}{\partial\lambda} \quad [441]$$

But $$\frac{\partial \ln u}{\partial\lambda} = \frac{\partial \ln u}{\partial u} \cdot \frac{\partial u}{\partial\lambda} = \frac{1}{u}\frac{\partial u}{\partial\lambda} \quad [442]$$

therefore:

$$\frac{\partial}{\partial\lambda}(r_w - A_{wj})^{l_{wi}} = u^v \ln u \frac{\partial V}{\partial\lambda} + v u^{v-1} \frac{\partial u}{\partial\lambda} \quad [443]$$

Substituting for u and v gives,

$$\frac{\partial}{\partial\lambda}(r_w - A_{wj})^{l_{wi}} = (r_w - A_{wj})^{l_{wi}} \ln(r_w - A_{wj}) \frac{d\,l_{wi}}{\partial\lambda}$$

$$+ \; l_{wi}(r_w - A_{wj})^{l_{wi}-1} \frac{d}{\partial\lambda}(r_w - A_{wj}) \qquad [444]$$

Alternatively we may write after performing the differetiation of the last term:

$$\frac{\partial}{\partial\lambda}(r_w - A_{wj})^{l_{wi}} = (r_w - A_{wj})^{l_{wi}} \ln(r_w - A_{wj}) \left[\frac{d\,l_{wi}}{\partial\lambda}\right] +$$

$$+ \; l_{wi}\,(r_w - A_{wj})^{l_{wi}-1} \left[\frac{d\,r_w}{\partial\lambda}\right]$$

$$- \; l_{wi}\,(r_w - A_{wj})^{l_{wi}-1} \left[\frac{dA_{wi}}{\partial\lambda}\right] \qquad [445]$$

for w = x,y or z

Once $\partial l_{wi}/\partial\lambda$, $\partial r_w/\partial\lambda$, and $\partial A_{wj}/\partial\lambda$ are known (w = x,y,x) then $\partial P/\partial\lambda$ of equation [421] can be evaluated from equations [423] and [435].

EVALUATION OF $\partial E/\partial\lambda$

Using [420] we may set out to differentiate E against λ as follows:

$$\frac{\partial E}{\partial\lambda}[A_j, \alpha_{ik}, r\,]$$

$$= \frac{\partial}{\partial \lambda} \left[e^{-\alpha_{ik}(r_x - A_{xj})^2} e^{-\alpha_{ik}(r_y - A_{yj})^2} e^{-\alpha_{ik}(r_z - A_{zj})^2} \right] \qquad [446]$$

in details we may obtain:

$$\frac{\partial E}{\partial \lambda} \frac{\partial E[A_j, \alpha_{ik}, r]}{\partial \lambda} =$$

$$\left[\frac{\partial}{\partial \lambda} e^{-\alpha_{ik}(r_x - A_{xj})^2} \right] e^{-\alpha_{ik}(r_y - A_{yj})^2} e^{-\alpha_{ik}(r_z - A_{zj})^2}$$

$$+ \; e^{-\alpha_{ik}(r_x - A_{xj})^2} \left[\frac{\partial}{\partial \lambda} e^{-\alpha_{ik}(r_y - A_{yj})^2} \right] e^{-\alpha_{ik}(r_z - A_{zj})^2}$$

$$+ \; e^{-\alpha_{ik}(r_x - A_{xj})^2} e^{-\alpha_{ik}(r_y - A_{yj})^2} \left[\frac{\partial}{\partial \lambda} e^{-\alpha_{ik}(r_z - A_{zj})^2} \right] \qquad [447]$$

Each term in [437] involves derivatives of the form,

$$\frac{\partial}{\partial \lambda} e^{-\alpha_{ik}(r_w - A_{wj})^2} \qquad [448]$$

with $w = x, y,$ or z.

Let $u = -\alpha_{ik}(r_w - A_{wj})^2$ so that equation [438] becomes,

$$\frac{de^u}{\partial\lambda} = \frac{de^u}{\partial u} \cdot \frac{du}{\partial\lambda} = e^u \frac{du}{\partial\lambda}$$

$$= e^{-\alpha_{ik}(r_w - A_{wj})^2} \frac{\partial}{\partial\lambda}(-\alpha_{ik}(r_w - A_{wj})^2 \qquad [449]$$

But

$$\frac{\partial}{\partial\lambda}(-\alpha_{ik}(r_w - A_{wj})^2 = (r_w - A_{wj})^2 \frac{\partial}{\partial\lambda}(-\alpha_{ik}) - \alpha_{ik}\frac{\partial}{\partial\lambda}(r_w - A_{wj})^2 =$$

$$= -(r_w - A_{wj})^2 \frac{\partial}{\partial\lambda}(-\alpha_{ik}) - 2\,\alpha_{ik}(r_w - A_{wj})\frac{\partial}{\partial\lambda}(r_w - A_{wj}) \qquad [450]$$

Therefore [439] becomes:

$$\frac{\partial}{\partial\lambda} e^{-\alpha_{ik}(r_w - A_{wj})^2} = -(r_w - A_{wj})^2\, e^{-\alpha_{ik}(r_w - A_{wj})^2}\left[\frac{\partial\alpha_{ik}}{\partial\lambda}\right]$$

$$-2\,\alpha_{ik}(r_w - A_{wk})\; e^{-\alpha_{ik}(r_w - A_{wj})^2}\left[\frac{\partial\alpha_w}{\partial\lambda}\right]$$

$$+2\,\alpha_{ik}(r_w - A_{wj})\; e^{-\alpha_{ik}(r_w - A_{wj})^2}\left[\frac{\partial A_{wj}}{\partial\lambda}\right] \qquad [451]$$

with w = x,y or z.

Once $\partial\alpha_{ik}/\partial\lambda$, $\partial r_w/\partial\lambda$, and $\partial A_{wj}/\partial\lambda$ are known (for w = x,y, and z), then $\partial E/\partial\lambda$ of equation [421] can be evaluated from equations [437] and [441].

Therefore, we may evaluate $\partial\eta_i/\partial\lambda$ of equations [421] [423], [435], [437] and [441]

$$\left[\frac{\partial d_{ik}}{\partial\lambda}\right], \left[\frac{\partial l_{wi}}{\partial\lambda}\right], \left[\frac{\partial r_w}{\partial\lambda}\right], \left[\frac{\partial A_{wj}}{\partial\lambda}\right], \text{ and } \left[\frac{\partial\alpha_{ik}}{\partial\lambda}\right] \qquad [452]$$

(i.e. the effect of λ on d_{ik}, l_{wi}, r_w, A_{wj} and x_{ik} respectively).

4.4.3 ATOMIC DERIVATIVES IN GEOMETRY OPTIMIZATION

In geometry optimization calculations, within all ab initio SCF. calculations, AO. types (i.e. the l_i vectors) are held constant. Therefore, any parameter λ has no effect on l_i.

$$\frac{\partial l_{wi}}{\partial\lambda} = 0 \qquad [453]$$

As well, if the contraction coefficients and the exponents of the basis set are held constant (i.e. no optimization of the basis set), then,

$$\frac{\partial d_{ik}}{\partial\lambda} = 0 \quad \text{and} \quad \frac{\partial\alpha_{ik}}{\partial\lambda} = 0 \qquad [454]$$

since the centers for the AO correspond to nuclear positions (when no floating Gaussians are used), then optimizing the geometry is equivalent to minimizing the total energy with respect to A_p (with p = 1 to N and N ≡ total number of nuclei).

Hence, $$\partial/\partial\lambda \equiv \partial/\partial A_{vp} \quad (v = x, y, \text{ or } z) \qquad [455]$$

We can find $\partial\eta_i/\partial A_{vp}$ under conditions [443] and [444] if we know

$$\frac{\partial r_w}{\partial A_{vp}} \equiv \text{the change in } r_w \text{ with } A_{vp} \qquad [456]$$

$$\frac{\partial A_{wj}}{\partial A_{vp}} \equiv \text{the change in } A_{wj} \text{ with } A_{vp} \qquad [457]$$

However, the vector r depends on the coordinate reference frame used in the calculation. Considering the reference frame to be independent of A_{vp} gives,

$$\frac{\partial r_w}{\partial A_{vp}} = 0 \qquad [458]$$

We can evaluate $\partial A_{wj}/\partial A_{vp}$ by considering the nuclei to be uncoupled (i.e. the jth nucleus is not coupled to the pth nucleus, unless $j = p$). Then gives,

$$\frac{\partial A_{wj}}{\partial A_{vp}} = \delta_{jp} \frac{\partial A_{wj}}{\partial A_{vj}} \qquad [459]$$

Considering the x,y,z components of A_j to be independent gives,

$$\frac{\partial A_{wj}}{\partial A_{vj}} = \delta_{wv} \frac{\partial A_{wj}}{\partial A_{wj}} = \delta_{wv} \qquad [460]$$

Since $\partial A_{wj}/\partial A_{wj} = 1$ (as $\partial x/\partial x = 1$)

Therefore,

$$\frac{\partial A_{wj}}{\partial A_{vp}} = \delta_{jp}\,\delta_{wv} \qquad [461]$$

Using [443], [448], and [451] gives for [435]

$$\frac{\partial (r_w - A_{wj})^{l_{wi}}}{\partial A_{vp}} = -\,\delta_{jp}\delta_{wv} l_{wi} (r_w - A_{wj})^{l_{wi}-1} \qquad [462]$$

and using [444], [448], and [451] gives for [441],

$$\frac{\partial e^{-\alpha_{ik}(r_w - A_{wj})^2}}{\partial A_{vp}} = 2\,\alpha_{ik}\,\delta_{jp}\delta_{wv}\,(r_w - A_{wj})\,e^{-\alpha_{ik}(r_w - A_{wj})^2} \qquad [463]$$

According to [452],

$$\frac{\partial (r_x - A_{xj})^{l_{xi}}}{\partial A_{xp}} = -\,\delta_{jp} l_{xi}\,(r_x - A_{xj})^{l_{xi}-1} \qquad [464a]$$

$$\frac{\partial (r_y - A_{yj})^{l_{yi}}}{\partial A_{xp}} = 0 \qquad [464b]$$

$$\frac{\partial (r_z - A_{zj})^{l_{zi}}}{\partial A_{xp}} = 0 \qquad [464c]$$

Using these, equation [423] becomes,

$$\frac{\partial P}{\partial A_{xp}}[b_i,A_j,r] = \delta_{jp}\, l_{xi}(r_x - A_{xj})^{l_{xi}-1}$$

$$(r_y - A_{yj})^{l_{yi}}(r_z - A_{zj})^{l_{zi}} \qquad [465]$$

or

$$\frac{\partial P[\, l_{xi},l_{yi},l_{zi}, A_j,r]}{\partial A_{xp}} = -\delta_{jp} l_{xi}\ P[l_{xi}-1,l_{yi},l_{zi}, A_j,r] \qquad [466]$$

According to [453]

$$\frac{\partial e^{-\alpha_{ik}(r_x - A_{xj})^2}}{\partial A_{xp}} = 2\ \delta_{jp}\ \alpha_{ik}\ (r_x - A_{xj})\ e^{-\alpha_{ik}(r_x - A_{xj})^2} \qquad [467a]$$

$$\frac{\partial e^{-\alpha_{ik}(r_y - A_{yj})^2}}{\partial A_{xp}} = 0 \qquad [467b]$$

$$\frac{\partial e^{-\alpha_{ik}(r_z - A_{zj})^2}}{\partial A_{xp}} = 0 \qquad [467c]$$

This gives for equation, [437]

$$\frac{\partial E[A_j,\alpha_{ik},r]}{\partial A_{x_p}} = 2\delta_{jp}\alpha_{ik}(r_x - A_{xj})e^{-\alpha_{ik}(r_x - A_{xj})^2}$$

$$e^{-\alpha_{ik}(r_y - A_{yj})^2}\, e^{-\alpha_{ik}(r_z - A_{zj})^2}$$

$$\frac{\partial E[A_j,\alpha_{ik},r]}{\partial A_{xp}} = 2\delta_{jp}\alpha_{ik}(r_x - A_{xj})\, E[A_j,\alpha_{ik},r] \qquad [468]$$

Finally, since $\partial d_{ik}/\partial A_{vp} = 0$ according to [444], then from equation [456] and [459] equation [421] becomes,

$$\frac{\partial \eta_i[l_i,A_j,\{d_{ik}\},\{\alpha_{ik}\},r]}{\partial A_{xp}} =$$

$$\sum_{k=1}^{a_i}\Big[-d_{ik}\delta_{jp}l_{xi}P[l_{xi}-1,l_{yi},l_{zi},A_j,r]\,E[A_j,\alpha_{ik},r] +$$

$$+ 2\,d_{ik}\delta_{jp}\alpha_{ik}(r_x - A_{xj})\,P[l_{xi},l_{yi},l_{zi},A_j,r]\,E[A_j,\alpha_{ik},r]\Big] \qquad [469a]$$

$$= \delta_{jp}l_{xi}\sum_{k=1}^{a_i} d_{ik}\; P[l_{xi}-1,\, l_{yi},l_{zi},A_j,r]\,E[A_j,\,\alpha_{ik},\,r]$$

$$+ \ \delta_{jp} \sum_{k=1}^{a_i} 2\, d_{ik}\, \alpha_{ik}\, P[l_{xi}+1, l_{yi}, l_{zi}, A_j, r]\, E[A_j, \alpha_{ik}, r] \qquad [469b]$$

$$= \delta_{jp}\, \eta_i\, [l_{xi}+1, l_{yi}, l_{zi}, A_j, \{2d_{ik}\, \alpha_{ik}\}, \{\alpha_{ik}\}, r]$$

$$-\delta_{jp} l_{xi} \eta_i [l_{xi}-1, l_{yi}, l_{zi}, A_j, \{d_{ik}\}, \{d_{ik}\}, r] \qquad [469c]$$

$$\frac{\partial \eta_i}{\partial A_{xp}} [l_{xi}, l_{yi}, l_{zi}, A_j, \{d_{ik}\}] = \delta_{jp} \Big[\eta_i [l_{xi}, l_{yi}, l_{zi}, A_j, \{2\alpha_{ik} d_{ik}\}]$$

$$- l_{xi} \eta_i [l_{xi}-1, l_{yi}, l_{zi}, A_j, \{d_{ik}\}] \Big] \qquad [470a]$$

Similarly,

$$\frac{\partial \eta_i}{\partial A_{yp}} [l_{xi}, l_{yi}, l_{zi}, A_j, \{d_{ik}\}] = \delta_{jp} \Big[\eta_i [l_{xi}, l_{yi}+1, l_{zi}, A_j, \{2\alpha_{ik} d_{ik}\}]$$

$$- l_{yi} \eta_i [l_{xi}, l_{yi}-1, l_{zi}, A_j, \{d_{ik}\}] \Big] \qquad [470b]$$

and,

$$\frac{\partial \eta_i}{\partial A_{zp}} [l_{xi}, l_{yi}, l_{zi}, A_j, \{d_{ik}\}] = \delta_{jp} \Big[\eta_i [l_{xi}, l_{yi}, l_{zi}+1, A_j, \{2\alpha_{ik} d_{ik}\}] \ -$$

$$- l_{zi} \eta_i [l_{xi}, l_{yi}, l_{zi}-1, A_j, \{d_{ik}\}] \Big] \qquad [470c]$$

Examples for s and p type orbitals are given in Table 5

Table 5. The change of orbital angular type as the consequence of differentiation

$\eta_i[\vec{l}_i]$	$\partial\eta_i[\vec{l}_i]/\partial A_{xp}$		$\partial\eta_i[\vec{l}_i]/\partial A_{yp}$		$\partial\eta_i[\vec{l}_i]/\partial A_{zp}$	
(l_{xi}, l_{yi}, l_{zi})	$(l_{xi}+1, l_{yi}, l_{zi})$	$(l_{xi}-1, l_{yi}, l_{zi})$	$(l_{xi}, l_{yi}+1, l_{zi})$	$(l_{xi}, l_{yi}-1, l_{zi})$	$(l_{xi}, l_{yi}, l_{zi}+1)$	$(l_{xi}, l_{yi}, l_{zi}-1)$
(0,0,0) s-type	(1,0,0) p_x-type		(0,1,0) p_y-type		(0,0,1) p_z-type	
(1 0 0) p_x-type	(2,0,0) d_{x^2} type	(0,0,0) s-type	(1,1,0) d_{xy}-type		(1,0,1) d_{xz}-type	
(0,1,0) p_y-type	(1,1,0) d_{xy}-type		(0,2,0) d_{y^2} type	(0,0,0) s-type	(0,1,1) d_{yz}-type	
(0,0,1) p_z-type	(1,0,1) d_{xz}-type		(0,1,1) d_{yz}-type		(0,0,2) d_{z^2}-type	(0,0,0) s-type

5. PROBLEMS

1 Calculation of the SCF energy and Mulliken population analysis for the He–H(+) system, which has only 2 electrons and therefore only 1 (doubly) occupied MO. The smallest possible basis set is used: 1 gaussian function (uncontracted) on each atom with the following orbital exponents (α):

He exponent: 0.77385975

H exponent: 0.41661243

The SCF for a different bond length may be calculated and thereby allowing the potential energy curve for the dissociation of He–H(+) to He + H(+) to be plotted. Initial MO coefficients are given for use, so that only a few cycles of the SCF must be done to achieve convergence. All calculations can be performed on a calculator, although writing a short computer program to handle some of the repeated operations is encouraged.

(*i*) Convert the bond length (given to you in Bohr) to Angstroms so that you have a feel for the internuclear separation.

(*ii*) Calculate the overlap matrix: S, normalized for the basis set:

$$\mathbf{S} = \begin{bmatrix} 1 & S_{12} \\ S_{21} & 1 \end{bmatrix}$$

Where $S_{21} = S_{12}$. Note that S_{12} (off–diagonal element of S) will be on the order of 0.4 at 1.75 Bohr (larger for short distances, smaller for long distances).

(*iii*) Calculate $V = S^{**}(-1/2)$, which orthogonalizes the AO basis, using the Jacobi diagonalization technique. (The smaller eigenvalue of S should match the value given below.)

(*iv*) Calculate the nuclear repulsion energy $E_n = V_{nn} = \sum_i \sum_j Z_i Z_j / R_{ij}$

(*v*) From the given approximate MO coefficients for the occupied MO (the virtual MO is immaterial), calculate the initial density matrix D.

(*vi*) From the given core hamiltonian h, the 2 electron integrals and D, calculate the initial Fock matrix, F, the electronic energy, E_e, and the total energy E^{tot}.

$$E^{tot} = E_e + E_n$$

(*vii*) Complete the SCF iterations (stop when the density is stable in the fifth decimal place, the energy stable in the sixth decimal place – this should take 2 iterations only).

(*viii*) Perform the Mulliken population analysis on your wavafunction, including the population matrix P, the gross atomic orbital charges, the "atom by atom" population matrix R, and the net atomic charges. Calculate the "bond order" or "overlap population" for the He–H(+) bond at your bond length. Note that in a 2 orbital –2 atom system **P** = **R**:

$$\mathbf{P} = \mathbf{R} = 2\begin{bmatrix} D_{11}S_{11} & D_{12}S_{12} \\ D_{21}S_{21} & D_{22}S_{22} \end{bmatrix} = \begin{bmatrix} 2D_{11} & 2D_{12}S_{12} \\ 2D_{12}S_{12} & 2D_{22} \end{bmatrix}$$

Overlap population: $2D_{12}\ S_{12}$

Atomic population:

$$\text{He:}\quad N_1 = 2D_{11} + D_{12}S_{12}$$

$$\text{H :}\quad N_2 = 2D_{22}\ \ D_{12}S_{12}$$

Net charge

$$\text{He:}\quad Q_1 = 2\text{–}N_1$$

$$\text{H :}\quad Q_2 = 1\text{–}N_2$$

Table 6. Input data for SCF calculation on protonated helium: He–H(+).

R(bohr)		1.05		1.15		1.35	
SMALLEST EIGENVALUE OF OVERLAP		3.0881SE-01		3.48796E-01		4.31265E-01	
CORE HAMILTONIAN (H=T+V)		1	2	1	2	1	2
	HE 1S 1	-2.537550	-1.881884	-2.478921	-1.757833	-2.374532	-1.599853
	H 1S 2	-1.881884	-1.975989	-1.757832	-1.904794	-1.509853	-1.766000
TWO ELECTRON INTEGRALS		I J K L	INTEGRAL	I J K L	INTEGRAL	I J K L	INTEGRAL
		1 1 1 1	9.926272E-01	1 1 1 1	9.926272E-01	1 1 1 1	9.926272E-0
		2 1 1 1	6.209076E-01	2 1 1 1	5.815627E-01	2 1 1 1	5.012196E-0
		2 1 2 1	4.158957E-01	2 1 2 1	3.691767E-01	2 1 2 1	2.815919E-0
		2 2 1 1	6.909870E-01	2 2 1 1	6.684023E-01	2 2 1 1	6.222243E-0
		2 2 2 1	5.071658E-01	2 2 2 1	4.710959E-01	2 2 2 1	3.986301E-0
		2 2 2 2	7.283182E-01	2 2 2 2	7.283182E-01	2 2 2 2	7.283182E-0
Guessed Coefficients	C_{11}	0.7829		0.7803		0.7880	
	C_{21}	0.2834		0.2976		0.3134	

2. Harmonic vibration, as may be examplefied by a spring and a pair of balls is described by the quadratic potential usually referred to a Hooke's law:

$$E = E_m + \frac{1}{2}k(x - x_m)^2$$

or

$$\Delta E = \frac{1}{2}k\Delta x^2$$

Figure 25 depicts schematically such a relationship:

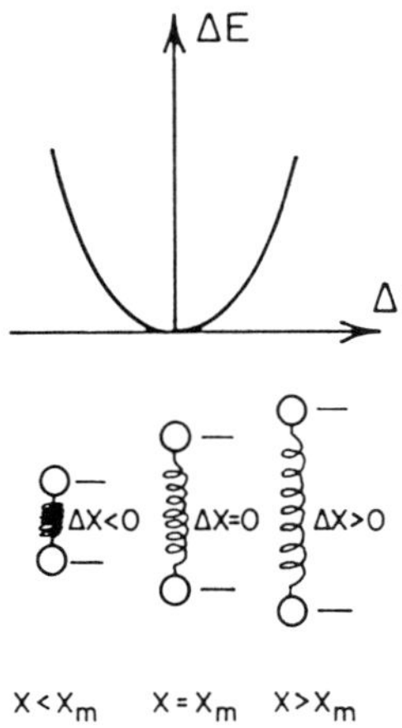

Figure 25. A schematic representation of Hooke's law. Central spring ($x=x_m$ or $\Delta x = 0$) represents the equilibrium position (i.e. minimum energy) the right hand state ($x>x_m$, i.e $\Delta x>0$) represents a stretched spring while the left hand state ($x<x_m$ i.e. $\Delta x<0$) represents a compressed spring.

In the above equation, k is the force constant. The chemical bond may be modeled by a spring. On the same set of axes, make plots of ΔE vs Δx for $k = 5 \times 10^5$, 10×10^5, and 15×10^5 dyne/cm (the approximate force constants for single, double, and triple bonds). What trend can you see in the potential energy of the spring as the bond gets tighter? What characteristic of the potential energy curve appear to be controlled by the magnitude of the force constant?

With the aid of the above force constants and with the data given below.

	X_m(Å)	E (Keal/mol)
C – C	1.536	84
C = C	1.339	167
C ≡ C	1.208	228

Plot the three full parabolas E versus x on the same scale (Note that the three sets of data k, x_m and E_m are all given in non–complementary units).

3. The following quadratic functional form will be used

$$F(x,y) = Ax^2 + By^2 + Cxy + Dx + Ey + F$$

where the parameters A, B, C, D, E, and the starting coordinates (x,y) may be chosen at will.

Show that the Optimally Conditioned algorithm Hessian update formula

$$H(new) = H - \frac{H|y><y|}{a} + \frac{|\sigma><\sigma|}{b} + \phi a \left[\frac{|\sigma>}{b} - \frac{H|y>}{a}\right] \left[\frac{<\sigma|}{a} - \frac{<y|H}{b}\right]$$

where $a = <y|H|y>$ and $b = <y|\sigma>$, reduces to the BFGS algorithm. Hessian update formula for $\phi = 1$.

4. Optimize the functional form above using the BFGS algorithm, starting at the given initial coordinates. As an aid to convergence, you should use an initial Hessian inverse guess (H) calculated as the inverse of a diagonal matrix whose elements are the exact second derivatives of the function for your set of constant (A–F). Your optimization should converge at iteration 6 (consider the function to be minimized when each gradient component is less than 1. 0D–10). Evaluate the function value at the minimum.

5. Optimize the functional form above using the Newton–Raphson technique (using analytic second derivatives). Why did the optimization converge in 1 step? Compare the G inverse matrix from this step with the last H matrix from the BFGS optimization – explain the result using NO equations.

6. One electron operators such as the dipole moment operator (μ) consist of nuclear and electronic components. For example,

$$\hat{\mu} = \hat{\mu}_n + \hat{\mu}_e$$

The dipole moment operator is given by

$$\hat{\mu} = \sum_{p}^{N} Z_p (R_p - Q) - \sum_{i}^{2M} (r_i - Q)$$

where

Z_p: nuclear charge on atom p,

R_P: coordinates of nucleus p,

r_i: coordinates of electron i,

the sums are over all nuclei (p) and electrons (i), and Q is some arbitrary origin. Show that the computed total dipole moment of a NEUTRAL molecule does not depend on the origin Q.

7. Consider a 6 electron system such as CH(+). Assume that the electronic wavefunction may be represented in the first approximation as a simple Hartree product wavefunction:

$$\psi(1,2,3,4,5,6) = \phi_1(1)\alpha(1)\phi_1(2)\beta(2)\oslash_2(3)\alpha\phi_2(4)\beta(4)\phi_3(5)\alpha\phi_3(6)\beta(6)$$

or more rigorously as an antisymmetrized Hartree–Fock product (Slater determinant) wavefunction:

$$\psi_{HF}(1,2,3,4,5,6) = \frac{1}{\sqrt{N!}} \sum_{v}^{N!} (-1)^P \hat{P}_v \psi(1,2,3,4,5,6)$$

where p is the number of interchanges and P_v is a permutation operator. Derive the electronic component of the dipole moment value (μ) for the simple Hartree wavefunction assuming that the electronic component of the dipole moment operator for the 6 electron system has the form:

$$\hat{\mu}(1,2,3,4,5,6) = \sum_{i=1}^{6} \hat{\mu}(i)$$

Also derive the electronic component of the dipole moment value (μ) for the anti–symmetrized Hartree–Fock wavefunction.

Do You Have SCF Stability and Convergence Problems?

H.B. Schlegel
Department of Chemistry
Wayne State University
Detroit, Michigan 48202, USA.

J.J.W. McDouall
Department of Chemistry
University of Manchester
Manchester M13 9PL, U.K.

ABSTRACT Because the Hartree-Fock (HF) equations are non-linear in the molecular orbitals (MO's), they must be solved iteratively to obtain a self-consistent set of MO's (i.e. the SCF method). When iterations of the SCF process converge, the energy is stationary with respect to infinitesmal variations in the orbitals. Whether this stationary point is stable (a local minimum) or unstable (a maximum in one or more directions) depends on the second derivatives of the energy with respect to the orbital variations. The internal and external stabilities or instabilities of various constrained HF wavefunctions (real and complex, restricted, unrestricted and general HF) are discussed. A simple iterative scheme does not always lead to convergence of the SCF equations. The connection between SCF stability and convergence is examined and a number of methods for improving SCF convergence are discussed, including damping, level shifting, extrapolation, univariate search, direct inversion of iterative subspace (DIIS) and quadratically convergent SCF. Techniques for constructing better initial guess MO's are also considered.

1. Introduction.

Most quantum chemical calculations using molecular orbital methods determine the orbitals by some iterative method. Typically, for each electron the orbital is optimized in the field of all the other electrons. This process is repeated until convergence is achieved and each orbital is optimal in the self consistent field (SCF) of all the other electrons. Hartree-Fock theory is the result when this approach is used with a single determinantal wavefunction; alternatively, this approach applied to a multi-configurational wavefunction yields an MC-SCF calculation. Because of the iterative nature of the SCF procedure, convergence problems are sometimes encountered. Furthermore, when convergence is achieved, there is often no guarantee that the solution to the SCF problem is stable with respect to all permitted changes in the wavefunction.

C. Ögretir and I. G. Csizmadia (eds.),
Computational Advances in Organic Chemistry: Molecular Structure and Reactivity, 167–185.

In this Chapter we will address some of the issues related to SCF convergence and stability within the single determinantal, Hartree-Fock (HF) picture. Problems with stability and convergence in MCSCF calculations are beyond the scope of this Chapter. Section 2 briefly outlines the Hartree-Fock approach for molecular orbitals expressed as linear combinations of atomic orbitals (LCAO-MO), i. e. the Roothaan-Hall equations. The question of the stability of the SCF wavefunction under various constraints is discussed in Section 3. The next Section deals with algorithms for obtaining SCF convergence. Section 5 covers various methods for obtaining initial estimates of the molecular orbitals to start the SCF process.

2. Hartree-Fock Theory and SCF Calculations

In the Hartree-Fock model, the electronic wavefunction for an n-electron system is represented by a single Slater determinant, $|\Psi_0\rangle$, of n spin orbitals $\{\Phi_i\}$

$$|\Psi_0\rangle = |\Phi_1\ \Phi_2 \cdots \Phi_n\rangle \tag{1}$$

The orbitals are optimized according to the variation principle such that the energy functional

$$E_0 = \langle\Psi_0|\ \hat{H}\ |\Psi_0\rangle \,/\, \langle\Psi_0|\Psi_0\rangle \tag{2}$$

is minimized. In eq 2, $\hat{H}$, represents the full electronic hamiltonian. The derivation of the Hartree-Fock equation, which determines the optimal spin orbitals, may be found in several standard texts[1-3] and will not be given here. We shall simply note that the Hartree-Fock equation is an eigenvalue equation of the form

$$\hat{F}(i)\ \Phi(\tau_i) = \varepsilon\ \Phi(\tau_i) \tag{3}$$

where $\hat{F}(i)$ is an effective one-electron operator (for the ith electron) called the Fock operator and is given by

$$\hat{F}(i) = \hat{h}(i) + \hat{V}^{HF}(i) \tag{4}$$

where $\hat{h}(i) = -\frac{1}{2}\nabla_i^2 - \sum_A^{nuclei} \frac{Z_A}{r_{iA}}$, $\quad \hat{V}^{HF}(i) = \sum_j \hat{J}_j(i) - \hat{K}_j(i)$

and

$$\hat{J}_j(1)\ \Phi_i(1) = \int \Phi_j(2) \frac{1}{r_{12}}\ \Phi_j(2)\ \Phi_i(1)\ d\tau_2$$

$$\hat{K}_j(1)\ \Phi_i(1) = \int \Phi_j(2) \frac{1}{r_{12}}\ \Phi_i(2)\ \Phi_j(1)\ d\tau_2\ ,$$

in which, ∇_i^2 is the Laplacian kinetic energy operator for electron i, Z_A is the charge on nucleus A, r_{iA} is the distance between electron i and nucleus A and $\hat{V}^{HF}(i)$ is the average potential experienced by electron i due to all other electrons. Hence, eq 3 replaces the instantaneous inter-electron repulsion by an averaged field. This is the central idea in the Hartree-Fock approximation.

To cast eq 3 into a form suitable for computation, we write the spin orbitals $\{\Phi_i\}$ as a linear combination of N one-electron basis functions $\{\chi_v\}$

$$(5) \qquad \Phi_i = \sum_{v}^{N} C_{vi}\, \chi_v$$

This approach was first applied to the case of closed-shell systems by Roothaan and Hall[4] and is commonly referred to as the restricted Hartree-Fock (RHF) method because all doubly occupied orbitals are constrained to have the same spatial form. The method was later extended to open-shell systems by Pople and Nesbet[5], who incorporated the additional flexibility of allowing a different spatial form for electrons of α and β spin. This method is known as the unrestricted Hartree-Fock (UHF) approach. These ideas are expanded further in Section 3; for the moment, no distinction is made between RHF and UHF determinants. Substituting eq 5 into eq 3 yields

$$(6) \qquad \sum_{v}^{N} \hat{F}\, \chi_v\, C_{vi} = \varepsilon_i \sum_{v}^{N} \chi_v\, C_{vi}$$

premultiplying eq 6 by χ_μ and integrating gives

$$(7) \qquad \sum_{v}^{n} \{F_{\mu v} - \varepsilon_i S_{\mu v}\}\, C_{vi} = 0 \qquad\qquad i=1,2 \cdots N$$

where $\qquad F_{\mu v} = \langle \chi_\mu | \hat{F} | \chi_v \rangle$ and $S_{\mu v} = \langle \chi_\mu | \chi_v \rangle$

The Fock matrix, in turn can be written in terms of the density matrix, **P**, and the one and two electron integrals ($h_{\mu v}$ and $(\mu\lambda||v\sigma)$, respectively).

$$(8) \qquad F_{\mu v} = h_{\mu v} + \sum_{\lambda\sigma} (\mu\lambda||v\sigma)\, P_{\lambda\sigma}$$

$$(9) \qquad P_{\lambda\sigma} = \sum_{i} C_{\lambda i}^{*}\, C_{\sigma i}\, , \qquad h_{\mu v} = \int \chi_\mu(1)^*\, \hat{h}(i)\, \chi_v(1)\, d\tau_1$$

$$(\mu\lambda||v\sigma) = \int \chi_\mu(1)^*\, \chi_\lambda(2)^*\, \frac{1}{r_{12}}\, (\, \chi_v(1)\, \chi_\sigma(2) - \chi_\sigma(1)\, \chi_v(2)\,)\, d\tau_1\, d\tau_2$$

Thus the basis set expansion has turned the integro-differential Hartree-Fock equation into a matrix equation,

$$(10) \qquad \mathbf{F\, C = S\, C\, \varepsilon}$$

where **F** is the matrix representation of the Fock operator, **C** is the matrix of molecular orbital coefficients, $\{C_{\nu i}\}$ (eq 5), **S** is the overalp matrix of basis functions and ε is a diagonal matrix containing the one-electron orbital energies. Equation 10 is nonlinear in the MO coefficients, since $\hat{J}$ and $\hat{K}$ in $\hat{F}$ depend on **C**. It is therefore necessary to solve eq 10 iteratively. The Fock matrix is built using an initial guess of the MO coefficients, $\{C_{\nu i}\}$. Equation 10 is solved for a new set of $\{C_{\nu i}\}$ which are then used to recompute the Fock matrix. This process is iterated until the change in the molecular orbitals and the orbital energies falls below some chosen threshold. The orbitals obtained in this manner are said to be self-consistent and the average potential that they describe for each electron is termed the self-consistent field (SCF).

Because of the nonlinearity of eq 10, the simple iterative procedure described above does not always converge. It may oscillate or even diverge, depending on the initial guess and the nature of the electronic structure being calculated. Even when convergence is obtained, eq 10 simply ensures that the energy is stationary with respect to the first-order variation of the $\{C_{\nu i}\}$. To determine the nature of the stationary point (i.e. whether the energy is a maximum or a minimum), the second-order variation of the energy with respect to the $\{C_{\nu i}\}$ must be examined[6]. This is the subject of the next section.

3. SCF Constraints and Stability.

Each member of the set of spin orbitals $\{\Phi_i\}$ is a function of the coordinates (τ) of a single electron. The electronic coordinate, τ, may be separated into a spatial component, which typically depends on the Cartesian coordinates, $\vec{r}$, of the electron and a spin coordinate, σ. In solving the SCF equations, it is common practice to impose certain constraints on the nature of the spin orbitals and carry out the variation of Ψ_0 in a restricted space. Typical examples of constraints include representing each Φ_i as a product of a function of Cartesian coordinates and a function of spin coordinates and restricting the functions to be real. Hence, to establish that the solution to eq 10 (within the constraints imposed) does lead to a local minimum in the energy (eq 2), it is necessary to examine its second-order variation and show that it is positive. Once a constrained minimum has been located, the consequences of removing one or more of the constraints can be investigated. The possible outcomes of removing constraints on Ψ_0 are: (a) $E(\Psi_0)$ remains at a local minimum, or (b) the first-order variation of Ψ_0 is zero but $E(\Psi_0)$ is no longer at local minimum, or (c) $E(\Psi_0)$ is no longer stationary. The stability conditions were originally derived by Thouless[6]; Cizek and Paldus[7] presented the stability problem form a chemical point of view. Systematic discussion of constraints and stability conditions were presented later by Ostlund[8], Fukutome[9] and Seeger and Pople[10]. The latter approach[10] will be followed in the next section.

<u>General Hartree-Fock (GHF)</u> In general, the spin orbital $\Phi(r,\sigma)$ is a complex function, usually taken to be part of an orthonormal set, i.e.

(11) $$\int \Phi_i^* \Phi_j \, d\tau = \delta_{ij}$$

Expanding Φ_i in spin space over the two spin basis functions $\alpha(\sigma)$ and $\beta(\sigma)$ gives

$$(12) \quad \Phi_i(r,\sigma) = \phi_i^\alpha(r)\ \alpha(\sigma) + \phi_i^\beta(r)\ \beta(\sigma)$$

where ϕ_i^α and ϕ_i^β are functions of the Cartesian coordinates alone (spatial orbitals). In general ϕ_i^α and ϕ_i^β are complex and may be separated into their real and imaginary components,

$$(13) \quad \Phi_i^\alpha = \mathrm{Re}(\phi_i^\alpha) + i\ \mathrm{Im}(\phi_i^\alpha)\ , \qquad \Phi_i^\beta = \mathrm{Re}(\phi_i^\beta) + i\ \mathrm{Im}(\phi_i^\beta)$$

which leads to the following general form for the spin orbitals in terms of the N spatial basis functions $\{\chi\}$

$$(14) \quad \Phi_i(r,\sigma) = \sum_{\mu}^{N} [\ C_{\mu i}^\alpha\ \chi_\mu(r)\ \alpha(\sigma) + C_{\mu i}^\beta\ \chi_\mu(r)\ \beta(\sigma)\ ] \qquad \text{(GHF orbitals)}$$

where $C_{\mu i}^\alpha$ and $C_{\mu i}^\beta$ are complex. The use of the full expression in eq 14, defining the spin orbitals in terms of 4N real numbers (where N is the number of spatial basis functions), leads to what has been termed general Hartree-Fock theory (GHF). Note that the GHF wavefunctions are not eigenfunctions of the spin operators $\hat{S}^2$ and $\hat{S}_z$. The spatial functions $\{\chi\}$ and the MO coefficients can be constrained to be real, yielding the real-GHF approximation. Other levels of Hartree-Fock theory can be obtained by applying other constraints to eq 14.

<u>Unrestricted Hartree-Fock</u> *(UHF)* The most commonly applied constraint is to write each spin orbital as a product of the form $\chi_\mu(r)\ \alpha(\sigma)$ or $\chi_\mu(r)\ \beta(\sigma)$ rather than the general linear combination of eq 14 (i.e. the UHF spin orbitals are either pure α or pure β rather than mixed). The α and β orbitals are independent complex functions and this approximation is usually referred to as unrestricted Hartree-Fock theory (UHF)

$$(15) \quad \Phi_i(r,\alpha) = \sum_\mu C_{\mu i}^\alpha\ \chi_\mu(r)\ \alpha(\sigma) \quad \text{and} \quad \Phi_i(r,\beta) = \sum_\mu C_{\mu i}^\beta\ \chi_\mu(r)\ \beta(\sigma) \qquad \text{(UHF orbitals)}$$

The UHF wavefunctions are eigenfunctions of the $\hat{S}_z$ but not of $\hat{S}^2$, the total spin operator[11]. At this level, each spin orbital may be specified by N complex coefficients $C_{\mu i}^\alpha$ or $C_{\mu i}^\beta$. The additional constraint of requiring the spatial functions $\{\chi\}$ and the MO coefficients to be real leads to the real-UHF approximation.

<u>Restricted Hartree-Fock</u> *(RHF)* The next constraint relates the spatial parts of orbitals of different spin. If the numbers of occupied α and β orbitals are n_α and n_β, where $n_\alpha \geq n_\beta$, then n_β of the α orbitals may be restricted to have the same spatial functions as the β orbitals. These n_β orbitals are then "doubly occupied" and the remaining $(n_\alpha - n_\beta)$ α orbitals are "singly occupied"; this approximation is referred to as restricted Hartree-Fock theory (RHF). All doubly occupied orbitals (i.e. closed-shells with $n_\alpha = n_\beta$) is an important special case in the RHF scheme.

$$(16)\quad \Phi_i(r,\sigma) = \sum_\mu C_{\mu i}\, \chi_\mu(r)\, \alpha(\sigma) \text{ and } \bar{\Phi}_i(r,\sigma) = \sum_\mu C_{\mu i}\, \chi_\mu(r)\, \beta(\sigma) \qquad \text{(RHF orbitals)}$$

All RHF wavefunctions (whether closed- or open-shell) have the advantage that they are eigenfunctions of the spin operators $\hat{S}^2$ and $\hat{S}_z$, and so represent pure spin states. For the examination of the stability of RHF wavefunctions, only closed shell RHF will be considered, since open shell RHF (ROHF) wavefunctions are always unstable with respect UHF.

Stability of HF Wavefunctions A Hartree-Fock wavefunction is considered stable if none of the possible changes in the one electron orbitals leads to a lower energy. It can be shown readily that the energy is invariant with respect to mixing the occupied orbitals in a set with themselves (e.g. doubly occupied with doubly occupied for RHF, α with α for UHF, etc.). In addition to the n occupied orbitals, there are N-n virtual orbitals ($\Phi_{n+1} \cdots \Phi_N$) which arise from the choice of the one-electron basis functions in eq 5. Thus, it is necessary to examine how the energy changes when these virtual orbitals are allowed to mix with occupied orbitals[13]. Let an occupied orbital change according to:

$$(17)\quad \Phi_i \rightarrow \Phi_i' = \Phi_i + \sum_a D_{ia}\, \Phi_a$$

where the sum extends over all virtual orbitals. The standard notation of indices i, j, k, etc. for occupied orbitals, indices a, b, c, etc. for virtual orbitals and indices p, q, r, etc. for any orbital (occupied or virtual) is used throughout. When the occupied orbitals Φ_i in eq 1 are replaced by $\Phi_{i'}$ and the wavefunction is expanded, the result can be written in terms of the Hartree-Fock ground state determinant and a sum over excited determinants obtained by replacing all possible combinations of occupied spin orbitals Φ_i ($1 \le i \le n$) with all possible combinations of virtual spin orbitals Φ_a ($n+1 \le a \le N$):

$$(18)\quad |\Psi\rangle = D_0\, |\Psi_0\rangle + \sum_{i\,a} D_{ia}\, |\Psi_i^a\rangle + \frac{1}{2!} \sum_{ij\,ab} D_{ia} D_{jb}\, |\Psi_{ij}^{ab}\rangle + \frac{1}{3!} \sum_{ijk\,abc} D_{ia} D_{jb} D_{ck}\, |\Psi_{ijk}^{abc}\rangle + \cdots$$

where

$$(19)\quad |\Psi_0\rangle = |\cdots \Phi_i\, \Phi_j\, \Phi_k \cdots\rangle,\quad |\Psi_i^a\rangle = |\cdots \Phi_a\, \Phi_j\, \Phi_k \cdots\rangle,\quad |\Psi_{ij}^{ab}\rangle = |\cdots \Phi_a\, \Phi_b\, \Phi_k \cdots\rangle, \text{ etc.}$$

and D_0 is introduced to allow for renormalization. Likewise, the energy, eq 2, can be expanded:

$$(20)\quad E = \langle\Psi|\, \hat{H}\, |\Psi\rangle \,/\, \langle\Psi|\Psi\rangle$$

$$= \Big\{ D_0^* D_0 E_0 + \sum_{ia} [\, D_0^* D_{ia} \langle\Psi_0|\, \hat{H}\, |\Psi_i^a\rangle + D_{ia}^* D_0 \langle\Psi_i^a|\, \hat{H}\, |\Psi_0\rangle\,]$$

$$+ \sum_{ij\,ab} [\, \tfrac{1}{2} D_0^* D_{ia} D_{jb} \langle\Psi_0|\, \hat{H}\, |\Psi_{ij}^{ab}\rangle + \tfrac{1}{2} D_{ia}^* D_{jb}^* D_0 \langle\Psi_{ij}^{ab}|\, \hat{H}\, |\Psi_0\rangle$$

$$+ D_{ia}^* D_{jb} \langle\Psi_i^a|\, \hat{H}\, |\Psi_j^b\rangle\,] + \cdots \Big\}\, (D_0^* D_0 + \sum_{kc} D_{kc}^* D_{kc})^{-1}$$

For the stability analysis, it is sufficient to adopt intermediate normalization, i.e. $D_0 = 1$. Since Ψ_0 is a converged Hartree-Fock wavefunction, the matrix elements involving the ground state and single excitations, e.g. $\langle\Psi_0|\hat{H}|\Psi_i^a\rangle$, are zero by Brillouin's theorem. If the expansion is truncated at second order, the energy can be written in a convenient matrix form[7]:

$$(21) \quad E = E_0 + \frac{1}{2} \sum_{ij\,ab} [\, D_{ia} D_{jb} \langle\Psi_0|\hat{H}|\Psi_{ij}^{ab}\rangle + D_{ia}^* D_{jb}^* \langle\Psi_{ij}^{ab}|\hat{H}|\Psi_0\rangle + 2\, D_{ia}^* D_{jb} \langle\Psi_i^a|\hat{H} - E_0|\Psi_j^b\rangle \,]$$

$$= E_0 + \frac{1}{2} \begin{bmatrix} \mathbf{D} \\ \mathbf{D}^* \end{bmatrix}^\dagger \begin{bmatrix} \mathbf{A} & \mathbf{B} \\ \mathbf{B}^* & \mathbf{A}^* \end{bmatrix} \begin{bmatrix} \mathbf{D} \\ \mathbf{D}^* \end{bmatrix} = E_0 + \frac{1}{2} \tilde{\mathbf{D}}^\dagger\, \mathbf{H}\, \tilde{\mathbf{D}}$$

where $\mathbf{D}^*$ and $\mathbf{D}^\dagger$ are the complex and hermitian conjugates of $\mathbf{D}$, the matrix of the coefficients D_{ia}. The matrix $\mathbf{H}$ is composed of submatrices $\mathbf{A}$ and $\mathbf{B}$, which are given by

$$(22) \quad A_{ia\,jb} = \langle\Psi_i^a|\hat{H} - E_0|\Psi_j^b\rangle = (\varepsilon_a - \varepsilon_i)\, \delta_{ab}\, \delta_{ij} + (aj||ib),$$

$$B_{ia\,jb} = \langle\Psi_{ij}^{ab}|\hat{H}|\Psi_0\rangle = (ab||ij)$$

Note that $\mathbf{A}$ is hermitian and $\mathbf{B}$ is symmetric. For E to be a minimum with respect to the variations $\tilde{\mathbf{D}}$, the matrix $\mathbf{H}$ must be positive semidefinite, i.e. no negative eigenvalues. For open shell systems, a zero eigenvalue is permissible, since the direction of spin quantization is indeterminate. For general Hartree-Fock theory, all variations in $\tilde{\mathbf{D}}$ are permitted. For the other, constrained HF wavefunctions, only certain selected variations are allowed and hence only specific portions of $\mathbf{H}$ need to be positive semidefinite to ensure stability.

Stability of Constrained Hartree-Fock Wavefunctions Figure 1 summarizes the types of stabilities that will be considered. A wavefunction can be tested for stability within the constraint (internal stability) or for stability with respect to a less constrained wavefunction (external stability). In the Figure, releasing constraints corresponds to moving down and/or to the right.

Real RHF	→	**Real UHF**	→	**Real GHF**
↓		↓		↓
Complex RHF	→	**Complex UHF**	→	**Complex GHF**

Figure 1. Hartree-Fock wavefunctions and the types stability considered.

One constraint that can be applied to RHF, UHF and GHF wavefunctions is that the wavefunction remain real. In this case, the matrices $\mathbf{A}$ and $\mathbf{B}$ in eq 21 are real and the energy can be written as:

(23) $E = E_0 + Re(\mathbf{D}^\dagger)\,(\mathbf{A} + \mathbf{B})\,Re(\mathbf{D}) + Im(\mathbf{D}^\dagger)\,(\mathbf{A} - \mathbf{B})\,Im(\mathbf{D})$

where $Re(\mathbf{D}) = 1/2\,(\mathbf{D} + \mathbf{D}^*)$ and $Im(\mathbf{D}) = 1/2\,(\mathbf{D} - \mathbf{D}^*)$. For internal stability, **D** must be real and the matrix **A** + **B** must be positive semidefinite. A non-zero imaginary component in **D** will result in a complex wavefunction. Thus if **A** - **B** has a negative eigenvalue, the energy will be lowered by making the wavefunction complex, i.e. the wavefunction is not stable with respect to becoming complex.

In a UHF wavefunction, the α spin orbitals cannot be mixed with the β spin orbitals. The excitations in **D** can be divided into two groups: spin-conserving $\alpha \rightarrow \alpha$ and $\beta \rightarrow \beta$ excitations, **D'**, and spin-unconserving $\alpha \rightarrow \beta$ and $\beta \rightarrow \alpha$ excitations, **D"**. Similarly, the matrices **A**, **B** and **H** can be divided into spin-conserving and spin-unconserving components, designated by ' and ", respectively[14]. Only spin conserving excitations, **D'**, maintain the constraints on a UHF wavefunction; hence, **H'** must be positive semidefinite for a UHF wavefunction to be internally stable. Spin-unconserving excitations, D", mix the α and β spin orbitals, changing the UHF wavefunction into a GHF wavefunction. Thus, if **H"** is not positive semidefinite, then the UHF wavefunction is GHF unstable. For UHF wavefunctions constrained to be real, **A'** + **B'** and **A"** + **B"** have to be positive semidefinite for internal real UHF and external real GHF stability, respectively. Real UHF versus complex UHF stability is determined by **A'** - **B'**.

For a closed shell RHF wavefunction, all of the α spin orbitals must remain identical to the β spin orbitals. The excitations in **D'** can be separated into 2 groups: $\mathbf{D}^{\alpha\alpha}$, the $\alpha \rightarrow \alpha$ excitations, and $\mathbf{D}^{\beta\beta}$, the $\beta \rightarrow \beta$ excitations in **D'**. A linear combination of these 2 terms yield a singlet component ${}^1\mathbf{D}' = 1/\sqrt{2}\,(\mathbf{D}^{\alpha\alpha} + \mathbf{D}^{\beta\beta})$ for which the $\alpha \rightarrow \alpha$ and $\beta \rightarrow \beta$ excitations are the same and a triplet component ${}^3\mathbf{D}' = 1/\sqrt{2}\,(\mathbf{D}^{\alpha\alpha} - \mathbf{D}^{\beta\beta})$ for which the $\alpha \rightarrow \alpha$ and $\beta \rightarrow \beta$ excitations are opposite in sign; the singlet excitations maintain the constraints of the RHF wavefunction and the triplet excitations change the RHF wavefunction into a UHF wavefunction. Similarly, the matrices **A'**, **B'** and **H'** can be divided into singlet and triplet components. Thus, for RHF internal stability, ${}^1\mathbf{H}'$ must be positive semidefinite; likewise, ${}^3\mathbf{H}'$ not positive semidefinite signals an instability in changing the RHF wavefunction to UHF. For a real RHF wavefunction, internal stability is dertemined by ${}^1\mathbf{A}' + {}^1\mathbf{B}'$, external stability with respect to a complex RHF wavefunction is determined by ${}^1\mathbf{A}' - {}^1\mathbf{B}'$.

The conditions for stabitity of the various Hartree-Fock wavefunctions are summarized in Table 1. The explicit formulas for the different **A**'s, **B**'s and **H**'s are given in references 10.

Table 2 summarizes the results for nitrogen molecule, showing the various forms of instability which occur as the bond is stretched. The ground state electronic configuration of N_2 is shown in Figure 2. At bond lengths less than the equilibrium value, no instabilities occur; hence, all types of wavefunctions (RHF, UHF, GHF) lead to the same

Table 1. Matrices required to be positive semidefinite in order for an SCF solution of type **X** to be stable in a space of type **Y**

Y =	real RHF	complex RHF	real UHF	complex UHF	real GHF	complex GHF
X =						
real RHF	$^1A'+^1B'$	$^1A'-^1B'$	$^3A'+^3B'$			
complex RHF		$^1H'$		$^3H'$		
real UHF			A'+B'	A'-B'	A"+B"	
complex UHF				H'		H"
real GHF					A+B	A-B
complex GHF						H

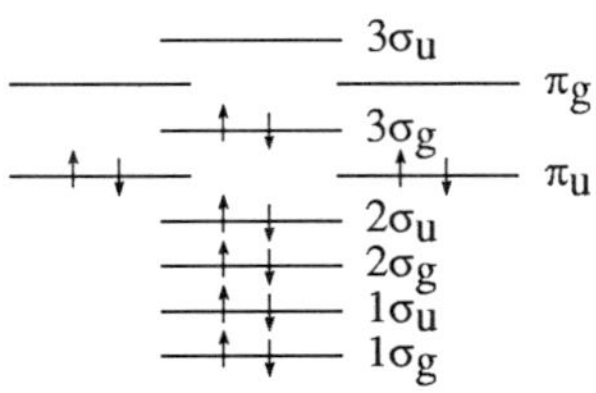

Figure 2. Electronic configuration of N_2 near R_{eq}.

Table 2. Total energies and instabilities that occur as the bond in N_2 is stretched.

R	real RHF	Types of instability	real UHF	Types of instability.
0.8	-106.680804	none	-106.680804	none
1.0	-107.419532	none	-107.419532	none
1.2	-107.487783	real UHF	-107.501203	GHF
1.4	-107.357815	complex RHF	-107.396565[b]	internal UHF
		real UHF		GHF
1.6	-107.184846[c]	internal RHF	-107.287416[d]	internal UHF
		complex RHF		complex UHF, GHF

[a] calculated with the STO-3G basis, R_{eq} = 1.1339 Å, total energies in hartree.

[b] E = -107.412014 after re-optimization for internal UHF instability.

[c] E = -107.225669 after re-optimization for internal RHF instability.

[d] E = -107.445187 after re-optimization for internal UHF instability.

energy. As the bond is stretched, various instabilities arise from the mixing of the occupied π_u and the virtual π_g orbitals. At R = 1.2 Å, the RHF wavefunction is UHF unstable; the anti-bonding π_g mixes with the bonding π_u to give a lower energy real UHF wavefunction. In turn, the UHF wavefunction is GHF unstable. A complex instability sets in near R = 1.4 Å for the RHF case and near 1.6 Å for the UHF case. Although the initial UHF wavefunctions obtained for R = 1.4 and 1.6 Å have lower energies lower than their RHF counterparts, the UHF solution themselves are internally unstable and can be optimized to give lower energies.

4. SCF Convergence

In general, the Hartree-Fock equations must be solved iteratively. In a simple SCF procedure, one cycle of the iteration consists of (a) constructing a new Fock matrix, $\mathbf{F}^{k+1}$, using the current approximation to the MO coefficients, $\mathbf{C}^k$, or the density matrix, $\mathbf{P}^k$, and (b) obtaining the new (hopefully, improved) coefficient matrix $\mathbf{C}^{k+1}$, by diagonalizing $\mathbf{F}^{k+1}$.

(24) $$F^{k+1}_{\mu\nu} = h_{\mu\nu} + \sum_{\lambda\sigma} (\mu\lambda||\nu\sigma)\, P^k_{\lambda\sigma}; \quad P^k_{\lambda\sigma} = \sum_j C^{k\,*}_{\lambda j}\, C^k_{\sigma j}$$

(25) $$\sum_\nu F^{k+1}_{\mu\nu}\, C^{k+1}_{\nu p} = \sum_\nu S_{\mu\nu}\, C^{k+1}_{\nu p}\, \varepsilon_p$$

This cycle is repeated until the changes in **C** or **P** are below a pre-set threshold. Unfortunately, this simple iteration scheme does not always converge; furthermore, the convergence difficulties are often more severe for open shell systems than for closed shells. A number of techniques for controlling and accelerating the convergence of SCF calculations are discussed in this section.

Damping One of the simplest and earliest used methods for forcing SCF convergence is damping[16], whereby a weighted average of the old and the new matrices is used,

(26) $$\mathbf{M}^{k+1\,'} = (1-\alpha)\, \mathbf{M}^{k+1} + \alpha\, \mathbf{M}^k \qquad 0 \leq \alpha < 1$$

where α is a constant. This can be applied to either the MO coefficient matrix, the density matrix or the Fock matrix (for MO coefficients, some care must be taken with phasing and mixing of degenerate orbitals). Damping has the effect of reducing the size of the step in the iteration and limits the tendency for oscillation. Typically a fixed value of α is chosen until the SCF has converged partially and α = 0 (no damping) is used until final convergence is reached (the case where α is optimized is treated below under Univariate Search). Alternatively the damping factor can be adjusted dynamically[17-19]. One of the drawbacks of damping is that it does not preserve the orthonormality of the MO's or the idempotency of the density matrix[20]. For the early stages of the SCF process, this is not a serious problem since the later (undamped) steps properly maintain the orthonormality of the MO's. Secondly, damping slows down the rate of convergence substantially;

however, in the absence of better techniques, slow convergence is better than no convergence.

Level Shifting[21] This is an alternative to damping in controlling the size of the step taken in an SCF cycle. The energies of the virtual orbitals are increased by a constant amount (i.e. a constant is added to the diagonal elements of the virtual-virtual block of the Fock matrix in the MO basis). The effect of this can be seen by using perturbation theory to estimate the changes in the MO coefficients for an SCF cycle. Parallel to eq 17, let the new coefficients for the occupied orbitals, $C^{k+1}_{\nu i}$, be written in terms of the old coefficients plus a contribution from the virtual orbitals.

$$(27) \qquad C^{k+1}_{\nu i} = C^{k}_{\nu i} + \sum_{a} D^{k+1}_{ia} C^{k}_{\nu a}$$

Insertion of eq 27 into eq 25 and pre-multiplying by $C^{k\,*}_{\mu a}$ leads to

$$(28) \qquad D^{k+1}_{ia} = - \sum_{\mu\nu} C^{k\,*}_{\mu a} F^{k}_{\mu\nu} C^{k}_{\nu i} / (\varepsilon_a - \varepsilon_i) = - F^{k}_{ai} / (\varepsilon_a - \varepsilon_i)$$

This estimate of the change in the MO coefficients is appropriate for the simple SCF procedure. If level shifting is used, ε_a is replaced by $\varepsilon_a + b$ where $b > 0$; consequently, the changes in the MO coefficients are smaller than without level shifting, and the tendency for the SCF to oscillate is reduced. The level shift can be fixed or adjusted dynamically[21,22]. While damping reduces the step by the same amount for each orbital, level shifting reduces the occupied-virtual mixing more for those pairs with small energy differences.

Extrapolation Just as damping can stabilize the early phase of an SCF calculation, extrapolation can speed up the final convergence. The Aitken extrapolation[23] assumes the matrix elements form a geometric series (e.g. $(M^{k+1} - M^{\infty}) / (M^{k} - M^{\infty})$ is a constant). Consideration of 3 consecutive points and elimination of the constant leads to the following formula for the extrapolation of individual matrix elements:

$$(29) \qquad M^{k+1'} = M^{k+1} + \Delta M^{k+1} / (\Delta M^{k} - \Delta M^{k+1}), \qquad \Delta M^{k} = M^{k} - M^{k-1}$$

This formula can be applied to the MO coefficients, density matrix or Fock matrix to improve convergence in the later stages of an SCF calculation. In addition, the Aitken extrapolation has been shown to overcome oscillatory divergence in the SCF[24]. A variety of other extrapolation schemes have also been used[25-29] As with the damping, orthonormality and idempotency are not maintained[20]; hence the final iteration must be carried out without extrapolation to obtain a variational energy and orthonormal MO's.

Univariate Search A simple SCF cycle provides a direction in which the orbital coefficient matrix or the density matix is to be changed, and a step length in that direction. Instead of using the simple SCF step length, better results can be obtained by minimizing the energy with respect to the step size. Care must be taken to ensure that the MO's remain orthonormal as the step size is varied. If the minimum in the univariate search is to be located accurately[30-33], the energy can be calculated at several points along the search path until the minimum is bracketed. A minimum is then found by fitting a polynomial to the energies along the search path. An approximate location of the minimum in the search direction can be found with much less effort[18,19,34] (in fact, without any additional energy or Fock matrix evaluations). A cubic polynomial can be fit to the energies, E^k and E^{k+1}, and to the derivatives $dE^k/d\Delta\mathbf{C}$ and $dE^{k+1}/d\Delta\mathbf{C}$ which can be obtained from the Fock matrices $\mathbf{F}^k$ and $\mathbf{F}^{k+1}$. The new matrix (MO, density or Fock) at the minimum in the cubic polynomial can then be found by interpolation. Optimizing the damping factor in eq 26 is a special case of this approach[18,19]. Experience with general unconstrained non-linear optimization problems has pointed out the importance of the univariate search step in obtaining rapid convergence in quasi-Newton minimizations[35]. The same appears to be true for the constrained optimization problem posed by the Roothaan-Hall equations.

<u>DIIS</u> The DIIS approach[36] is a general and efficient method for obtaining a solution to a set of equations that must be solved iteratively. Its roots are linked to the Lánczos method[37] for eigenvalue problems. An iterative solution of a set of equations produces a series of vectors that approach the desired result. Initially, these vectors span a smaller space than the full set of equations. The equations can be projected into this small space and inverted directly to yield an approximate solution that is used as a better input for the next iteration. Specifically, at the kth iteration, a better approximation to the solution is obtained that is of the form $\mathbf{p} = \Sigma\, c_i\, \mathbf{p}^i$ subject to the constraint $\Sigma\, c_i = 1$ such that the error $\Sigma\, c_i\, \mathbf{e}^i$ is minimized in a least squares sense. This leads to the following set of equations:

$$(30)\qquad \begin{bmatrix} 0 & -\mathbf{1} \\ -\mathbf{1}^T & \mathbf{b} \end{bmatrix} \begin{bmatrix} -\lambda \\ \mathbf{c} \end{bmatrix} = \begin{bmatrix} -1 \\ \mathbf{0} \end{bmatrix}$$

where $\mathbf{1} = (1,1,\cdots)$, $b_{ij} = \mathrm{tr}\, \mathbf{e}^{i\dagger}\, \mathbf{e}^j$ and $\mathbf{e}^i$ is an estimate of the error at the ith iteration. In one form or another, this approach has been used for geometry optimization[38], CI calculations[39], CPHF equations[40] and SCF iterations[36,41]. For SCF calculations, the occupied-virtual block of the Fock operator in the MO basis could be used as the error vector[36a]. A better choice for the error vector has been found to be[36b]

$$(31)\qquad \mathbf{e}^k = (\,\mathbf{S}^{-1/2}\,)^T\,(\,\mathbf{F}^k\,\mathbf{D}^k - \mathbf{D}^k\,\mathbf{F}^k\,)\,\mathbf{S}^{-1/2}$$

The linear combination coefficients, c_i, are determined by eq 30 (if the equation becomes ill conditioned, results from the earliest iterations are discarded one at a time until the equations are well behaved again). An improved Fock matrix is constructed according to $\mathbf{F} = \Sigma\, c_i\, \mathbf{F}^i$, which is diagonalized to obtain the coefficients for the next iteration.

The DIIS approach can be thought of as a multivariate search and hence should be significantly better than the univariate search methods. Applications indicate that DIIS overcomes most of the convergence problems of simple SCF[36]. Damping or level shifting can be used with DIIS to overcome most of the remaining convergence problems. The DIIS approach is often a factor of 2 or 3 faster than simple SCF[36] and, in a number of cases[36b], has been shown to be twice as fast as quadratically convergent SCF[42].

Quadratically Convergent SCF (QC-SCF)[42] One source of the difficulties with the simple SCF iteration scheme is the fact that the Fock matrix depends on the MO coefficients. A quadratically convergent SCF procedure remedies this problem, in part, by taking this dependency into account through first order. As discussed in the Stability section, the Hartree-Fock wavefunction is invariant with respect to the occupied-occupied and virtual-virtual orbital mixing. Hence only the effect of occupied-virtual mixing on the energy needs to be considered. As in the stability calculation, small contributions from the virtual orbitals are added to the occupied orbitals, eq 17. The wavefunction and the total energy are expressed in terms of the occupied-virtual mixing coefficients, D_{ia}, shown in eq 17 - 20. Since the mixing coefficients are no longer infinitesimal, the denominator must be retained through second order. The variational energy can then be writen as:

$$(32) \quad E = E_0 + \left\{ \begin{bmatrix} \mathbf{D} \\ \mathbf{D}^* \end{bmatrix}^\dagger \begin{bmatrix} D_0 \boldsymbol{F} \\ D_0^* \boldsymbol{F}^* \end{bmatrix} + \frac{1}{2} \begin{bmatrix} \mathbf{D} \\ \mathbf{D}^* \end{bmatrix}^\dagger \begin{bmatrix} \mathbf{A} & D_0 \mathbf{B} \\ D_0^* \mathbf{B}^* & \mathbf{A}^* \end{bmatrix} \begin{bmatrix} \mathbf{D} \\ \mathbf{D}^* \end{bmatrix} \right\} \{ D_0^* D_0 + \mathbf{D}^\dagger \mathbf{D} \}^{-1}$$

where $\boldsymbol{F} = \langle \Psi_i^a | \hat{H} | \Psi_0 \rangle$, the occupied-virtual block of the Fock matrix. Since $D_0 \approx 1$ near convergence, this expression can be simplified by dropping the D_0 dependence of the terms in $\mathbf{B}$ to obtain an equation that is quadratic in the D's. The energy can now be written as

$$(33) \quad E = E_0 + \{ \tilde{\mathbf{D}}^\dagger \tilde{\boldsymbol{F}} \tilde{\mathbf{D}}_0 + \tfrac{1}{2} \tilde{\mathbf{D}}^\dagger \mathbf{H} \tilde{\mathbf{D}} \} \{ \tfrac{1}{2} \tilde{\mathbf{D}}_0^\dagger \tilde{\mathbf{D}}_0 + \tfrac{1}{2} \tilde{\mathbf{D}}^\dagger \tilde{\mathbf{D}} \}^{-1}$$

where $\tilde{\mathbf{D}}_0 = \begin{bmatrix} D_0 \\ D_0^* \end{bmatrix}$ and $\tilde{\boldsymbol{F}} = \begin{bmatrix} \boldsymbol{F} & 0 \\ 0 & \boldsymbol{F}^* \end{bmatrix}$

The energy must be stationary with respect to variations in the D's. Hence

$$(34a) \quad \frac{dE}{d\tilde{\mathbf{D}}_0} = \{ \tilde{\mathbf{D}}^\dagger \tilde{\boldsymbol{F}} - (E - E_0)\, \tilde{\mathbf{D}}_0^\dagger \} \{ 1/2\, \tilde{\mathbf{D}}_0^\dagger \tilde{\mathbf{D}}_0 + 1/2\, \tilde{\mathbf{D}}^\dagger \tilde{\mathbf{D}} \}^{-1} = 0$$

$$(34b) \quad \frac{dE}{d\tilde{\mathbf{D}}^\dagger} = \{ \tilde{\boldsymbol{F}} \tilde{\mathbf{D}}_0 + \mathbf{H} \tilde{\mathbf{D}} - (E - E_0)\, \tilde{\mathbf{D}} \} \{ 1/2\, \tilde{\mathbf{D}}_0^\dagger \tilde{\mathbf{D}}_0 + 1/2\, \tilde{\mathbf{D}}^\dagger \tilde{\mathbf{D}} \}^{-1} = 0$$

This can be written conveniently in matrix form as an eigenvalue problem:

$$(35) \quad \begin{bmatrix} E_0 \mathbf{I} & \tilde{\boldsymbol{F}}^\dagger \\ \tilde{\boldsymbol{F}} & E_0 \mathbf{I} + \mathbf{H} \end{bmatrix} \begin{bmatrix} \tilde{\mathbf{D}}_0 \\ \tilde{\mathbf{D}} \end{bmatrix} = E \begin{bmatrix} \tilde{\mathbf{D}}_0 \\ \tilde{\mathbf{D}} \end{bmatrix}$$

with the eigenvector normalized such that $\{ 1/2\, \tilde{\mathbf{D}}_0^\dagger \mathbf{D}_0 + 1/2\, \tilde{\mathbf{D}}^\dagger \tilde{\mathbf{D}} \} = 1$. In the case of a wavefunction constrained to be real, eq 35 reduces to the equation presented by Bacskay[42].

$$(36) \quad \begin{bmatrix} E_0 & \boldsymbol{F}^\dagger \\ \boldsymbol{F} & E_0 \mathbf{I} + \mathbf{A} + \mathbf{B} \end{bmatrix} \begin{bmatrix} D_0 \\ \mathbf{D} \end{bmatrix} = E \begin{bmatrix} D_0 \\ \mathbf{D} \end{bmatrix}; \qquad D_0^2 + \mathbf{D}^T \mathbf{D} = 1$$

If $\mathbf{B}$ and the two electron part of $\mathbf{A}$ are omitted, the QC-SCF method is the same as the simple SCF[42].

At first glance, eq 35 and 36 would seem to be very expensive to solve, since each iteration involves a 4 index transformation of the 2 electron integrals and requires a large matrix to be diagonalized to obtain the D's. However, Bacskay[42] has shown that the eigenvalue problem can be solved iteratively and efficiently by avoiding the explicit 4 index transformation. The products $\mathbf{H\tilde{D}}$ or $\mathbf{AD}$ and $\mathbf{BD}$ can be constructed from the untransformed integerals[42]. Thus one micro-iteration of the solution of the eigenvalue problem requires about the same amount of work as one iteration of the conventional SCF. Far fewer macro-iterations are needed to converge the QC-SCF, with the net result that convergence is reached in significantly less cpu time than the simple SCF. Although DIIS is generally faster[36b] than QC-SCF, the latter may have some advantages for difficult to converge cases or where very tight convergence is needed. An alternative approach to QC-SCF has been presented by Douady, Elinger, Subra and Levy[43] based on exponential transformation of the MO coefficients.

<u>Relation Between Stability and Convergence</u> Whether or not a simple SCF calculation converges is closely related to the stability conditions[44-46]. An SCF calculation will converge if the changes in the MO's decrease from one cycle to the next. Let the changes be written as

(37) $$\Delta\mathbf{M}^{k+1} = \mathbf{Q}\,\Delta\mathbf{M}^{k}$$

The iterative process will converge only if the absolute values of the eigenvalues of Q are less than unity. As before (eq 17 or 27), it is sufficient to consider only the occupied-virtual orbital mixing in determining the changes in the orbitals. Let $C^{k}_{\mu p}$ be the eigenvectors of $\mathbf{F}^{k}$. These eigenvectors can be used to construct F^{k+1} according to eq 24. The coefficients and density matrix can be related to the corresponding quantities from the previous iteration:

(38) $$C^{k}_{\nu i} = C^{k-1}_{\nu i} + \sum_{a} D^{k}_{ia}\, C^{k}_{\nu a}$$

(39) $$P^{k}_{\lambda\sigma} = P^{k-1}_{\lambda\sigma} + \sum_{ia} [\, D^{k\,*}_{ia}\, C^{k\,*}_{\lambda a}\, C^{k}_{\sigma i} + D^{k}_{ia}\, C^{k\,*}_{\lambda i}\, C^{k}_{\sigma a}\,]$$

where only the first order terms in D have been retained. Since F^{k} was built from $\mathbf{P}^{k-1}$, only the contribution from $\mathbf{P}^{k}$ - $\mathbf{P}^{k-1}$ needs to be added to $\mathbf{F}^{k}$ to obtain F^{k+1}. Inserting eq 39 into eq 24 and transforming to the $C^{k}_{\mu p}$ basis yields

(40) $$F^{k+1}_{pq} = F^{k}_{pq} + \sum_{jb} [\, (pj||qb)\, D^{k}_{jb} + (pb||qj)\, D^{k*}_{jb}\,]$$

$$= \varepsilon_p\, \delta_{pq} + \sum_{jb} [\, (pj||qb)\, D^{k}_{jb} + (pb||qj)\, D^{k*}_{jb}\,]$$

since the $C^{k}_{\mu p}$ are the the eigenvectors of F^{k}. With the definitions in eq 21 and 22, the occupied-virtual block of F^{K+1} can be written as

(41) $$\boldsymbol{F}^{k+1} = (\mathbf{A} - \boldsymbol{\Delta})\, D^{k} + \mathbf{B}\, D^{k*}$$

where $\Delta_{ia\,jb} = (\varepsilon_a - \varepsilon_i)\, \delta_{ij}\, \delta_{ab}$. The new coefficients, $C^{k+1}_{\mu i}$, (eq 24) can be obtained by diagonalizing $\mathbf{F}^{k+1}$ (see eq 25); alternatively the D's can be computed by perturbation theory (see eq 28):

$$(42)\quad D^{k+1}_{ia} = - F^{k+1}_{ai} / (\varepsilon_a - \varepsilon_i) = - \sum_{jb} [\, (aj||ib)\, D^{k}_{jb} + (ab||ij)\, D^{k*}_{jb}\,] / (\varepsilon_a - \varepsilon_i)$$

$$\text{or}\qquad \tilde{\mathbf{D}}^{k+1} = - \tilde{\mathbf{\Delta}}^{-1} (\tilde{H} - \tilde{\Delta})\, \tilde{\mathbf{D}}^{k}\,; \qquad \tilde{\mathbf{\Delta}} = \begin{bmatrix} \mathbf{\Delta} & \mathbf{0} \\ \mathbf{0} & \mathbf{\Delta} \end{bmatrix}$$

After a bit of rearrangment, eq 42 can be put in the form of eq 37.

$$(43)\quad \tilde{\mathbf{\Delta}}^{-1/2}\, \tilde{\mathbf{D}}^{k+1} = (1 - \tilde{\mathbf{\Delta}}^{-1/2}\, \mathbf{H}\, \tilde{\mathbf{\Delta}}^{-1/2})\, \tilde{\mathbf{\Delta}}^{1/2}\, \tilde{\mathbf{D}}^{k}$$

For the iteration to converge, the absolute values of the eigenvalues of $\mathbf{Q}=(\mathbf{1}-\tilde{\mathbf{\Delta}}^{-1/2}\mathbf{H}\tilde{\mathbf{\Delta}}^{-1/2})$ must be less than 1. Hence, the eigenvalues of $\tilde{\mathbf{\Delta}}^{-1/2}\, \mathbf{H}\, \tilde{\mathbf{\Delta}}^{-1/2}$ must be between 0 and 2[47]. Since $\tilde{\mathbf{\Delta}}$ is diagonal and positive definite, **H** must at least be positive definite for the simple SCF to converge. But if **H** is positive semidefinite, then the wavefunction is stable. Thus, if it converges, the simple SCF procedure (without techniques such as damping, level shifting, etc.) yields a stable wavefunction. However, the converse is not necessarily true. If a wavefunction is stable, **H** is positive semidefinite; however $\tilde{\mathbf{\Delta}}^{-1/2}\, \mathbf{H}\, \tilde{\mathbf{\Delta}}^{-1/2}$ may have eigenvalues larger than 2 and the simple SCF does not converge. If any of the eigenvalues of $\tilde{\mathbf{\Delta}}^{-1/2}\, \mathbf{H}\, \tilde{\mathbf{\Delta}}^{-1/2}$ are near 0 or 2, convergence will be slow. An eigenvalue near zero signals a near degeneracy between two states. Large eigenvalues of $\tilde{\mathbf{\Delta}}^{-1/2}\, \mathbf{H}\, \tilde{\mathbf{\Delta}}^{-1/2}$ can occur when an energy difference between occupied and virtual orbitals is small relative to the 2 electron contributions to **H**. This gives rise to large changes in the Fock matrix from step to step and results in oscillatory divergence if any of the eigenvalues of $\tilde{\mathbf{\Delta}}^{-1/2}\, \mathbf{H}\, \tilde{\mathbf{\Delta}}^{-1/2}$ are greater than 2. This analysis also illustrates how level shifting affects the convergence of a conventional SCF calculation. Since level shifting increases all of the occupied-virtual orbital energy differences (i.e. all of the elements of $\mathbf{\Delta}$ become larger), it reduces the eigenvalues of $\mathbf{1} - \tilde{\mathbf{\Delta}}^{-1/2}\, \mathbf{H}\, \tilde{\mathbf{\Delta}}^{-1/2}$ and forces convergence, even in cases that would otherwise be divergent.

Examples The behavior of some of the algorithms discussed above is shown in Table 3. The convergence is judged by the root mean square change in the density matrix (or the rms orbital rotation gradient in the case of QCSCF). The number of Fock matrix evaluations (and number of micro-iterations to solve QCSCF equations) is used as the iteration count. Since most of the work in the SCF for a larger molecule is in the formation of the Fock matrix, this method of counting iterations gives a fair comparison of the work required to achieve convergence (a micro-iteration of the QCSCF equations requires about the same amount of work as a Fock matrix evaluation[42]). For the simple SCF, the energy was constant to 10 significant figures after 10 iterations, but the density matrix continued to change. The SCF seemed to be heading toward convergence on the density matrix for the first 18 cycles; however, it then started to diverge. Extrapolation did not alter this behavior significantly. Level shifting slowed the rate of convergence, but did not prevent divergence. A univariate search with accurate location of the minimum, the SCFDM approach detailed in reference 32, did lead to convegence after 9 searches which required a total of 65 Fock matrix evaluations. By contrast, the DIIS algorithm converged very rapidly, in only 8 cycles. The QCSCF method started with a univariate search that required 7 points to find the minimum. This was followed by 2 QCSCF macro-iterations, each

Table 3. Convergence behavior of various SCF algorithms for O_2[a]

Iteration	Simple SCF[b,c]	Level Shift[b,d]	SCFDM[e]	DIIS	QCSCF[f]
1					6.7(-1)
2	4.9(-4)	4.6(-4)		3.0(-4)	
3	1.7(-4)	1.8(-4)		1.8(-4)	
4	7.4(-5)	8.2(-5)		1.2(-4)	
5	3.5(-5)	4.1(-5)	4.1(-3)	2.0(-5)	
6	1.7(-5)	2.2(-5)		9.6(-9)	
7	8.5(-6)	1.1(-5)			
8	4.2(-6)	6.0(-6)	6.6(-4)	1.9(-11)	4.5(-3)
9	2.1(-6)	3.2(-6)			
10	1.0(-6)	1.7(-6)			
11	5.1(-7)	9.0(-7)	5.3(-4)		1.6(-6)
12	2.5(-7)	4.8(-7)			
13	1.3(-7)	2.5(-7)			
14	6.2(-8)	1.3(-7)			2.5(-10)
15	3.1(-8)	7.1(-8)			
16	1.5(-8)	3.8(-8)			
17	8.5(-9)	2.0(-8)			
18	8.2(-9)	1.1(-8)	3.9(-4)		
19	1.4(-8)	5.9(-9)			
31	3.5(-5)	1.4(-6)	5.5(-6)		

[a] UHF triplet at R = 1.7Å; convergence on root mean square change in the density matrices; the iteration count indicates the number of Fock matrix evaluations (and micro-iterations required to solve the QCSCF equations)

[b] failed to converge in 32 cycles.

[c] 3 and 4 point extrapolation did not alter convergence significantly.

[d] virtual orbital energies shifted by 0.1 au.

[e] ref 32, univariate search with accurate location of the minimum; converged after 65 Fock matrix evaluations

[f] values quoted in this column are rms orbital rotation gradients rather than rms density matrix changes.

requiring 3 micro-iterations to solve the QCSCF equations, and a final energy and rotation gradient calculation, for a total equivalent to 14 Fock matrix evaluations. An accurate univariate search, both in QCSCF and SCFDM, imposes a large penalty in terms of the total number of Fock matrix evaluations. This suggests that an approximate search might be more cost effective; however, this may risk non-convergence. The DIIS method is clearly the best algorithm for this example; similar results have been demonstrated for other cases[36]. QCSCF also performs quite well and should be used if DIIS fails. Simple SCF is not satisfactory for this example, even with various convergence controlling techniques.

5. Initial Guess.

Whether or not an SCF calculation converges and how many iterations it takes to converge often depends on the quality of the initial guess for the MO's or the density matrix. Some of the approaches used to obtain a suitable initial guess are listed below.

(a) Eigenvectors of the one electron hamiltonian ($h_{\mu\nu}$ in eq 9). This is equivalent to choosing a zero density matrix in building the Fock matrix for the zeroth iteration.

(b) Semi-empirical wavefunction. Extended Hückel (EHT), CNDO, MNDO etc. provide reasonable MO's; the empirical parameters in the EHT are sometimes adjusted until more suitable orbitals are obtained.

(c) Smaller basis set ab initio calculations. These can be projected onto larger basis sets to yield good initial guesses.

(d) Calculations from near by points. This is useful for geometry optimizations, numerical differentiations and potential surface scans.

The above methods are roughly in order of increasing quality. For problematic cases, a number of additional techniques and tricks can be taken to obtain convergence:

(e) Alter the orbital occupancy. Selected occupied and virtual orbitals can be switched until a combination is found that leads to convergence for the desired state. Symmetry and a close inspection of the orbitals usually indicates which orbitals need to be switched.

(f) Higher symmetry. Often, higher symmetry structures are better behaved, since symmetry reduces the orbital mixing that occurs. The converged orbitals can serve as an initial guess for near by lower symmetry structures.

(g) Localized orbitals. It may be easier to choose the correct occupancy if the orbitals are not delocalized over the entire molecule.

(h) Changing the geometry. Elongated bonds, strained structures and strongly interacting fragments sometimes cause stability and convergence problems. It may be possible to converge the SCF calculation for a structure with shorter bonds, a more relaxed geometry or non-interacting fragments, respectively (however, care should be taken not to change the symmetry of the system or its alginment with the coordinate axes). These orbitals can then be used as a guess for the problematic case.

(i) Stability analysis. If a stability calculation indicates that a converged wavefunction is unstable, the wavefunction can be displaced along the offending eigenvector and optimized by a univariate search, to give an initial guess for a lower energy wavefunction.

(j) Changing the charge. It may be possible to converge a calculation with fewer electrons (but the same number of unpaired electrons) or with greater charges on the nuclei. The electrons in a cation are more strongly bound and hence more likely to converge. The converged orbitals may be suitable as an initial guess for the unaltered molecule.

There are, of course, cases for which the best initial guess will not lead to SCF convergence with a given SCF procedure. These can only be overcome by using a better algorithm. With a combination of improvements in the initial guess and the use of more powerful algorithms, most SCF convergence problems can be overcome.

6. Summary.

Stability SCF calculations don't always converge to the most stable wavefunction. Molecules with unusual geometric or electronic structure (e.g. strained geometry, transition states, hypervalency, open shells etc.) can have internal and/or external instabilities in the wavefunction, and therefore should be tested for stability. If necessary, a lower energy wavefunction can be found by changing the MO's according to the offending eigenvector obtained in the stability analysis.

Convergence Getting a converged SCF wavefunction can sometimes be a problem. The methods discussed above can be summarized as follows:

(a) If at all possible, avoid the simple SCF iteration scheme. DIIS is more stable, requires fewer iterations and costs no more per iteration than simple SCF.

(b) Use a good initial guess - either from a reliable semi-empirical calculation, a lower level ab initio calculation or a near by structure. Sometimes occupied and virtual orbitals have to be switched or mixed to get a suitable initial guess.

(c) Damping may be useful in getting a partially converged wavefunction, but level shifting is usually more efficient. Univariate search is even better but may be more expensive.

(d) Extrapolation can speed up convergence during the latter part of a simple SCF calculation, but may cause problems if applied too early (DIIS normally does not need additional extrapolation).

(e) Quadratically convergent SCF is usually more expensive than DIIS, but may be the method of choice if DIIS fails; however, the initial guess must be suitably close to the desired wavefunction (see (c) to get close enough).

References.

1. A. Szabo and N.S. Ostlund, *Modern Quantum Chemistry: Introduction to Advanced Electronic Structure Theory*, Macmillan, New York, 1982.
2. R. McWeeny and B.T. Sutcliffe, *Methods of Molecular Quantum Mechanics*, Academic Press, London, 1969.
3. J.P Lowe, *Quantum Chemistry*, Academic Press, New York, 1978.
4. C. C. J. Roothaan, Rev. Mod. Phys., **23**, 69 (1951); G. G. Hall, Proc. Roy. Soc. (London), **A205**, 541 (1951)
5. J. A. Pople and R. K. Nesbet, J. Chem. Phys. **22**, 571 (1954).
6. D. J. Thouless, *The Quantum Mechanics of Many-Body Systems*, Academic, NY, 1961.
7. J. Cizek and J. Paldus, J. Chem. Phys. **47**, 3976 (1967), Phys. Rev A, **3**, 525 (1971); J. Paldus and J. Cizek, Chem. Phys. Lett., **3**, 1 (1969), J. Chem. Phys., 52, 2919 (1970).
8. N. S. Ostlund, J. Chem. Phys. **57**, 2994 (1972).
9. H. Fukutome, Int. J. Quantum Chem. **20**, 955 (1981).
10. R. Seeger and J.A. Pople, J. Chem. Phys., **66**, 3045, (1977).
11. UHF wavefunctions can be projected[56] to make them eigenfunctions of $\hat{S}^2$ but the result i multideterminantal and, hence, outside the scope of this article.
12. P. -O. Löwdin, Phys. Rev., **97**, 1509 (1955); for a more recent review see I. Mayer, Adv. Quantum Chem., 12, 189 (1980).

13. For ROHF wavefunctions, singly and doubly occupied orbitals must also be mixed.
14. The stabilities of projected Hartree-Fock methods are discussed in reference 15
15. Y. G. Smeyers and G. Delgado-Barrio, Phys. Rev. A, **15**, 1359 (1977); P. Karadakov and O. Castano, Int. J. Quantum Chem., **22**, 1033 (1982); O. Castano and P. Karadakov, Chem. Phys. Lett., **130**, 123 (1986).
16. D. R. Hartree, *The Calculation of Atomic Structures*, Wiley, New York, 1957; C. C. J. Roothaan and P. S. Bagus, in *Methods in Computational Physics*, Vol 2, B. Adler, S. Fernbach and M. Rotenberg, eds., Academic Press, New York, 1963.
17. M.C. Zerner and M. Hehenberger, Chem. Phys. Lett., **62**, 550, (1979).
18. G. Karlström, Chem. Phys. Lett., **67**, 348, (1979).
19. P. de Montgolfier and A. Hoareau, J. Chem. Phys. **65**, 2477 (1976).
20. For methods of maintaining idempotency while changing the density matrix see ref. 30.
21. V. R. Saunders and I. H. Hillier, Int. J. Quantum Chem. **7**, 699 (1973); M. F. Guest and V. R. Saunders, Mol. Phys., **28**, 819 (1974).
22. A. V. Mitin, J. Comput. Chem. **9**, 107 (1988).
23. For a discussion of Aitken extrapolation and related methods see W. H. Press, B. P. Flannery, S. A. Teukolsky and W. T. Vetterling, *Numerical Recipes. The Art of Scientific Computing*, Cambridge Univ. Press, Cambridge, 1986, or any similar volume on numerical analysis.
24. J. Cioslowski, J. Chem. Phys., **89**, 2126 (1988).
25. W. B. Neilsen, Chem. Phys. Lett., **18**, 225 (1973).
26. I. Röeggen, Chem. Phys. Lett., **22**, 140 (1973).
27. M. J. S. Dewar and P. K. Weiner, Comput. Chem. **2**, 31 (1978).
28. L. G. Ferreira, J. Comput. Phys., **36**, 198 (1980).
29. Z. Chen, Chem. Phys. Lett., **87**, 455 (1982).
30. R. McWeeny, Proc. Roy. Soc. (London) **A235**, 496 (1956).
31. I.H. Hillier and V.R. Saunders, Int. J. Quantum Chem., **4**, 503, (1970), Proc. Roy. Soc. (London), **A320**, 161 (1970).
32. R. Seeger and J. A. Pople, J. Chem. Phys. **65**, 265 (1976).
33. R. N. Camp and H. F. King, J. Chem. Phys. **75**, 268 (1981).
34. J. F. Rico, J. M. G. de la Vega, M. Paniagua and J. I. Fernández-Alonso, J. Chem. Phys. **79**, 4407 (1983).
35. L. E. Scales, *Introduction to Non-Linear Optimization*, Springer, New York, 1985.
36. P. Pulay, Chem. Phys. Lett. **73**, 393 (1980); J. Comput. Chem. 3, 556 (1982).
37. C. Lanczos, J. Res. Natl. Bur. Std. US, **45**, 255 (1955).
38. P. Császár and P. Pulay, J. Mol. Struct., **114**, 31 (1984).
39. E. R. Davidson, J. Comput. Phys. **17**, 87 (1975).
40. J. A. Pople, R. Krishnan, H. B. Schlegel and J. S. Binkley, Int. J. Quantum Chem., Quantum Chem. Symp., **13**, 225 (1979).
41. I. Bálint and M. I. Bán, Chem. Phys. Lett. **101**, 153 (1983).
42. G. B. Bacskay, Chem. Phys. **61**, 385 (1981), **65**, 383 (1982).
43. J. Douady, Y. Elinger, R. Subra and B. Levy, J. Chem. Phys. **72**, 1452 (1980).
44. R. E. Stanton, J. Chem. Phys. **75**, 3426, 5416 (1981).
45. J. C. Facelli and R. H. Contreras, J. Chem. Phys. **77**, 1076 (1982); **79**, 3421 (1983).
46. M. A. Natiello and G. E. Scuseria, Int. J. Quantum Chem., **26**, 1039 (1984).
47. with the appropriate change in notation, this result is the same as reference 45.

Post-SCF methods: Theory and Practice

Michael A. Robb
Department of Chemistry
King's College London
Strand
London WC2R 2LS
United Kingdom

Joseph J.W. McDouall
Department of Chemistry
The University of Manchester
Manchester M13 9PL
United Kingdom

Abstract.

The many post-SCF methods for the computation of electron correlation effects will be discussed from a unified but practical point of view. The lectures are divided into roughly two parts.

In the first part, we shall discuss the physical and theoretical basis of dynamic and non-dynamic electron correlation so that a sound foundation can be established for deciding which method is likely to be reliable for a given chemical problem. For this purpose we shall use the MCSCF+CI formulation of the electron correlation problem and the Löwdin partitioning method (Bloch equations) to provide a theoretical basis for this discussion, since methods such as single or multi-reference CI, Møller-Plesset perturbation theory, and coupled-cluster approaches can be derived from this common starting point.

In the second part, the various single reference and multi-reference methods will be discussed from a practical point of view. Our objective here is to enable the organic chemist to make a sensible choice amongst the extensive standard software for the particular problem under consideration, given the accuracy/cost dilemmas involved. The choice between single and multi-reference methods at the SCF (ie SCF vs. MCSCF) and post SCF levels will be considered carefully. We shall also discuss size extensivity problems as they apply to all methods.

C. Ögretir and I. G. Csizmadia (eds.),
Computational Advances in Organic Chemistry: Molecular Structure and Reactivity, 187–227.

Introduction.

The combination of dramatic developments in computer hardware over the past 10 years and the availability of standard computer software for electronic structure computations has meant that one may begin to use the techniques of quantum chemistry in the manner of a numerical spectrometer. The typical user of quantum chemistry software for electronic structure computations may have little theoretical or computational experience and is interested mainly in the confidence that he may have in the accuracy of the computed results relative to the cost of performing the computation. However, while the choice of basis set and the reliability of SCF computations is now well understood from this point of view, the use of post-SCF methods for the treatment of electron correlation is not so well understood. Thus the user of a standard quantum chemistry code such as GAUSSIAN (at the G88 level) is presented with the options of using Møller-Plesset perturbation theory, CI, coupled-cluster or a combination of coupled-cluster and perturbation theory at the post-SCF level, but has only to choose between restricted and unrestricted methods at the SCF level. Further, the list of post-SCF methods to be found in this code is being extended with each successive edition. How is the user who is interested in the chemistry to make a sensible choice? The choice of say SCF + MP4, as a good default position, would simply be quite wrong in many situations.

Thus the purpose of this article is twofold. In Part I, we shall try to establish some physical models for electron correlation to discuss these models within a common theoretical framework that encompasses the most common post-SCF methods . In Part II, we will discuss the CI and coupled-cluster/ MP methods in rather more detail. We hope this discussion will guide the user of standard post-SCF software in the selection of the correct method for the problem at hand.

I. Physical and Theoretical Basis for the Computation of Dynamic and Non-Dynamic Electron Correlation.

i)<u>Dynamic versus non-dynamic Electron Correlation</u>

We shall begin our discussion with a discussion of the physical nature of the electron correlation problem taking H_2 as a simple example. Electron correlation is usually taken to be the energy difference between the Restricted Hartree-Fock energy and the exact non-relativistic energy assuming that the Restricted Hartree-Fock energy is a good zeroth-order wavefunction. Let us assume this for the time being and examine the origin of electron correlation for a pair of electrons in the same orbital (eg. the σ_g orbital of H_2). The following argument is adapted from a much more precise argument given by Sinanoglu [2]. If Fig. 1 below we depict the physical situation where each electron is surrounded by a $1/r_{12}$ potential into which the other electron cannot penetrate

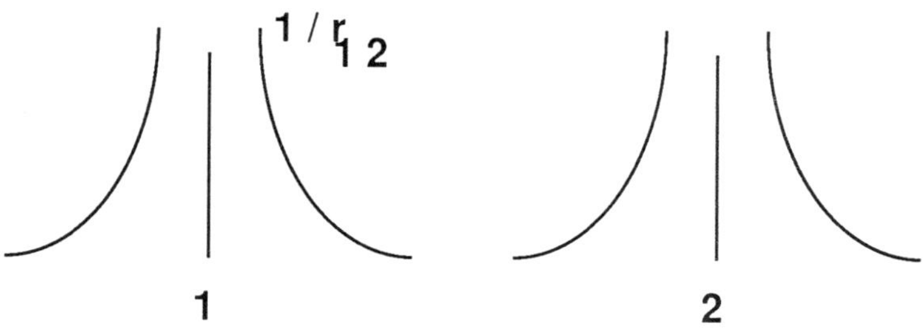

Fig.1 Coulomb Hole around each electron

However, since the orbital is doubly occupied, each electron in fact sees an average coulomb potential arising from the electron density ϕ^2 of the form $\int \phi^2(2)/r_{12}\, dr_2$ as illustrated in Fig. 2

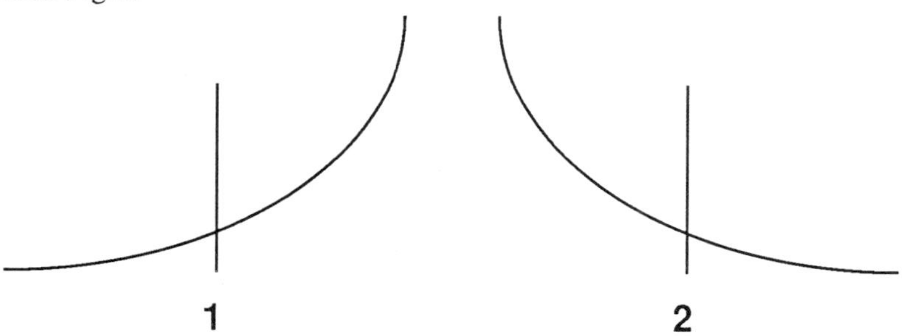

Fig. 2 Average SCF potential

Thus the motions of the electrons are not correlated. Electron 2 can get close to electron 1 and the repulsion is too high. We know that CI corrects such a problem. Let us remind ourselves why. If we take a CI wavefunction for H_2 of the form

$$\Psi = C_1\, \sigma_g{}^2(1s) + C_2\, \sigma_u{}^2(1s) + C_3\, \pi_u{}^2(2p) + C_4\, \pi_g{}^2(2p) + \ldots\ldots \quad (1)$$

where the orbitals have the form illustrated schematically in fig. 3

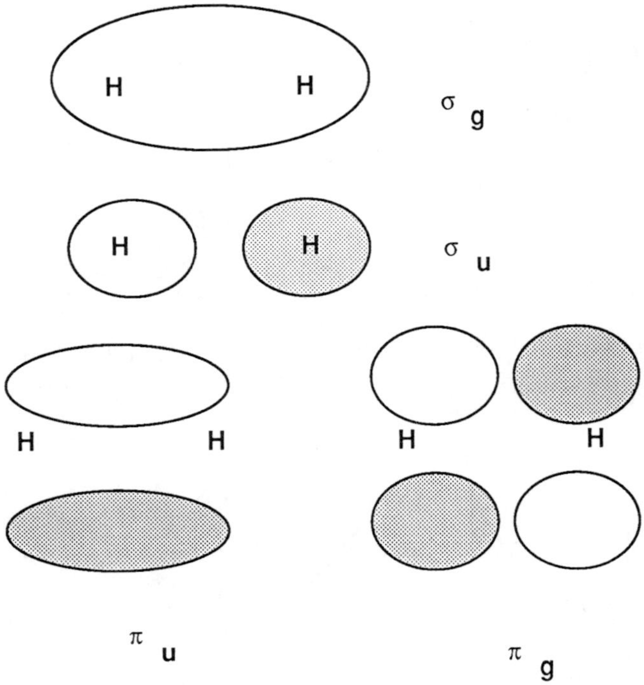

Fig. 3 Excited orbitals for H_2

Adding a $\sigma_u^2(1s)$ configuration to the $\sigma_g^2(1s)$ SCF configuration builds in a left-right node and keeps the two electrons apart thus lowering the energy through the reduction of the electron-electron repulsion energy. Similarly, the $\pi^2(2p)$ configurations build in transverse nodes. Thus it is apparent that one must go on adding configurations of this form until the energy converges. Also the excited orbitals should be localized in the same region of space as the occupied ones in order for efficient interaction. Thus CI is correcting the poor treatment of the instantaneous electron-electron repulsion at the SCF level. This is usually called *Dynamic* or *all external* electron correlation. It is simple to think about physically, since it depends essentially on the number of electrons and the electron density. For a molecular system one gets an order of magnitude estimate taking the sum of atomic correlations and adding a correction for the number of new electron pairs in the molecule.

There is a second type of correlation referred to as *non-dynamic* or *internal* correlation that is much more subtle, yet is familiar. Near dissociation the σ_g^2 configuration cannot describe two H atoms and the MO method fails completely. As is well known the σ_g^2 and σ_u^2 configurations can be expanded in terms of electronic configurations of electrons in atomic orbitals ($1s_a$ on one centre and $1s_b$ on the other centre)

$$\sigma_g^2 = 1s_a(1)1s_a(2) + 1s_b(1)1s_b(2) + 1s_a(1)1s_b(2) + 1s_b(1)1s_a(2)$$
$$= \quad \text{ionic} \quad + \quad \text{covalent} \tag{2a}$$

$$\sigma_u^2 = 1s_a(1)1s_a(2) + 1s_b(1)1s_b(2) - 1s_a(1)1s_b(2) + 1s_b(1)1s_a(2) \tag{2b}$$
$$= \quad \text{ionic} \quad - \quad \text{covalent}$$

Thus the correct wavefunction for dissociation into two H atoms is $\sigma_g^2 - \lambda\, \sigma_u^2$ which eliminates the unwanted ionic terms. The difference in energy obtained with the restricted HF wavefunction and that obtained with a wavefunction that dissociates correctly is the *non-dynamic* electron correlation. It is clearly of a completely different physical nature than the rather simple dynamic correlation.

We are thus lead to the conclusion that the reference wavefunction is perhaps the most important concept to be grasped from the outset. For H_2 , except in the region of the equilibrium bond-length, we need a two-dimensional reference space consisting of the σ_g^2 and σ_u^2 configurations in order to discuss the electron correlation problem. Thus our reference point for a discussion of dynamic correlation is a Multi-configuration SCF (MCSCF) rather than SCF. The effect of the σ_u^2 configuration is no longer assigned to dynamic electron correlation.

At this stage it turns out to be expedient to use the configurations of the electrons in atomic orbitals in order to get some physical insight into the nature of dynamic correlation with a multi-reference wavefunction. Thus let us divide the orbitals into two types: valence orbitals $1s_a$ and $1s_b$ and unoccupied orbitals 2p etc. We can write a CI wavefunction, that includes dynamic correlation approximately, as

$$\begin{aligned}\Psi = {} & A_1 \{1s_a(1)\ 1s_a(2) + {}^1C_3\, \pi_u^2(2p) + {}^1C_4\, \pi_g^2(2p)\} \\ & + A_2 \{ 1s_b(1)\ 1s_b(2) + {}^2C_3\, \pi_u^2(2p) + {}^2C_4\, \pi_g^2(2p)\} \\ & + A_3 \{ 1s_a(1)\ 1s_b(2) + {}^3C_3\, \pi_u^2(2p) + {}^3C_4\, \pi_g^2(2p)\} \\ & + A_4 \{1s_b(1)\ 1s_a(2) + {}^4C_3\, \pi_u^2(2p) + {}^4C_4\, \pi_g^2(2p)\}\end{aligned} \tag{3}$$

and the correlation energy becomes roughly

$$\begin{aligned}\Delta E = {} & (A_1)^2\, \Delta E\, \{1s_a(1)\ 1s_a(2)\} + (A_2)^2\, \Delta E\, \{1s_b(1)\ 1s_b(2)\} \\ & + (A_3)^2\, \Delta E\, \{1s_a(1)\ 1s_b(2)\} + (A_4)^2\, \Delta E\, \{1s_b(1)\ 1s_b(2)\}\end{aligned}$$

$$= 2\,(A_1)^2\ \Delta E(H^+H^-) + 2\,(A_3)^2\ \Delta E(H{\cdot}H{\cdot}) \tag{4}$$

However $\Delta E(H{\cdot}H{\cdot})$ and $\Delta E(H^+H^-)$ will be approximately equal to sums of atomic correlation and independent of distance. Since $\Delta E(H{\cdot})$ and $\Delta E(H^+)$ are zero, then

$$\Delta E = 2\,(A_1)^2\ \Delta E(H^-) \tag{5}$$

and the true dynamic correlation energy is a function of the electron correlation of H^- weighted by the coefficient of the ionic contribution to the wavefunction that goes to zero at dissociation. Thus the non-dynamic electron correlation that arises from the $\sigma_g^2 \rightarrow \sigma_u^2$ excitation is strongly structure dependent. However, the dynamic correlation is essentially atomic in character. Its structure dependence arises mainly because the ratio of ionic and covalent configurations is structure dependent.

To summarize, we wish to emphasize two points. Firstly, the choice of the reference space is critical. CI on top of RHF makes no sense if the RHF wavefunction is not a good zeroth-order wavefunction. Secondly, the dynamic correlation is essentially atomic in character and it is important to try to exploit this fact in computational methods. The interatomic correlation is really non-dynamic correlation and should be computed using a reference space of more than 1 dimension.

It is now our intention to present a rather general methodology for the computation of electron correlation effects. Most of the approximations commonly used for the treatment of electron correlation effects can be extracted from this formalism.

ii)A Unified Formalism for Post-SCF Methods:

Before we can discuss the various methods for treating electron correlation we need a theoretical formalism in which to work. Thus in this section we shall derive two formalisms that are very general, from which most currently used methods can be derived. These are the Multi-reference CI (MR-CI) formalism and the Multi-reference coupled-cluster (MR-CC) formalism. While the results may be quite unfamiliar to the reader, we shall start our development from a Configuration Interaction (CI) expansion [3] which will is known.

We begin this section by formulating the solution of the electronic Schrödinger equation as a Configuration Interaction (CI) expansion in configuration state functions (CSF). We will avoid completely the detailed definition of a CSF. We need only the fact that a CSF is in one to one correspondence with an electronic configuration of N electrons in n orbitals with some spin-coupling information. Let us assume that we have 3 types of orbitals: core orbitals (for which we use a running index *a, b, c, d*), valence orbitals (denoted by a running index μ,ν,η,χ) and unoccupied orbitals (with a running index *r,s,t,u*). The 3 types of orbital are distinguished by their population in what we shall call *reference configurations*. In a reference configuration the *core* orbitals are doubly occupied, the *unoccupied* orbitals have zero occupancy and the *valence* orbitals can have occupancies of 2,1 or 0. The orbital occupation of a reference configuration has the form shown in Fig. 4 below

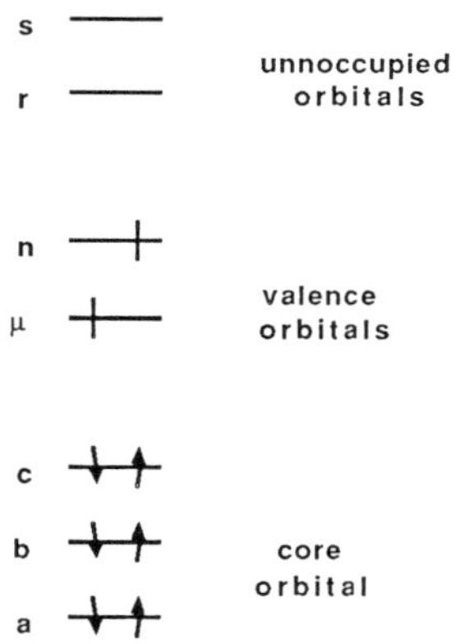

Fig. 4 The orbital occupation pattern of a *reference configuration*

We will subsequently see that with this definition we can define a CI expansion in terms of CSF that will converge quite rapidly for any electronic state, open- or closed-shell. For the moment we need only observe that if we have no *valence* orbitals then we have only one reference configuration and this could be the familiar Restricted Hartree-Fock (RHF) configuration. On the other hand, if all the valence orbitals were singly occupied with parallel spins, then we would also have a single reference *open-shell* configuration. However, the formalism is quite general and we could describe a wide variety of chemical situations by taking a linear combination of reference configurations with different *valence* orbital occupation numbers. We shall use this linear combination of *valence configurations* as the starting point for our discussion of electron correlation and post SCF methods. Thus we shall write a reference wavefunction Λ_0 as

$$\Lambda_0 = \Sigma_\kappa |\kappa\rangle A_\kappa \qquad (6)$$

where $|\kappa\rangle$, $|\lambda\rangle$ .... are possible *reference configurations* distinguished by different possible occupancies (and spin-coupling) of the *valence* orbitals. The A_κ are chosen to minimize the energy via the usual eigenvalue equation.

$$\mathbf{H_0}\,\mathbf{A} = E_0\,\mathbf{A} \qquad (7)$$

where

$$\mathbf{H_0} = \{\langle\kappa|\, H\, |\lambda\rangle\} \text{ ; } \kappa \text{ and } \lambda \text{ are reference configurations} \qquad (8)$$

A CI expansion that corresponds to the full solution of the electronic Schrödinger equation is obtained by forming single, double, etc. excitations from the reference space. The various terms are distinguished by the occupancies of the core, valence and unoccupied orbitals. We use N_{ch} to denote the number of *holes* created in the doubly occupied closed shell core and N_p to denote the occupancy of the unoccupied orbitals of the reference configurations (the number of *particles)*. The occupancy of the valence orbitals, N_v is expressed relative to *n* the number of valence electrons in the reference configurations. The possibilities for double excitation are enumerated below:

No. of core *holes* N_{ch}	No. of Valence *electrons* (n valence electrons in reference configurations) N_v	No. unoccupied *particles* N_p
2	n	2
0	n-2	2
2	n+2	0
1	n-1	2
1	n+1	1

The first three types of terms occur in such an expansion (for double excitation) , i) a core - unoccupied excitation, ii) a valence - unoccupied excitation and iii) a core - valence excitation are shown in Fig. 5 below.

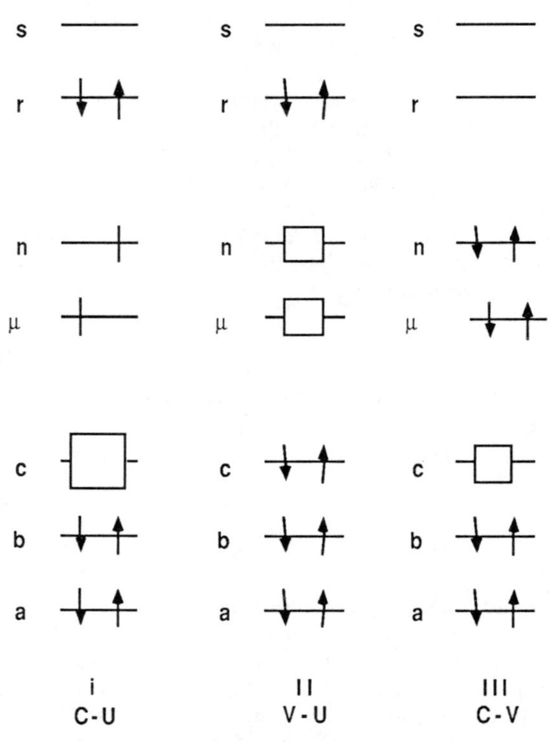

Fig. 5 Examples of Double Excitations that occur in Multi-reference CI

The possibilities are apparent by comparison with fig. 4. The square box indicates where electrons have been removed. In the limit of closed shell RHF only class *i* is possible. Note that for classes ii and iii the valence orbitals play the role of *holes* (ie electrons are excited *from* them $N_V < n$) or *particles* (ie electrons are excited *into* them, $N_V > n$).

The CI expansion can now be written in partitioned form as

$$\Psi = \Sigma_K |K\rangle A_K + \Sigma_R |R\rangle B_R \qquad (9)$$

where the configurations $|R\rangle$, $|S\rangle$ .... are the *secondary space* configurations obtained by C-U, V-U, and C-V orbital excitation. The corresponding eigenvalue problem has the form

$$\begin{pmatrix} H_0 & Z^\dagger \\ Z & W \end{pmatrix} \begin{pmatrix} A_\lambda \\ B_\lambda \end{pmatrix} = E_\lambda \begin{pmatrix} A_\lambda \\ B_\lambda \end{pmatrix} \qquad (10)$$

where

$$W = \{\langle R|H|S\rangle\}$$

are matrix elements between *secondary space* configurations,

$$Z = \{\langle K|H|R\rangle\}$$

are matrix elements between *reference* and *secondary* configurations and $\{A_\lambda|B_\lambda\}$ is the eigenvector with energy E_λ. Clearly in the limit where the reference space is one dimensional eq. 10 is just the ordinary CI problem. The preceding development assumes that we are interested in only a few solutions of the CI problem that are dominated by the reference configurations. In ideal cases the reference space will have only one configuration, the RHF closed shell determinant. However, we shall keep the discussion very general pointing out the obvious simplifications for the single reference case as we proceed.

In general we are interested in a small number of solutions that are dominated by the reference configurations. Let us now explore the solution of the eigenvalue problem that yields only these solutions. It is convenient to think of any method for solving the CI problem that finds a subset of the solutions as *perturbation theory* . To begin with, we shall not use any order by order summations but will assume that we can solve any equations exactly. Thus we are concerned with *infinite-order perturbation theory*. There are two types of *infinite-order perturbation theory* that will concern us: i) *perturb then diagonalize* and ii) *diagonalize then perturb* . In i) we have the aim of computing a subset of the eigenvectors of the CI problem with equal accuracy. In ii) we are interested in only one solution, usually the lowest. We shall explore two formulations of each method; one based upon CI the other based upon a cluster expansion.

Let us begin with the *perturb then diagonalize* methodology in a CI basis since it is the most general. We start by writing the eigenvalue problem of eq. 10 in a slightly different form (known as the Bloch equations [4], see the exhaustive review by Durand and Malrieu [5] for an up to date discussion and many applications)

$$\begin{bmatrix} H_0 & Z^{\dagger} \\ Z & W \end{bmatrix} \begin{bmatrix} U_{AA} & U_{AB} \\ U_{BA} & U_{BB} \end{bmatrix} = \begin{bmatrix} U_{AA} & U_{AB} \\ U_{BA} & U_{BB} \end{bmatrix} \begin{bmatrix} H_{eff} & 0 \\ 0 & W \end{bmatrix} \qquad (11)$$

Thus rather than require the CI hamiltonian to be diagonal (which would yield all possible solutions), we *block* diagonalize it so that a sub-set of the eigenvalues are obtained from

$$H_{eff}\, A_{\lambda} = E_{\lambda}\, A_{\lambda} \qquad (12)$$

By matrix multiplication of the blocks of eq. 11. it can easily be seen that the matrix **U** must satisfy

i) by multiplying the second row of the first matrix on each side with the first column on each side

$$Z\, U_{AA} + W\, U_{BA} = U_{BA} H_{eff} \qquad (13a)$$

ii)by multiplying the first row of the first matrix on each side with the first column on each side

$$H_0\, U_{AA} + Z^{\dagger}\, U_{BA} = U_{AA} H_{eff} \qquad (13b)$$

and a true upper bound to the energy is obtained by forming

$$H_{eff} = \{ U^{-1}\, H\, U \}_{AA} \qquad (13c)$$

However, without any loss in generality, we can choose U_{AA} to be the identity matrix and write **U** in the form

$$\begin{bmatrix} U_{AA} & U_{AB} \\ U_{BA} & U_{BB} \end{bmatrix} = 1 + X = 1 + \begin{bmatrix} 0 & -C^{\dagger} \\ C & 0 \end{bmatrix} \qquad (14)$$

where we make the identification

$$X = \begin{pmatrix} 0 & -C^{\dagger} \\ C & 0 \end{pmatrix} \tag{15}$$

We can now simplify eq. 13a/b to

$$\{Z\}_K + \{W - (H_{eff})_{K K} 1\} C_K - \Sigma_{\lambda \neq K} \{H_{eff}\}_{K \lambda} C_\lambda = 0 \tag{16a}$$

$$\{H_0\}_{K\lambda} + \{Z^{\dagger}\}_K C_\lambda = \{H_{eff}\}_{K\lambda} \tag{16b}$$

where we use the obvious notation where $\{Z\}_K$ $\{Z^{\dagger}\}_K$ is the Kth column or row of the matrix **Z**. Note that there is 1 equation for each reference function K. In the case where we have only one reference function then H_{eff} becomes a scalar E (the total energy) and the summation $\Sigma_{\lambda \neq K}$ does not occur.

In the equation system of eq.16a/16b, the subsequent diagonalization H_{eff} yields a variational upper bound to the energy only if the C_K satisfy eq. 16a exactly. In order by order perturbation theories this will not be the case, because we will solve eq. 16a only approximately. Thus let us now examine the alternative method of computation for H_{eff} based upon eq. 13c. Firstly, the matrix **U** is not unitary so we define

$$U^{\dagger}U = S \tag{17}$$

Since we must have

$$S^{-1}U^{\dagger}U = 1 \tag{18}$$

then

$$U^{-1} = S^{-1}U^{\dagger} \tag{19}$$

and we can write

$$H_{eff} = S^{-1}U^{\dagger}H\,U \tag{20a}$$

$$H_{eff} = S^{-1}(1 - X)\,H\,(1 + X) \tag{20b}$$

Given an approximate solution (for example via perturbation theory) of eq. 16a, then eq. 20 always gives a variational upper bound to the energy. In the case of a single reference configuration, **S** just becomes the normalization integral.

The preceding matrix manipulations should be quite obvious. Now we shall rewrite these equations in a more compact but abstract way that will prove essential for the discussion of MCSCF and coupled-cluster methods. We shall refer to this approach as

Multi-reference Coupled-Cluster (MR-CC). Let us begin with the introduction of some general excitation operators (often called cluster operators and hence the name MR-CC). We define an excitation operator (see for example [6])

$$E^{ijkl\ldots}{}_{mnop\ldots} = E^{[I]}{}_{[J]} \qquad (21)$$

where [I] is the set of integers i<j<k<l etc denoting the spin orbitals or just spatial orbitals. The operator $E^{[I]}{}_{[J]}$ annihilates the orbitals *mnop....* and replaces them with the orbitals *ijkl....* Thus for example

$$E^{\alpha\beta\gamma}{}_{\nu\mu\lambda} = X^{\dagger}_{\alpha}\, X^{\dagger}_{\beta}\, X^{\dagger}_{\gamma}\, X_{\nu} X_{\mu} X_{\lambda} \qquad (22)$$

where X_{λ} when it acts on a CSF deletes orbital λ if it is occupied in the CSF or gives 0 if it is unoccupied and $X^{\dagger}_{\alpha}$ when it acts on a CSF creates an occupied orbital α if it is unoccupied in the CSF or gives 0 if it is occupied. The X_{η} satisfy the usual anticommutation relationships

$$X_{\beta} X^{\dagger}_{\alpha} = -\delta^{\alpha}{}_{\beta}\, X^{\dagger}_{\alpha} X_{\beta} \qquad (23a)$$

$$X^{\dagger}_{\beta} X^{\dagger}_{\alpha} = -\, X^{\dagger}_{\alpha} X^{\dagger}_{\beta} \qquad (23b)$$

and we can manipulate these operators in a manner that is very similar to the way in which one manipulates determinants. We can now write the matrix **U** in operator form as

$$\boldsymbol{U} = \Sigma_{K}\ \exp\,(\mathbf{T}_{K})\ \mathbf{P}_{K} \qquad (24)$$

where $\mathbf{P}_{K}$ is a projector onto the reference configuration κ

$$\mathbf{P}_{K} = |\,\kappa\rangle\langle\,\kappa\,| \qquad (25)$$

The operator $\mathbf{T}_{K}$ is a combination of single, double etc excitation operators of the form defined in eq. 21

$$\mathbf{T}_{K} = \Sigma_{\alpha,r}\, {}^{\kappa}t^{r}{}_{\alpha}\, \{E^{r}{}_{\alpha} - E^{\alpha}{}_{r}\} + \Sigma_{\alpha\beta,r\geq s}\, {}^{\kappa}t^{rs}{}_{\alpha\beta}\, \{E^{rs}{}_{\alpha\beta} - E^{\alpha\beta}{}_{rs}\} + \,.. \qquad (26)$$

and the ${}^{\kappa}t^{r}{}_{\alpha}$ and ${}^{\kappa}t^{rs}{}_{\alpha\beta}$ are coefficients for reference state κ. In general $\alpha\beta$ are core and valence orbitals while *rs* are valence or unoccupied orbitals. At this stage an example may be helpful. The coefficient of a double excitation $|R\rangle = |\,{}^{rs}{}_{\alpha\beta}\rangle$ in $\mathbf{C}_{K}$ can be written in this notation as

$$\mathbf{C}_{K}(\,{}^{rs}{}_{\alpha\beta}\,) = 1/2\ [\ {}^{\kappa}t^{r}{}_{\alpha} \langle R\,|\{E^{r}{}_{\alpha} - E^{\alpha}{}_{r}\}\,|\,\kappa\rangle.\ \{\ {}^{\kappa}t^{s}{}_{\beta} \langle R\,|\{E^{s}{}_{\beta} - E^{\beta}{}_{s}\}\,|\,\kappa\rangle$$

$$\quad {}^{\kappa}t^{s}{}_{\alpha} \langle R\,|\{E^{s}{}_{\alpha} - E^{\alpha}{}_{s}\}\,|\,\kappa\rangle.\ \{\ {}^{\kappa}t^{r}{}_{\beta} \langle R\,|\{E^{r}{}_{\beta} - E^{\beta}{}_{r}\}\,|\,\kappa\rangle]$$

$$+\ {}^{\kappa}t^{rs}{}_{\alpha\beta}\ \langle R\,|\,\{E^{rs}{}_{\alpha\beta} - E^{\alpha\beta}{}_{rs}\}\,|\,\kappa\rangle \qquad (27)$$

only if orbitals α and β are occupied in κ. The product of single excitations in eq. 27 occurs as a result of the expansion of the exponential in eq. 24.

The utility of the exponential formalism is immediately apparent. One can see that the double excitation coefficient in eq. 27 has a contribution from a product of single excitations. The usefulness of this arises from the fact that one can truncate T_K at say double excitations and one has, *automatically* a contribution to quadruple excitations arising from a product of two double excitations. However one need only determine the ${}^{\kappa}t^{rs}_{\alpha\beta}$ explicitly.

Let us recast eq. 16 using this formalism. We get immediately

$$H \exp(T_K) \mid \kappa > - \Sigma_\lambda \exp(T_\lambda) \mid \lambda > \{H_{eff}\}_{\kappa\lambda} = 0 \qquad (28)$$

where the summation restriction $\lambda \neq \kappa$ has been removed and $(H_{eff})_{\kappa\kappa}$ has been included in the second term of eq. 28. If we now premultiply by $\exp(-T_K)$ and project onto the secondary space $<R|$ we obtain an equation that corresponds to eq. 13 (one eq. for each R and κ)

$$< R \mid \exp(-T_K) H \exp(T_K) \mid \kappa >$$
$$- \Sigma_{\lambda \neq \kappa} < R \mid \exp(-T_K) \exp(T_\lambda) \mid \lambda > \{H_{eff}\}_{\kappa\lambda} = 0 \qquad (29a)$$

Note that the restriction $\lambda \neq \kappa$ appears again because $\exp(-T_K) \exp(T_K)$ is just the identity and $< R \mid \lambda > = 0$. The equation corresponding to eq.16b is just

$$\{H_{eff}\}_{\kappa\lambda} = < \kappa | H \exp(T_\lambda) \lambda > \qquad (29b)$$

Let us now briefly summarize what we have accomplished. In eq.16 and eq 29 we have two formalisms for the solution of the electronic Schrödinger equation for a subset of states whose projection on the reference space is large. If no truncations of the CI space in eq. 16 or the excitation operators in eq. 29 are introduced then both are exact. If truncations are made then the two formalisms lead to different results and we shall return to a discussion of this point subsequently. The usual single reference theory is obtained in each case when the $\Sigma_{\lambda \neq \kappa}$ is dropped from eq. 16a or 29a and H_{eff} is replaced by the scalar E corresponding to the ground state energy.

The formalism just discussed is very general. The diagonalization of H_{eff} yields a subset of the eigenvalues (those with the largest projection on the chosen reference space). We have only assumed that the orbital space can be divided into 3 types of orbitals and that this division defines the reference and secondary spaces. Let us now discuss a formulation of the problem that is intermediate between the multi-reference formalism just discussed and the limiting case of single reference CI: single reference CI/CC with a diagonal H_0 (eq.11). Clearly in eq. 16 and eq. 29 one "*perturbs then diagonalizies*". In other words in eq. 16 we solve the equation system for an effective hamiltonian H_{eff} as opposed to a single eigenvalue. We now explore the alternative situation corresponding to "*diagonalize then perturb*" where we begin with a diagonal H_0 and we solve for a single eigenvalue. We shall use the notation $| \kappa' >$ to denote the linear combinations of states $|\kappa>$ that satisfy

$$H_0 \mid \kappa' > = E_{\kappa'} \mid \kappa' > \qquad (30)$$

where

$$|\kappa'\rangle = \sum_{\kappa} |\kappa\rangle A_{\kappa\kappa'} \qquad (31)$$

We suppose that we are interested in the lowest energy solution $|0'\rangle$. Then eq. 16 and 29 remain valid with the simplification that

i) $\mathbf{H}_{\mathbf{eff}} \Rightarrow E_{0'}$

ii) the term involving $\Sigma_{\lambda \neq \kappa}$ does not occur

and

iii) $\kappa \Rightarrow |0'\rangle$.

iv) $\mathbf{C}_{\mathbf{K}} \Rightarrow \mathbf{C}_{0'}$ in eq. 16 is *partitioned* into

a)a contribution from the reference space configurations $\mathbf{C}^{a}{}_{0'} = \{C_{\lambda'0'}\}$

and

b)a contribution from the secondary space configurations

$\mathbf{C}^{b}{}_{0'} = \{C_{R0'}\}$.

The hamiltonian is now partitioned as

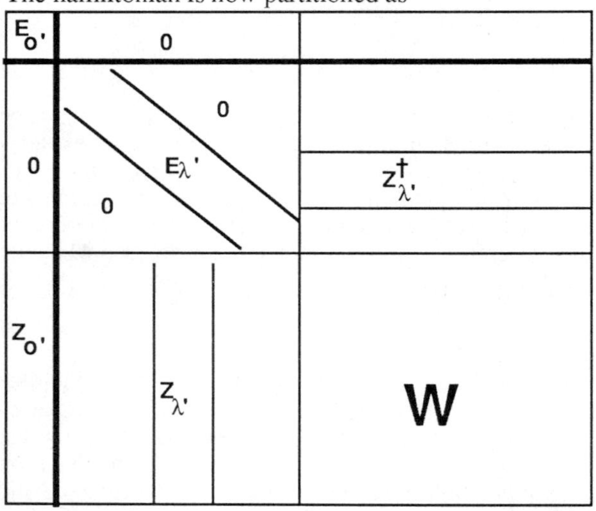

Thus eq. 16a becomes

$$\{ E_{\lambda'} - E_{0'} \} C_{\lambda'0'} + \Sigma_R Z^{\dagger}{}_{\lambda' R} C_{R0'} = 0 \qquad (32a)$$

(one equation for each λ')

$$Z_{0'R} + \Sigma_{\lambda'} Z_{R\lambda'} C_{\lambda'0'} + \Sigma_S \{ W_{SR} - E_{0'}\delta_{RS}\} C_{R0'} = 0 \qquad (32b)$$

(one equation for each R)

Clearly this equation system reduces to the single reference case when the summations over λ are removed.

We can deal with eq. 29 in a similar fashion. We must redefine $\mathbf{T}_K \Rightarrow \mathbf{T}_{0'}$ as

$$\exp(\mathbf{T}_{0'}) = \exp(\mathbf{X})\exp(\mathbf{M}) \tag{33}$$

where

$$\mathbf{M} = \Sigma_{\lambda'} C_{\lambda'0'} \{ |0'\rangle\langle\lambda'| - |\lambda'\rangle\langle 0'| \} \tag{34}$$

In eq. 32 we define **X** to have the same form as in eq. 26 but omitting the dependence on κ.

$$\mathbf{X} = \Sigma_{\alpha,r} X^r_\alpha \{E^r_\alpha - E^\alpha_r\} + \Sigma_{\alpha\beta,r\geq s} X^{rs}_{\alpha\beta} \{E^{rs}_{\alpha\beta} - E^{\alpha\beta}_{rs}\} + .. \tag{35}$$

Eq. 29a now becomes (one eq. for each R)

$$\langle R| e^{-\mathbf{M}} e^{-\mathbf{X}} H e^{\mathbf{X}} e^{\mathbf{M}} |0'\rangle = 0 \tag{36}$$

It is perhaps helpful to write eq. 36 in matrix form. Using the expansion of the exponential ($e^{aG} = 1+aG+a^2G^2/2!+...$) we have

$$H^* = e^{-\mathbf{G}} H e^{\mathbf{G}} = H + [H,\mathbf{G}] + 1/2 \; [[H,\mathbf{G}],\mathbf{G}] + ... \tag{37}$$

where

$$e^{\mathbf{G}} = e^{\mathbf{X}} e^{\mathbf{M}} \tag{38}$$

and

$$H^* = \{H + [H,\mathbf{X}] + 1/2 \; [[H,\mathbf{X}],\mathbf{X}] + [H,\mathbf{M}] + 1/2[[H,\mathbf{M}],\mathbf{M}] + 1/2 \; [[H,\mathbf{X}],\mathbf{M}] + ... \} \tag{39}$$

Thus eq. 36 becomes

$$\begin{pmatrix} E^X \\ E^M \end{pmatrix} + \begin{pmatrix} E^{XX} & E^{XM} \\ E^{MX} & E^{MM} \end{pmatrix} \begin{pmatrix} \boldsymbol{X} \\ \boldsymbol{M} \end{pmatrix} = 0 \tag{40}$$

with $\boldsymbol{X}$ a vector containing the coefficients X^r_α, $X^{rs}_{\alpha\beta}$ ... etc and $\boldsymbol{M}$ contains the $C_{\lambda'0'}$. The arrays E^X etc contain matrix elements of the excitation operators, for example (with $R = |^r_\alpha\rangle$ and $S = |^s_\beta\rangle$)

$$\{E^{XX}\}_{r\alpha,s\beta} = \langle R| [H, \{E^r_\alpha - E^\alpha_r\}], \{E^s_\beta - E^\beta_s\}] |S\rangle \tag{41}$$

and

$$\{E^{XM}\}_{r\alpha,\lambda} = \langle R| [H, E^r_\alpha] | \lambda\rangle \tag{42}$$

and similar expressions involving higher-order commutators.

To summarize, we have demonstrated a formalism for both a *perturb then diagonalize* scheme (eq. 16 and eq. 29) and a *diagonalize then perturb* scheme (eq. 36 and 40). Clearly, the latter scheme, although simpler, is only valid for the ground state while the former can yield excited states as well. Both schemes remain valid when the desired solution cannot be represented by a single reference configuration. We can state the main features of all post-SCF methods in this formalism before looking at the details of implementation in part II.

iii) <u>The Choice of the Reference Space:</u>
<u>A Unified treatment of Single and Multi-Configuration SCF theory</u>

We now look at methods that transcend the single reference SCF approach from a rather general point of view. Clearly our objective is the computation of the non-dynamic electron correlation. This involves a selection of the reference space and some sort of SCF process (ie MCSCF). We shall begin by demonstrating that SCF theories all correspond to a truncation of the excitation operators in MR-CC to single excitations. We then turn our attention to the choice of reference space that is necessary for the computation of the dynamic correlation energy.

We can develop SCF theories in an immediate fashion from MR-CC where only the linear excitation operators (E^r_α) are retained. This is possible because there is a simple relationship between orbital rotations (in SCF) and single excitations generated by the E^r_α.

We now give this in a simple (but not very rigorous) way. Consider the set of n by n unitary matrices representing unitary transformations of the orthonormal orbitals $\{|i>\}$. The E^i_j in the basis of the orthonormal orbitals $\{ |i> \}$ have a simple matrix form

$$<i| E^m_n |j> = \delta_{im}\, \delta_{nj} \tag{43}$$

and any unitary matrix can be written as

$$\mathbf{U} = \exp \mathbf{A} \tag{44}$$

where

$$\mathbf{A} = \Sigma_{i>j}\, a_{ij}(E^i_j - E^j_i) + i\ \Sigma_{i<j}\ a_{ij}\ (E^i_j + E^j_i) \ + i\ a_{ii}(E^i_i) \tag{45}$$

and the a_{ij} are parameters and are anti-hermitian ($a_{ij}=-a_{ji}^*$). Obviously if we consider that the orbitals are real and we do not allow transformations where they become complex we can omit the last two terms from eq. 45. We shall adopt this convention from this point onwards. (A more general discussion is to be found in ref.[7]) The E^i_j are the "generators" of the unitary group where the term generators is used because they satisfy a commutation relation

$$[E^i_j, E^k_l] = E^i_j E^k_l - E^k_l E^i_j = \delta_{jk} E^i_l - \delta_{li} E^k_j \tag{46}$$

The CSF $\{| K >\}$ must also carry a representation of this same group of unitary matrices. Thus the generators $\{<K| E^i_j |L> \}$ will satisfy the commutation relation eq. 46 as well.

Further, the "representation" of an n x n unitary matrix **U** in the space of the CSF {| K >} is just (restricting ourselves to real transformations as discussed following eq. 45)

$$\mathbf{U} = \exp \mathbf{A} \qquad (47)$$

where

$$\mathbf{A} = \Sigma_{i>j}\, a_{ij} <K|\; E^i_j - E^j_i\; |L> \qquad (48)$$

Thus, a unitary transformation in the orbital space **U** = exp **A** *induces* the corresponding transformation **U**^= exp **A**^ in the space of the CSF. They are completely determined by the same set of parameters a_{ij}. But note that the matrix elements $<K|E^i_j-E^j_i|L>$ that occur in eq. 48 are exactly the same as the ones that occur in eq. 29 or eq. 36. Thus we can now state a remarkable fact that follows immediately.

1) If the operators T_λ *in eq. 26 or* **X** *in eq. 35 are truncated at single excitations*

(ie. $T_K = \Sigma_{\alpha,r}{}^K t^r_\alpha \{E^r_\alpha - E^\alpha_r\}$, $\mathbf{X} = \Sigma_{\alpha,r} X^r_\alpha \{E^r_\alpha - E^\alpha_r\}$),

then the corresponding cluster amplitudes ${}^K t^r_\alpha$ and X^r_α *found in the solution of eq. 29 or eq. 36 can be taken as the parameters of a unitary t transformation of the orbitals.*

2) The solution of the CC equations 29 or 36 ***is exactly equivalent to SCF theory.***

Let us now demonstrate **2)** explicitly for the case of *diagonalize then perturb* with a simple derivation of the MCSCF equations. In MCSCF theory we wish to minimize the energy of the reference wavefunction

$$\Psi = |0'> + \Sigma_{\lambda'}\; C_{\lambda'0'}\; |\,\lambda'> \qquad (49)$$

with respect to orbitals Φ which are obtained from the starting orthogonal set $\{\Phi_0\}$ by unitary (in practise orthogonal) transformation as

$$\Phi = \mathbf{U}\,\Phi_0 \qquad (50)$$

where the unitary transformation **U** is written in terms of its parameters as **U**=exp**X** (where the expansion in eq. 35 is truncated at single excitations) and the CI wavefunction is written as Ψ = exp **M** |0'> (eq 34). Since the MCSCF energy is a function of the parameters x_{ij} and $C_{\lambda'0'}$ (which can be varied without constraint) we can write it as a Taylor expansion in terms of the derivatives of parameters x_{ij} (E^X etc) and $C_{\lambda'0'}$ (E^M etc) as

$$E_{MC\text{-}SCF} = E_0 + E^X \boldsymbol{X} + \boldsymbol{X}^t E^{XX} \boldsymbol{X} + \boldsymbol{M}^t E^{MM} \boldsymbol{M} + \boldsymbol{M}^t E^{MX} \boldsymbol{X} + \boldsymbol{X}^t E^{XM} \boldsymbol{M} \qquad (51)$$

where the arrays ***X*** and ***M*** collect the orbital rotation parameters x_{ij} and CI coefficients $C_{\lambda'0'}$. The quantities E^X etc. denote the first and second derivatives of the energy with respect to orbital rotations or CI expansion coefficient variation. Thus for example,

$$E^M = ||\partial E/\partial C_{oK}\,|| \qquad (52)$$

$$E^X = \| \partial E/\partial X_{ij} \| \tag{53}$$

Differentiation of the energy expression (eq.51) immediately gives the usual Newton-Raphson equations of Eq. 40 and the derivatives can be identified with terms of the form given in eq. 41 and 42. Thus in practise one solves equations of the form given in eq. 40 for $\boldsymbol{X}$ and $\boldsymbol{M}$ and then transforms the orbitals by $\mathbf{U}=\exp\mathbf{X}$ repeating the process until convergence ($E^X=0$).

The preceding discussion at first sight appears rather different that the conventional textbook derivation of SCF theory. We have carried out the derivation in this fashion in order to remain very general. However perhaps we should briefly digress to extract conventional closed shell SCF theory from this formalism in order to orientate the reader. For the closed-shell RHF case we have only a single reference so the blocks E^{MM} , E^{MX}etc are omitted from the energy expression (eq.51) and eq. 40. Thus the equation system reduces to

$$E^X + E^{XX} \boldsymbol{X} = 0 \tag{54a}$$

which has the formal solution

$$\boldsymbol{X} = -(E^{XX})^{-1} E^X \tag{54b}$$

If we take E^{XX} to be diagonal and approximate it as

$$(E^{XX})_{ij,ij} = \varepsilon_i - \varepsilon_j \tag{55}$$

Then the X_{ij} are found as

$$X_{ij} = -(E^X)_{ij}/(\varepsilon_i - \varepsilon_j) \tag{56}$$

However, provided X_{ij} is small enough so that $\tan 2X_{ij} \approx X_{ij}$, then we can recognize eq. 56 as the formula for a 2X2 Jacobi plane rotation. Then we can determine $\mathbf{U}$ rather than $\mathbf{X}$ by solution (diagonalization) of

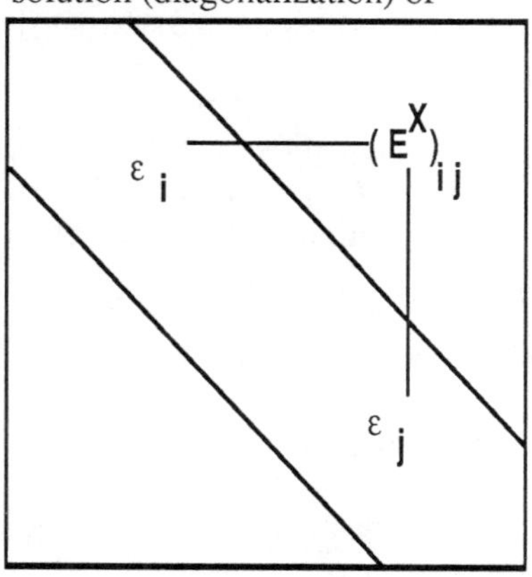

$$\mathbf{U} = \varepsilon\, \mathbf{U} \tag{57}$$

When one evaluates the expressions for the $(E^X)_{ij}$ one simply obtains the usual Hartree-Fock operator matrix elements. The ε_i are arbitrary but are usually taken to be the diagonal elements of the Hartree-Fock operator. It should be apparent that eq. 54 is just the second-order approximation and the usual Hartree-Fock equations (eq. 57) are a first-order approximation. The only difference from the conventional textbook formulation at this stage is that we have formulated the problem in an orthogonal basis rather than the usual AO basis.

We now have the apparatus to discuss methods for the computation of non-dynamic electron correlation from a very general point of view. The central point of the discussion is related to the choice of reference space. Let us begin by reminding the reader of our division of the orbital space into core, valence and occupied orbitals. Remember, that the reference configurations are charaterized by the fact that the *core* orbitals are doubly occupied, the *unoccupied* orbitals have zero occupancy and the *valence* orbitals can have occupancies of 2,1 or 0. If the reference space contains all possible configurations of the valence electrons in the valence orbitals then we have a *Complete Active Space* (CAS) reference. If one uses a CAS space then the energy is invariant to a transformation of the *valence* orbitals and one need only consider *core-valence* and *valence* -unoccupied orbital rotations in addition to the *core-unoccupied* rotations that occur in single reference SCF theory. (The energy will always be invariant to *core-core* and *unocupied-unoccupied* orbital rotations). Clearly, the choice of the reference space is critical for the determination of the non-dynamic correlation and its choice involves two considerations: i) the choice of the valence orbitals and ii) (if a CAS reference is not used) the selection of the reference configurations. However, one cannot make an arbitrary choice of reference configurations. The rule is simple.

i) For a CAS reference space the energy is invariant to a transformation of the valence orbitals so that the valence-valence single excitations are all included in the reference space and valence-valence orbital rotations do not need to be considered

ii) For a reference space which is not of the CAS type the energy must be optimized with respect to valence-valence orbital rotations. As a consequence, no two reference configurations can be related by a valence-valence single excitation. In other words we must choose reference states so that

$$<\kappa|\ E^{\mu}{}_{\nu} - E^{\nu}{}_{\mu}\ |\lambda> = 0 \text{ for all } \kappa, \lambda$$

where $\mu\nu$ are valence orbitals.I n other words, we must not include valence-valence excitations in the reference space since these terms are included in the orbital optimization. Otherwise one has a linear dependence between orbital rotations and CI coefficients.

There are two obvious sets of reference configurations that satisfy this criterion: i) all valence orbitals are either *doubly* occupied or *empty* in every reference configuration and ii) all valence orbitals are *singly* occupied in every reference configuration. The various forms of GVB[8] wavefunctions (viz. $H_2\ \sigma_g{}^2 - \lambda\ \sigma_u{}^2$ as a trivial example) satisfy criterion *i* while a perfect pairing VB (PP-VB) wavefunction (viz. H·H. ,built from orthogonal orbitals) satisfies criterion *ii.* However, while one may argue that GVB or PP-VB reference spaces may recover much of the non-dynamic correlation in specific cases, the CAS reference is much more general. If one uses a CAS reference then one needs only choose the *valence* orbitals for a particular chemical problem. We now give an argument to justify this.

Consider a reaction that involves breaking two bonds 1-2 and 3-4 and forming 2 new ones 1-3 and 2-4. We can imagine that the total energy E arises from the mixing of two diabatic surfaces α_R and α_P according to the resonance β as shown schematically below.

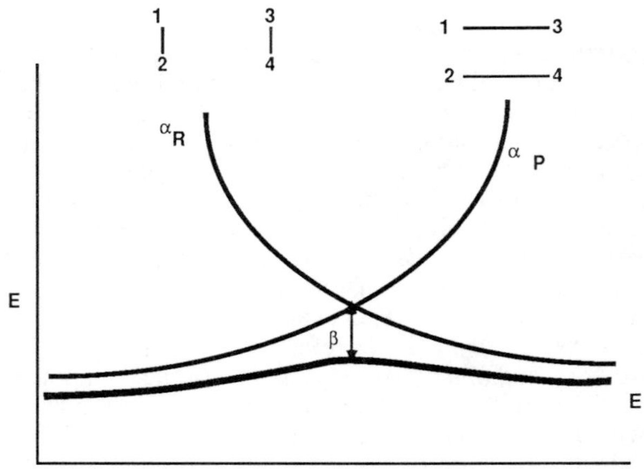

Fig. 6 Schematic representation of the diabatic surfaces for the exchange of two bonds.

A GVB wavefunction of the form

$$\Psi = \Lambda_{13}(1,2)\Lambda_{24}(3,4)$$

$$\Lambda_{13}(1,2) = (\sigma_{13})^2 - \lambda(\sigma^*{}_{13})^2$$

where σ_{13} is some localized 2 centre bonding orbital can describe bond breaking/forming between centres 13/24. However, unless one includes a second "pairing" (viz. $\Lambda_{12}(1,2)\Lambda_{34}(3,4)$) as well then one can only describe one possible diabat. But we can have at most two doubly occupied and two unoccupied orbitals for this problem. Thus the effect of $\Lambda_{12}(1,2)\Lambda_{34}(3,4)$ must be included by adding terms of the form $(\sigma_{13})(\sigma_{24}) + \omega(\sigma^*{}_{13})(\sigma^*{}_{24})$ and we are lead to the CAS reference space. Hence we are lead to the following conclusion:

> *The valence orbitals must be chosen so that all diabatic surfaces which contribute to the total energy are represented with equal accuracy. Thus one needs 2 active orbitals for each bond that is being made/broken in a chemical reaction. The reference space must include all possible spin couplings. This effectively implies a CAS space.*

Finally, we must now give a brief discussion of the type of SCF procedure to be used for the computation of non-dynamic electron correlation and the physical interpretation of each procedure. Clearly, the very general form of SCF which involves the solution of the MR-CC equations eq. 29 with a different T_K ($T_K=\Sigma_{\alpha,r}{}^{K}t^{r}{}_{\alpha}\{E^{r}{}_{\alpha}-E^{\alpha}{}_{r}\}$) for each reference κ, will be the most accurate. It is obvious that one has a different set of optimized orbitals (via the orbital rotation parameters ${}^{K}t^{r}{}_{\alpha}$) for each reference state κ and each reference state is computed with equal accuracy. This is to be compared with the equation system of eq. 36 where one determines only a single set of X^r_α for a reference that is a fixed linear combination (eq. 31) given by the prior diagonalization of $\mathbf{H}_0$. The difference in energy has been called semi-internal correlation by Sinanoglu and is very important from a chemical point of view. A simple example illustrates the main ideas. In fig. 7 below we show how a *formally* singly excited secondary space configuration can be derived in two different ways from two different reference configurations.

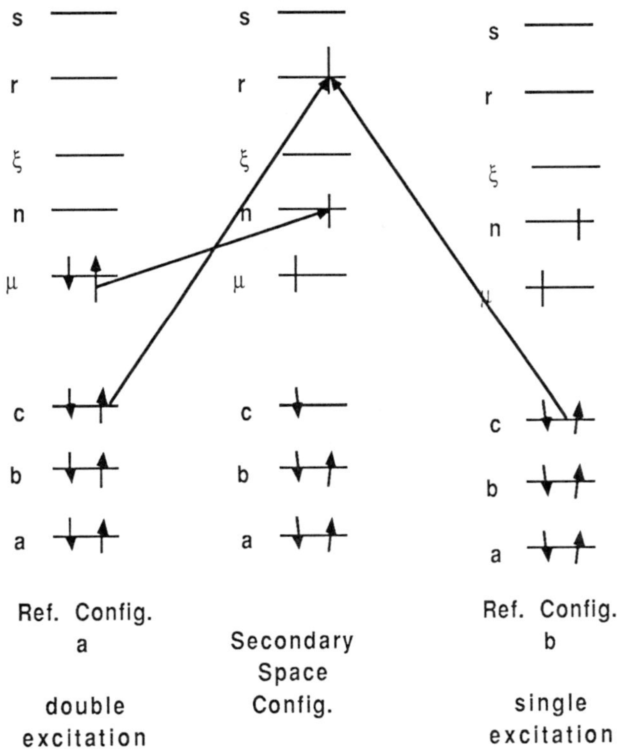

Fig. 7a A secondary space configuration (middle) derived by double excitation (valence-valence single coupled with core-unoccupied single) from reference configuration a or derived by core-unoccupied single excitation from reference configuration b.

The middle configuration is doubly excited with respect to reference configuration *a* and thus is really *dynamic* correlation. The effect of this configuration occurs only as $t^r_a\, t^\mu_\alpha$. However this same configuration is singly excited with respect to configuration *b* and contributes to the orbital rotation X^r_c. In contrast the same orbital optimization rotation for reference configuration a arises as shown in fig. 7b

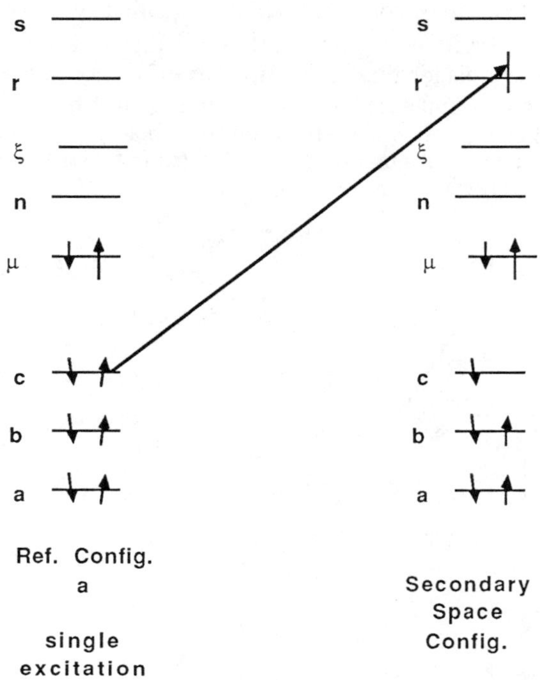

Fig.7b Configuration giving rise to $X^r{}_c$ for configuration *a*

It should now be obvious that in the general equation scheme of eq. 29 there will be two different $^\kappa t^r{}_c$ (κ=a and b) and in the equation scheme of eq. 36 there will be only one parameter $X^r{}_c$ to describe both situations. Further, the dynamic correlation effect of the semi-internal correlation (Fig. 7a double excitation of reference configuration a) is not described at all in the scheme of eq. 36. This semi-internal correlation would require terms of the form $^\kappa t^{r\mu}{}_{a\alpha}\ \{E^{r\mu}{}_{a\alpha} - E^{a\alpha}{}_{r\mu}\}$

Thus the point to be emphasised at this stage is that the usual MCSCF scheme that corresponds to eq. 36 with only the single excitation clusters has limitations:

1) Semi-internal correlation is not treated

2)The orbitals are optimized for a fixed linear combination of the reference configurations as opposed to the situation where each reference configuration has its own set of fully optimized orbitals

3)While one may obtain approximations to excited states using the equation scheme of eq. 36 one must optimize each excited state independently and variational collapse is possible. However, if the more general scheme of eq. 29 is used, then all excited states are obtained with the same accuracy.

At this moment in time many MCSCF codes exist corresponding to eq. 36 with single excitation operators. However, MCSCF schemes based upon eq. 29 have not yet been implemented. However, even in this approach, the important effects of semi-internal correlation effects are not treated. The so-called POL-CI or First-order CI method [9-11] does remedy this defect in the context of CI introduced via eq. 16 in that it includes both the effect of allowing for the fact that each reference state should have an independently optimized set of orbitals and includes semi-internal correlation effects as well. One merely uses a CAS reference and generates all CI configurations with at most one electron in an unoccupied orbital and one vacancy in a core orbital. The only deficiency in such a wavefunction is that he higher-order cluster excitations ${}^{\kappa}t^{r}{}_{\alpha}{}^{\lambda}t^{s}{}_{\beta}$ are omitted but the semi-internal effects ${}^{\kappa}t^{r\mu}{}_{a\alpha}$ are explicitly included. Thus the point to be emphasised at this stage is that the *single excitations* from an MCSCF wavefunction do not vanish to first-order. Rather they account for an important physical effect: *semi-internal* electron correlation.

iv) A general Discussion of the methodology for the Computation of Dynamic Correlation.

In part II of these notes it is our intention to give a rather detailed discussion of the most commonly used (and readily available in standard software) methods for the treatment of the dynamic correlation. In this section we will try to give a more general discussion based upon the formalism just presented.

As we have tried to indicate, from a practical point of view there are only two methods available to us for the computation of dynamic correlation, Multi-Reference CI (MR-CI) (eq. 16) and Multi-reference Coupled Cluster (MR-CC) (eq. 29). We assume the *diagonalize then perturb* approach (eq.32/36) as a limiting case. Both of these approaches reduce to the single reference situation as the number of reference configurations goes to one (and the distinction between valence and core orbitals goes away). The various forms of perturbation theory merely correspond to the solution of eq.16 or 29 via series expansion and thus are just one example of an algorithm for solving these equations to which we return later. Of the two methods we have used MR-CC has an advantage because it is *size-extensive* and we begin by discussing this point.

From a practical point of view one must truncate MR-CI and MR-CC at double excitations. However they are no longer equivalent when truncated. Thus for truncation at double excitations, the MR-CC formalism retains a contribution from say quadruple excitations of the form $t^{rs}{}_{\alpha\beta}\,\{E^{rs}{}_{\alpha\beta} - E^{\alpha\beta}{}_{rs}\}\, t^{tu}{}_{\gamma\delta}\,\{E^{tu}{}_{\gamma\delta} - E^{\gamma\delta}{}_{tu}\}$ but CI does not. A simple example illustrates the consequences. Suppose we have two subsystems X and Y. The wavefunction of X-Y must dissociate correctly into the product of wavefunctions of systems X and Y. Now, because of the invariance of the energy to transformations of core or valence orbitals among themselves, we can localize the orbitals onto the two subsystems. Thus for the wavefunction of X-Y to dissociate correctly into the product of wavefunctions of systems X and Y, we must include excitations of the form $E^{rs}{}_{\alpha\beta}|X\rangle \,.\, \{E^{tu}{}_{\gamma\delta}\,|Y\rangle$ where $\alpha\beta$ are occupied orbitals localized on subsystem X and $\gamma\delta$ are occupied orbitals localized on subsystem Y. Clearly, the limitation to double excitations alone in CI would not be correct and only CC will behave correctly. The numerical consequence of this effect obviously depends upon the number of electrons. For a large number of electrons the contribution from those terms that arise as products of double excitation operators will actually be much greater than the contribution from the double excitations themselves. While the exponential form used in MR-CC guarantees size extensivity in an obvious way, it is not obvious that any approximation to eq. 29 would be size-extensive (viz. the truncation of the expansion used in eq. 37). If fact, it is a particular property of the commutators that appear in eq. 37 that only connected diagrams would appear if a diagrammatic representation was used because the commutator or two second quantized operators can only contain connected terms. This was the motivation in the step from eq. 28 to eq. 29. With the pre-multiplication of eq. 28 by $\exp(-T_K)$ the term involving $\{H_{eff}\}_{\lambda\lambda}$ which could lead to disconnected diagrams is removed. Having said all this, it should be pointed out that the choice of the reference space has much to do with the numerical consequences of the preceding observations. If we choose a CAS reference space then the zeroth-order wavefunction is size extensive by construction. Thus for small molecules, the effects of the lack of size-extensivity in double-excitation CI may be negligible if a suitable CAS reference is chosen.

Up to this point we have not considered perturbation theory at all. Of course perturbation theory represents merely one algorithm for the solution of the equation systems of eq. 16,29,32 or 40. Nevertheless, perturbation methods can yield very cheap approximate solutions of these equations. Thus at this stage we shall consider perturbation theories from a very general point of view, returning to some details in part II.

The development of perturbation methods is most easily illustrated using eq. 16a. We can write the formal solution to eq. 16 as

$$C_K = -\{W - (H_{eff})_{KK}\,1\}^{-1}\,\{Z^*\}_K \qquad (58)$$

with

$$\{Z^*\}_K = -\Sigma_{\lambda\neq K}\{H_{eff}\}_{K\lambda}\,C_\lambda \qquad (59)$$

$$\{H_{eff}\}_{K\lambda} = \{H_0\}_{K\lambda} + \{Z^\dagger\}_K\,C_\lambda \qquad (60)$$

(The distinction between $\{Z^*\}_K$ and $\{Z\}_K$ vanishes for the single reference limit). We can now use the expansion

$$(A-B)^{-1} = A^{-1} + A^{-1}BA^{-1} + \ldots\ldots \qquad (61)$$

to develop a series expansion for the inverse that appears in eq. 58. Having developed a formal series expansion for C_K, the expansion for $\{H_{eff}\}_{K\lambda}$ and C_λ must be inserted. Clearly, a great deal of flexibility is possible in the choice of the partition of $\{W-(H_{eff})_{KK}1\}$ into $(A-B)$; however, one clearly requires **A** to be a diagonal matrix. The series will diverge if the matrix **A is not positive definite**. The rate of convergence will be very slow unless BA^{-1} is small (ie the eigenvalues of A must be large). If the expansion for $\{H_{eff}\}_{K\lambda}$ is not used then we have Brillouin-Wigner type perturbation theory. Expanding $\{H_{eff}\}_{K\lambda}$ from **A** we obtain Rayleigh-Schrödinger perturbation theories. The most useful perturbation methods are obtained by defining **A** to be differences of orbital energies obtained from some reference one-electron hamiltonian. Thus we partition as follows

$$\{W-(H_{eff})_{KK}1\} = \{ \Sigma_{i(R)}\, \varepsilon_i^* - \Sigma_{j(R)}\, \varepsilon_j \; \Sigma_R \; |R><R|\}$$
$$- [\{W-(H_{eff})_{KK}1\} - \{(\Sigma_{i(R)}\, \varepsilon_i^* - \Sigma_{j(R)}\, \varepsilon_j) \Sigma_R \; |R><R| \}] \quad (62)$$

taking

$$\mathbf{A} = \{ \Sigma_{i(R)}\, \varepsilon_i^* - \Sigma_{j(R)}\, \varepsilon_j \; \Sigma_R \; |R><R|\} \quad (63)$$

where we use ε_i^* to denote unoccupied orbital energies. Such theories are called Møller-Plesset perturbation theories or Many-Body perturbation theory. They are very powerful because when one replaces the *sums over states* |R> implied by the matrix multiplication by *sums over orbitals* one has simple expressions that are amenable to computation. There is much cancellation between the series expansion for H_{eff} and the original terms in the expansion of $(A-B)^{-1}$and one can use diagrammatic methods to enumerate a perturbation series with the cancelled terms removed [12]. In particular, one needs to be certain that the energy evaluated from perturbation theory at any-order will be size-extensive. If one starts from the MR-CC formalism then this is easy to demonstrate.

Clearly in perturbation methods, one is limited to the lowest orders of perturbation theory, otherwise it is more efficient to solve the CI problem itself. The choice of reference space is again critical. If the reference space does not describe the problem adequately, there will be large contributions from the secondary space and the perturbation expansion will be slowly convergent. From our derivation of perturbation theories it is obvious that they represent just an algorithm for the solution of the MR-CI or MR-CC problem. Thus a reference space that is adequate for MR-CC (an infinite-order perturbation theory) may fail completely in a perturbation expansion. The only way to tell for certain is to examine the rate of convergence of the lowest orders of perturbation theory.

II Practical Methods for the Computation of Electron Correlation:

This section is concerned with the more practical aspects of evaluating the contribution of the structure independent, dynamic correlation energy. We will concentrate in a little more detail on the traditional methods of CI, perturbation theory and coupled cluster methods as they are actually implemented in current molecular structure codes. Hence, we will assume that a suitable reference function (either of the SCF or MCSCF variety) has been obtained according to the guidelines of section 1.

Today, there are three principal methods which are used for treating the dynamic correlation problem. The first and conceptually the simplest is the use of configuration interaction (CI) wavefunctions. This method is usually applied with a well chosen truncation of the CI expansion, for example, all double substitutions from the reference space (CID) or all single and double substitutions (CISD). The second class of methods which have found wide application are based on perturbation theory (PT) (see for example reference 12b). There are a wide range of PT wavefunctions, each differentiated by the choice of the zeroth-order hamiltonian and the energy denominator (ie the partition A/B in eq. 61). Here we shall only be concerned with perturbation expansions based on the Rayleigh-Schrödinger theory (RSPT) (ie the partition used in eq. 63) since molecular energies obtained by these methods have the desirable attribute of scaling linearly with the number of electrons (hence they fulfil the size-extensivity requirement discussed in section 1). The practical form of RSPT has been termed many-body PT (MBPT) [12] and is concerned with the direct cancellation of non-size-extensive terms that arise in RSPT. Finally, there are the various methods based on the cluster development of the wavefunction. These form an infinite-order extension of MBPT. Rather than employing an order-by-order expansion in the perturbation (as in MBPT), the exact wavefunction is derived by applying an exponential electron cluster operator (eq. 24) to the reference function (including up to N-electron clusters for an N-electron system). This is the basis of the coupled-cluster (CC) approach. Various approximations then result by restricting the exponential operator to include only certain low-order clusters. The simplest of these involve restriction to two-electron clusters (CCD) or one- and two-electron clusters (CCSD). Further extensions [13] of the CC approach have been developed recently which treat four-electron clusters and also subsets of three-electron clusters.

i)Configuration Interaction (CI) Wavefunctions.

It is useful to begin the discussion of CI methods by considering the full CI wavefunction, which we will denote by $|\Psi_{FCI}>$. The full CI gives the exact result with a given orbital basis and number of electrons and thus provides a *benchmark* for testing all approximate methods (see the review paper of Handy [13] for several interesting examples).

For full CI we can choose any single determinant $|\Phi_0>$ as a reference and thus $|\Psi_{FCI}>$ for an N-electron system may be written as

$$|\Psi_{FCI}> = |\Phi_0> + \Sigma_{n=1,N}\, \omega_n\, |\Phi_0> \qquad (64)$$

where ω_n is an operator which induces n-fold excitations from the reference function, $|\Phi_0>$. ω_n can be written in terms of replacements from a suitable set of orthonormal orbitals $\{\chi\}$. In all that follows we shall denote orbitals occupied in $|\Phi_0>$ by $\alpha,\beta,\gamma..$ and unoccupied or virtual orbitals by r,s,t,u...

$$\omega_1 = \Sigma_{\alpha,r} w^r{}_\alpha\, \{E^r{}_\alpha\} \qquad (65a)$$

$$\omega_2 = \Sigma_{\alpha\beta,r\geq s}\, w^{rs}{}_{\alpha\beta}\, \{E^{rs}{}_{\alpha\beta}\} \qquad (65b)$$

etc ......

where $E^r{}_\alpha |\Phi_0> = |\Phi^r{}_\alpha>$, $E^{rs}{}_{\alpha\beta} |\Phi_0> = |\Phi^{rs}{}_{\alpha\beta}>$ etc.

hence $E^r{}_\alpha |\Phi_0>$ removes an electron from orbital $\chi_\alpha$and places an electron in orbital χ_r, yielding the excited determinant $|\Phi^r{}_\alpha>$. The $w^{rs}{}_{\alpha\beta}$ are the weights of the double excitation for example.

The full CI expansion increases very rapidly in size with the number of electrons and so is limited to small systems ($\approx$ 10 electrons). Hence it has found limited application because of the high computational cost. However, full CI does provide the most accurate upper bound to the exact energy within a given basis set. Furthermore,the energy is invariant to bound to the exact energy within a given basis set. Furthermore,the energy is invariant to the nature of the orbitals used (e.g. Hückel, SCF or MCSCF orbitals), such that any unitary transformation of the one-particle basis leaves the energy unchanged. Finally, full CI possesses the important property that it is size-extensive. Hence it is the best result we can obtain within our algebraic (basis set expansion) approximation and all other methods must attempt to achieve the full CI result, but at much lower computational cost. This introduces the question of accuracy versus computational cost, which is a central concern of the practising quantum chemist.

The simplest CI wavefunction utilizes a single reference function $|\Phi_0>$ and is restricted to less than all possible substitutions.

$$|\Psi_{CI}> = |\Phi_0> + \Sigma_{n(n<N)}\ \omega_n\ |\Phi_0> \tag{66}$$

Within this framework, the most commonly used approximations involve restricting eq.4 to double substitutions (CID) or restriction to single and double substitutions (CISD). Separate unitary transformations among the occupied orbitals and among the virtual orbitals still leave the energy invariant. However, many of the properties of the full CI approach mentioned above are lost when the CI expansion is restricted in this fashion. The most severe problem, assuming that a single reference is adequate, is the lack of size -extensivity.

While the severity of the size-extensivity error can be reduced by including higher levels of excitation or increasing the reference space, there exist various *approximate* formulas for estimating the contributions of higher level excitations, the most familiar of these is the so called Davidson correction [15] to CID, which attempts to account for the effects of certain quadruple excitations (specifically a subset of the simultaneous double excitations)

$$\Delta E_{Davidson} = (1 - C_0^2)\ E_{CID} \tag{68}$$

In eq.68, C_0 is the coefficient of the reference determinant in the normalized CID wavefunction. The corrected energy is evaluated from

$$E = E_{CID} + \Delta E_{Davidson} \tag{69}$$

As we have shown [16] this formula generalizes in the multi-reference case to

$$\Delta E_{Davidson} = (1 - A_\lambda^\dagger A_\lambda)\ A_\lambda^\dagger Z^\dagger A_\lambda\ //\ A_\lambda^\dagger A_\lambda \tag{70}$$

where $Z^\dagger$ is defined in eq. 10. It is derived by removing the renormalization of the energy due to double excitations from fourth-order onwards in perturbation theory, since it is assumed that this is cancelled by the quadruple excitations. As we show in ref [16] one can formulate a very general argument based upon eq. 16a using eq. 20b to compute $\mathbf{H_{eff}}$. In practise there are difficulties that should be remembered: i) the results will be meaningless unless C_0^2 or $A_\lambda^\dagger A_\lambda$ is close to 1 and ii)it is completely wrong for a two particle system.

In recent years a great deal of research has been directed to the problem of truncated CI expansions based on multi-configurational reference functions [17] which we have discussed in full generality in part I. These techniques have been slower to find their way into standard quantum chemistry software because it is only relatively recently that methodological and technical problems with implementation have been solved (see the reviews [18,19] for up to date accounts). The Saxe/Fox/Schaefer/Handy program [20] is contained in GAUSSIAN 88. Strictly speaking the energy still has all the shortcomings of a truncated CI expansion. But with a well chosen reference space the net effect of these errors is substantially reduced and such multi-configurational wavefunctions have been shown to yield energies which can approach the full CI result.

In building a CI expansion based on a multi-configurational reference function the choice of orbitals is an important consideration in determining how rapidly the expansion converges. While it is possible to build the expansion from SCF orbitals, experience has shown this to be a poor choice. The reason lies in the fact that the virtual orbitals determined in an SCF calculation are too diffuse (too high in energy relative to the occupied orbitals). A better choice is to use orbitals determined from an MCSCF calculation. In the MCSCF wavefunction certain low-lying virtual orbitals are allowed to have partial occupancy and so they are much less diffuse than the corresponding SCF orbitals and show a much more rapid convergence of the CI expansion. In fact it has been shown that the optimum orbitals from which to build a truncated CI expansion are the so called natural orbitals (NO's). The NO's are chosen such that they diagonalize the one-particle density matrix and may be defined for the SCF or MCSCF case. In practise, the one-particle density matrix is brought into diagonal form by a unitary transformation, subsequently the orbitals are subjected to the same transformation yielding the set of NO's.

To conclude our discussion of CI we shall discuss two *benchmarks:* full CI computations on CH_2 and H_2O which are collected in Tables 1 and 2.

Table 1 CI results for CH_2

Method	$^1A_1$	$^3B_1$	D (S-T)/kcal mol^{-1}
RHF/UHF	-38.8863[a]	-38.9326	29.05
MCSCF	-38.9076	-38.9276[b]	12.6
MCSCF-SDCI[c]	-39.0222	-39.0416[b]	12.2
EXACT (Full CI)[c]	-39.0272	-39.0463	11.97

a)Equivalent to RHF
b)The triplet 2 determinant (S_z=0) MCSCF is identical to a restricted open shell SCF (ROHF)
c)C. W. Bauschlicher and P. R. Taylor, J. Chem. Phys. **85**, 5510 (1986)

Table 2 CI results for H_2O

Method	r_e	$1.5\ r_e$	$2.0\ r_e$
SCF	-76.0098	-75.8035	-75.5952
SDCI [a]	-76.1500	-75.9921	-75.8448
Davidson correction	-76.1559	-76.0119	-75.9049
EXACT (full CI)[a]	-76.1579	-76.0145	-76.9052
4 in 4[b]			
MC-SCF[c]	-76.0651	-75.9273	-75.8284
MCSCF-SDCI[c]	-76.1559	-76.0125	-76.9033
6 in 6[d]			
MC-SCF[c]	-76.0993	-75.9559	-75.8456
MCSCF-SCDI[c]	-76.1575	-76.0125	-76.9033
8 in 8[f]			
MC-SCF[c]	-76.1326	-75.9820	-75.8662
MCSCF-SDCI[c]	-76.1578	-76.0143	-75.9051

a)N. C. Handy, Faraday Symp. Chem. Soc. **19** ,17 (1984)
b)active space consisted of *4* electrons in the *4* O-H (H2s and O 2p) bonding orbitals and all possible excitations of the core O 1s,2s as well
c) F. B. Brown, I. Shavitt, and R. Shepard, Chem. Phys. Lett. **105**, 363 (1984). In d)to the active space in b was added the additional 3p orbital to correlate the O 2p (lone pair)
f) to the active space in e was added the 3s orbital to correlate O 2s

In CH_2 the singlet state cannot be represented with a single configuration wavefunction because it has the quasi-open shell configuration σ^2 - C π^2. As can be seen from the table most of the error in the SCF result is recovered in the MCSCF and the multi-reference CI result is in good agreement with the full CI. In table 2 there are collected the full CI results in comparison with single and multi-reference CI for H_2O at r_e ,1.5 r_e and 2.0 r_e. The SCF and SDCI results are not parallel to the full CI. In contrast, a CI with a modest multi-reference wavefunction is in good agreement with the full CI as illustrated in Fig.8 below.

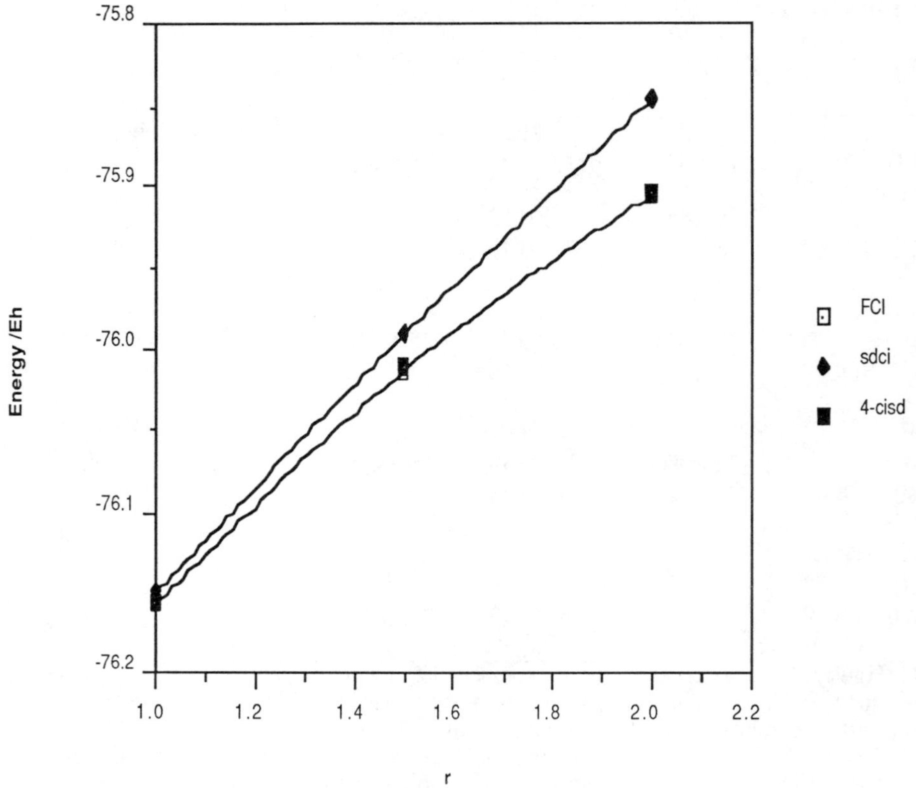

Fig. 8 SDCI, Multi-reference SDCI and Full CI for H_2O

Thus for structural problems in the region of equilibrium geometries single reference methods are fine. When bonds are partly cleaved multi-reference methods are essential.

ii)Perturbation Theory (PT) Wavefunctions.

We shall now give a more detailed discussion of perturbation methods that we have introduced briefly in part I. We believe that this is important because PT methods are without doubt the most cost effective methods for the computation of electron correlation effects. However, because convergence can be slow or non-existent, these methods must be used with great caution and it is important to understand the background theory.

The discussion of PT based methods falls naturally into two distinct but closely related parts. The first part deals with the general apparatus of Rayleigh-Schrödinger PT (RSPT). RSPT has the advantage of being a size-extensive theory. Bruckner [21] was the first to note that the RSPT energy shows a linear scaling with the size of system because of cancellation of terms which are non-linear in the number of electrons. Bruckner showed this for low orders of RSPT. Goldstone [22] extended Brueckner's result to the general case by showing that the cancellations occur at all orders of RSPT, this is the basis of the linked-cluster theorem [12]. The second part follows on from the linked-cluster theorem by attempting to develop a theory in which the terms which are non-linear in the number of electrons are explicitly removed; as opposed to RSPT where they are evaluated and numerically cancelled. This latter approach forms the basis of what has been termed many-body PT (MBPT).

We begin by reviewing the development of the general results of RSPT in a more concrete way than in part I. In part I we showed that perturbation methods can be regarded as a series expansion of the inverse of the coefficient matrix in the set of pseudo-linear equations 16. The central feature that distinguishes various perturbation theories is the partition into A and B in eq. 61 and it is obvious that an infinite variety is possible. In order to develop the various forms of PT methods in common use the partition A/B is usually formulated in terms of "model" hamiltonians (such as the Hartree-Fock hamiltonian). Thus we now proceed using an operator formalism (rather than matrix calculus as used in part I) where the distinctions between various types of PT expansions will appear less arbitrary.

We shall start with single reference PT. The total electronic hamiltonian, H, is decomposed into two parts

$$H = H_0 + V \qquad (71)$$

where H_0 is a model or reference hamiltonian and V is a perturbation. We assume that a complete set of eigenfunctions are known for H_0 and denote them by $\{\Phi_j\}$. The eigenfunctions of H are denoted by $\{\Psi_j\}$ and the state we are specifically interested in is state j. Hence,

$$H_0 |\Phi_j\rangle = E_j^0 |\Phi_j\rangle \qquad (72)$$

$$H |\Psi_j\rangle = (H_0 + V) |\Psi_j\rangle = E_j |\Psi_j\rangle \qquad (73)$$

Introducing an arbitrary parameter, ε, (which as will be seen, shall enable us to define various forms of PT) we rearrange eq.73 as

$$(\varepsilon - H_0) |\Psi_j\rangle = (\varepsilon - E_j + V) |\Psi_j\rangle \qquad (74)$$

The exact wavefunction is again assumed to obey intermediate normalization ($\langle\Phi_j|\Psi_j\rangle = 1$). Defining a projection operator, P, acting on the reference space

$$P = |\Phi_j\rangle \langle\Phi_j| \qquad (75)$$

such that

$$P |\Psi_j\rangle = |\Phi_j\rangle \langle\Phi_j|\Psi_j\rangle = |\Phi_j\rangle \qquad (76)$$

and its complement, Q,

$$Q = 1 - |\Phi_j\rangle \langle\Phi_j| = \Sigma_{k\neq j} |\Phi_k\rangle \langle\Phi_k| \qquad (77)$$

the total wavefunction $|\Psi_j\rangle$ may be written

$$|\Psi_j\rangle = |\Phi_j\rangle + Q\,|\Psi_j\rangle \tag{78}$$

Using the property that the projector Q commutes with the model hamiltonian, H_0, we can project eq.74 to obtain an expression for $Q|\Psi_j\rangle$

$$Q\,|\Psi_j\rangle = Q\,(\varepsilon - H_0)^{-1}(\varepsilon - E_j + V)\,|\Psi_j\rangle \tag{79}$$

Inserting eq.79 into eq.78 gives

$$|\Psi_j\rangle = |\Phi_j\rangle + Q\,(\varepsilon - H_0)^{-1}(\varepsilon - E_j + V)\,|\Psi_j\rangle \tag{80}$$

iterating the last step we obtain an order by order expansion of the exact wavefunction $|\Psi_j\rangle$

$$|\Psi j\rangle = \Sigma_{n=0,\infty}\,[Q\,(\varepsilon - H_0)^{-1}(\varepsilon - E_j + V)]^n\,|\Phi_j\rangle \tag{81}$$

In eq. 81 the operator $(\varepsilon - H_0)$ plays the role of -A in eq. 61 and $(\varepsilon - E_j + V)$ plays the role of B.A similar order by order expansion may be obtained for the exact energy by integrating eq.9.2 by preempt by $|\Phi_j\rangle$

$$E_j = E_j{}^0 + \langle\Phi_j|V|\Phi_j\rangle \tag{82}$$

and substituting for $|\Psi_j\rangle$

$$E_j - E_j{}^0 = \Delta E_j = \Sigma_{n=0,\infty}\,\langle\Phi_j|V[Q\,(\varepsilon - H_0)^{-1}(\varepsilon - E_j + V)]^n\,|\Phi_j\rangle \tag{83}$$

At this stage eqs.81and 83 are quite general and any type of PT expansion may be obtained by specifying ε and H_0. Two common perturbation series result by choosing ε to be either E_j (which yields the Brillouin-Wigner PT (BWPT) and leads to energies which are not size-extensive) or choosing $\varepsilon = E_j{}^0$ and inserting the series back on itself for each occurrence of E_j (which yields the RSPT in which the energies are size-extensive). Substituting $\varepsilon = E_j{}^0$ into eq.83, and we define

$$[1/b] = \Sigma_k\,\{\,|\Phi_k\rangle\langle\Phi_k|\}\,/\,(E_j{}^0 - E_k{}^0)$$

we obtain for the first few orders in RSPT

$$E_0 = E_j{}^0 \tag{84a}$$

$$E_1 = \langle\Phi_j|V|\Phi_j\rangle \tag{84b}$$

$$E_2 = \langle\Phi_j|\,V\,[1/b]V\,|\Phi_j\rangle \tag{84c}$$

$$E_3 = \langle\Phi_j|V\,[1/b]\,V\,[1/b]\,V|\Phi_j\rangle\; - E_1\,\langle\Phi_j|V\,[1/b]^2V\,|\Phi_j\rangle \tag{84d}$$

$$E_4 = \langle\Phi_j|V\,[1/b]\,V\,[1/b]\,V\,[1/b]\,V|\Phi_j\rangle - E_1 \langle\Phi_j|\,V\,[1/b]\,V\,[1/b]^2 V\,|\Phi_j\rangle$$
$$- E_1 \langle\Phi_j|\,V\,[1/b]^2\,V\,[1/b]\,V\,|\Phi_j\rangle + (E_1)^2 \langle\Phi_j|\,V\,[1/b]^3 V\,|\Phi_j\rangle$$
$$-E_2 \langle\Phi_j|V\,[1/b]^2 V\,|\Phi_j\rangle \qquad (84e)$$

The condition of size-extensivity has been mentioned several times and is an important consideration in developing practical methods for evaluating the dynamic correlation energy. It is a simple matter to show that eqs. 84 scale linearly with the size of system. For example (the reader can verify this for themselves), consider a minimal basis description of two hydrogen molecules at infinite separation. Since the integrals involving orbitals on distinct molecules will vanish, only contributions equivalent to those arising from the monomer units will be non-zero and there will be twice as many of these since there are two molecules, hence we have E(dimer) = 2 E(monomer).

What about eq.84d and 84e ? In eq. 84d there is a term which is multiplied by the first-order energy, E_1. If E_1 is itself size-extensive, then the factor in E_3 which is multiplied by E_1 might be expected to give rise to terms which are proportional to the square of the number of electrons ! This is In fact the case, but it is possible to show that all these terms cancel yielding size-extensive energies [23]. Further in eq. 84e, the term $E_2\langle\Phi_j|V[1/b]^2V|\Phi_j\rangle$ contains *renormalization* effects arising from the double excitation contribution to E_2. These are partly cancelled by certain terms in $\langle\Phi_j|V[1/b]V[1/b]V[1/b]V|\Phi_j\rangle$ that arise from quadruple excitations! (In fact this feature is used in the Davidson correction where the effect of the $E_2\langle\Phi_j|V\,[1/b]^2V\,|\Phi_j\rangle$ is removed to all orders of PT). However, it is not desirable to have a formalism in which unphysical terms arise and we rely on numerical precision to ensure proper cancelling. The aim of MBPT is to develop a formalism based on RSPT, in which the non-size-extensive terms are never computed.

Hence in MBPT the cancellation of the non-size-extensive terms is carried out explicitly and eq.83 is modified to

$$E_j - E_j^0 = \Delta E_j = \{\, \Sigma_{n=0,\infty} \langle\Phi_j|V[Q\,(E_j^0 - H_0)^{-1}V]^n\,|\Phi_j\rangle\}_L \qquad (85)$$

where the subscript L, indicates that only "linked" *diagram* contributions are included. In the diagrammatic formulation of PT, the terms which have a non-linear dependence on the number of electrons may be associated with "unlinked" diagrams (i.e. diagrams that may be separated into two or more distinct parts). Terms which are linear in the number of electrons may be associated with linked diagrams alone. This is essentially a statement of Goldstone's linked-cluster theorem. We shall not go into diagrammatic methods here and hence we shall not derive the linked-cluster theorem (which is best done in the diagrammatic language). For an introduction to diagrammatic methods the reader is referred to references [24,25] and for a derivation of the linked-cluster theorem to reference 12b (or see reference [26] for a detailed but pedestrian derivation).

It is useful to relate eq.84 to a commonly used form of MBPT known as Møller-Plesset PT (MPPT). In MPPT, H_0 is taken as the Hartree-Fock hamiltonian

$$H_0 = \Sigma_i f(i) = \Sigma_i\{h(i) + \Sigma_\gamma J_\gamma(i) - K_\gamma(i)\} = \Sigma_i h(i) + V^{HF}(i) \qquad (86)$$

where J_γ and K_γ are the usual coulomb and exchange operators and the perturbation, V, has the form

$$V = \Sigma_{i<j} 1/r_{ij} - \Sigma_i V^{HF}(i) \qquad (87)$$

hence

$$H_0 + V = \Sigma_i h(i) + \Sigma_{i<j} 1/r_{ij} \qquad (88)$$

and the energies of the reference states are simply the one-electron orbital energies $\{\varepsilon_j\}$, within the Hartree-Fock approximation. Using eq.87 in eq.84 yields the MPPT energies

$$E_0 = \Sigma_i \varepsilon_i \qquad (89a)$$

$$E_1 = -1/2\, \Sigma_{ij} <\gamma\delta \,||\, \gamma\delta> = -1/2\, \Sigma_{ij} <\gamma\delta \,|\, \gamma\delta> - <\gamma\delta \,|\, \delta\gamma> \qquad (89b)$$

$$E_2 = 1/4\, \Sigma_{\gamma\delta}\, \Sigma_{rs}\, |<\gamma\delta \,||\, rs>|^2\, (\Delta_{\gamma\delta}{}^{rs})^{-1} \qquad (89c)$$

$$\begin{aligned} E_3 = {} & 1/8\, \Sigma_{\gamma\delta}\, \Sigma_{rstu} < \gamma\delta \,||\, rs> <rs \,||\, tu> <tu \,||\, \gamma\delta > (\Delta_{\gamma\delta}{}^{rs}\, \Delta_{\gamma\delta}{}^{tu})^{-1} \\ & + \Sigma_{\eta\gamma\delta} \Sigma_{rst} <\eta\gamma \,||\, rs> <r\delta \,||\, \eta t> <st \,||\, \eta\delta> (\Delta_{\eta\gamma}{}^{rs}\, \Delta_{\eta\delta}{}^{st})^{-1} \\ & + 1/8\, \Sigma_{\eta\gamma\delta\chi}\, \Sigma_{rs} <\eta\gamma \,||\, rs> <tu \,||\, rs> <rs \,||\, \delta\chi> (\Delta_{\eta\gamma}{}^{rs}\, \Delta_{\delta\chi}{}^{rs})^{-1} \end{aligned} \qquad (89d)$$

where

$$\Delta_{\gamma\delta}{}^{rs} = (\varepsilon_\gamma + \varepsilon_\delta - \varepsilon_r - \varepsilon_s) \qquad (90)$$

and

$$<\gamma\delta|\, rs> = \int \chi_\gamma(1)\, \chi_\delta(2)\, 1/r_{ij}\, \chi_r(1)\, \chi_s(2)\, d\tau_1\, d\tau_2 \qquad (91)$$

In eq.89d in contrast to eq.84d, the cancellation of the terms multiplied by the first-order energy has already been carried out. It should be noted that both eqs.84 and 89 do not provide an upper bound to the exact energy (do not possess a variational condition) since the energies have been evaluated from the unsymmetric equation

$$E_{PT} = <\Phi_0|H|\Psi_{PT}> \qquad (92)$$

(see the general discussion in part one about eq. 16b leading to eq. 20b)

It is possible to obtain a variational result from a perturbational wavefunction by using a symmetric equation

$$E_{VPT} = <\Psi_{PT}|H|\Psi_{PT}> / <\Psi_{PT}|\Psi_{PT}> \qquad (93)$$

but in doing so one gains a variational bound at the expense of introducing unlinked energy diagrams and hence losing the size-extensivity of the energy expression. Indeed, except in the case of full CI, the requirements of size-extensive and variationally bound energies are mutually incompatible.

The central difficulty with MPPT is demonstrating convergence. A dramatic example given by Handy [14] illustrates the point. The reader will recall the behaviour of SDCI for H_2O at r_e $1.5r_e$ and $2.0\ r_e$ given in table 2. SDCI gives a very poor result at $2.0\ r_e$ because a multi-reference wavefunction is required. In Fig. 9 we show the behaviour of MPPT-n as demonstrated by Handy[14].

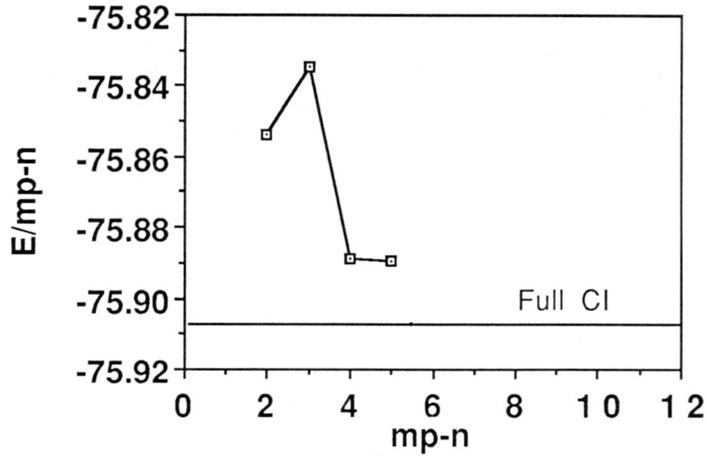

Fig. 9 MP-n for H_2O for $2.0\ r_e$ from ref [14]

From Fig. 9 we see the erratic convergence in the situation where a multi-reference wavefunction is required. In fact, as shown by Handy the series converges but not until MP-13 !

Since we have given a very general discussion of multi-reference perturbation methods in part I we need only a few brief remarks on implementation and applications at this stage. Now that MCSCF methods are being used quite routinely, the motivation for the development of multi-reference MP methods is very high. Unfortunately there are severe technical problems that are only slowly being overcome [28-34]. The central difficulty is that while a completely general MP or MBPT formalism for multi-reference perturbation theory [28] has existed for more than 20 years, there is simply no obvious choice of H_0 as in single reference PT. Also it is not obvious whether the *diagonalize then perturb* method (for a recent application see [32]) or the *perturb then diagonalize* method (see ref. 34 for an implementation of a hybrid approach) is best. The most important point to be made is that multi-reference MP of the *perturb then diagonalize* variety becomes incredibly complicated beyond third-order, yet multi-reference MP2 resembles the single reference case. The CIPSI method[35] is an example of a simple *diagonalize then perturb* method that has been routinely used for many years. At this stage we will give a simple second-order procedure (adapted from ref.34) that illustrates the main ideas.

In multi-reference MPPT one forms a PT expansion for each matrix element of $\mathbf{H_{eff}}$ as implied by the expansion of the inverse in eq. 58. Using the notation introduced for single reference PT we have

$$(\mathbf{H_{eff}})_{\kappa\lambda}^{(2)} = (\mathbf{H_0})_{\kappa\lambda} + 1/4\ \Sigma_{\gamma\delta}\ \Sigma_{rs}\ |\langle\gamma\delta\ ||\ rs\rangle|^2\ (\Delta_{\gamma\delta}{}^{rs})^{-1} \qquad (94)$$

where the following restrictions on ij ab are understood

i) $\gamma\delta$ are *core or valence* orbitals that are occupied in reference state λ

ii) rs are *valence or virtual* orbitals that are unoccupied in reference state λ

iii) terms that correspond to excitation that corresponds to generating another reference configuration are *explicitly excluded* (since this state is obviously included in the reference space itself.

The important fact to be appreciated is that the *valence* orbitals play a dual role: as holes (from which electrons are excited) and *particles* (into which electrons are excited). Thus the denominators $\Delta_{\gamma\delta}{}^{rs} = (\varepsilon_\gamma + \varepsilon_\delta - \varepsilon_r - \varepsilon_s)$ cause problems when any of $\gamma\delta$rs are valence orbitals since a given valence orbital may play the role of a virtual orbital γ for one reference configuration and the role of a hole r in another reference configuration. Thus if the same H_0 is used for each reference configuration, certain ε_i can be positive and ε_a can be negative for *valence* orbitals. This can lead to very small denominators and divergence. However, there is nothing that precludes using a different H_0 for each reference configuration (ie one uses a different A/B partition for each set of equations 58. This has been found to work well in the examples studied so far [34], but would be difficult to take beyond 2nd order.

iii)Coupled-Cluster (CC) Wavefunctions.

We now give a brief discussion of the coupled cluster method giving some of the detail that was omitted in part I. We limit our discussion to the single reference space for simplicity. Our objective is to reinforce the comparison with CI and PT methods

Recall that in eq.65 the excitation operators ω_n produced n-fold substitutions from the reference determinant, $|\Phi_0>$. Hence each ω_n refers to an n-electron "cluster". The CC method specifically introduces correlations among electrons within clusters as well as couplings between clusters. These interactions are introduced by applying the so called cluster operator, T defined in eq. 26 in its most general form, to some suitable reference function

$$|\Psi_{CC}> = e^T |\Phi_0> \quad (95)$$

T generates one-, two-......N-electron clusters; i.e.

$$T = T_1 + T_2 + T_3 \ldots\ldots + T_N \quad (96)$$

where, for example, T_1 is given as

$$T_1 = \Sigma_{\alpha,r} t^r{}_\alpha \{E^r{}_\alpha - E^\alpha{}_r\} \quad (97)$$

in which $\{E^r{}_\alpha - E^\alpha{}_r\}$ is a cluster excitation operator and $t^r{}_\alpha$ is the weight or amplitude of the $\alpha \to r$ excitation. The full CI wavefunction eq. 64, is obtained by expanding the operator in eq.95

$$e^T |\Phi_0> = \{1 + T_1 + T_2 + 1/2!\, T_1{}^2 + T_3 + 1/3!\, T_1{}^3 + T_1 T_2 + T_4$$

$$+ 1/4!\ T_1^4 + 1/2!\ T_2^2 + T_3 T_1 + 1/2!\ T_1^2 T_2 + ...\} |\Phi_0>$$ (98)

and grouping the terms for each total level of excitation

$$e^T |\Phi_0> = \{1 + \omega_1 + \omega_2 + \omega_3 + \omega_4 + ...\} |\Phi_0>$$ (99)

We then obtain the following relationships between the configuration excitation operators ω_n in eq. 64 and the individual cluster components T_n

$$\omega_1 = T_1$$ (100a)

$$\omega_2 = T_2 + 1/2!\ T_1^2$$ (100b)

$$\omega_3 = T_3 + 1/3!\ T_1^3 + T_1 T_2$$ (100c)

$$\omega_4 = T_4 + 1/4!\ T_1^4 + 1/2!\ T_2^2 + T_3 T_1 + 1/2!\ T_1^2 T_2$$ (100d)

Inspecting eqs.100 more closely we see that each order of excitation is composed of linked clusters (which describe the simultaneous correlation of n electrons) and unlinked clusters (which represent products of lower-order clusters). In the ω_4 term (eq.100d) it has been shown that the T_2^2 term is dominant, in that it represents the simultaneous interaction of two distinct electron pairs. The T_4 term is expected to be much smaller in magnitude since it represents the simultaneous interaction of four electrons. The remaining terms involving T_1 can be made small by using MCSCF orbitals. Because of the exponential form of the operator in eq. 24 or 98, even low-order truncations of T lead to a wavefunction that contains all the unlinked higher-order clusters, which serve to make the energy size-extensive.

The cluster amplitudes, t_α^r, $t_{\alpha\beta}^{rs}$ ...... are determined (see discussion surrounding eq. 28 for the general case) by requiring that eq.94 satisfies the Schrödinger equation

$$H e^T |\Phi_0> = E e^T |\Phi_0>$$ (101)

premultiply by e^{-T} gives

$$e^{-T} H e^T |\Phi_0> = E |\Phi_0>$$ (101)

$e^{-T} H e^T$ can be expanded in the well known series

$$\{H + [H,T] + 1/2!\ [[H,T],T] + 1/3!\ [[[H,T],T],T] + 1/4!\ [[[[H,T],T],T],T]\} |\Phi_0> = E |\Phi_0>$$ (102)

This series truncates exactly at four commutators regardless of the level of truncation of T_n. This is because H does not contain higher than two-electron integrals (hence no more than four orbitals may be substituted). The CC energy is therefore a quartic function of the cluster amplitudes. Again the energy is evaluated from (as in eq. 29b)

$$E_{CC} = \langle\Phi_0|H|\Psi_{CC}\rangle \qquad (103)$$

which is not variationally bound (unless all clusters up to N are included) but is size-extensive.

We conclude by considering in detail the case where the T operator is truncated at two-electron clusters (i.e. $T \approx T_2$), which is commonly known as the CC doubles (CCD) model. The motivation for this choice relies on the known importance of pair interactions. The hamiltonian is separated as in eq.71. Eq.101 is then projected onto $|\Phi_0\rangle$ to yield

$$E_{CCD} = \langle\Phi_0|e^{-T2}\, H\, e^{T2}|\Phi_0\rangle$$

$$= \langle\Phi_0|(1 - T_2 + 1/2\, T_2^2)\, H\, (1 + T_2 + 1/2\, T_2^2)\, |\Phi_0\rangle$$

$$= \langle\Phi_0|H|\Phi_0\rangle + \langle\Phi_0|HT_2|\Phi_0\rangle$$

$$= E_{HF} + \Sigma_{\gamma<\delta}\, \Sigma_{r<s} \langle\gamma\delta \,||\, rs\rangle\, t_{\gamma\delta}^{rs} \qquad (104)$$

where we have used the fact that $\langle\Phi_0|T_2 = 0$ and $\langle\Phi_0|HT_2T_2|\Phi_0\rangle = 0$ (quadruple excitations do not interact with $|\Phi_0\rangle$).

The amplitudes $t_{\gamma\delta}^{rs}$ are determined by projecting eq.101 against all doubly excited determinants $|\Phi_{\gamma\delta}^{rs}\rangle$

$$\langle\Phi_{\gamma\delta}^{rs}|e^{-T2}\, H\, e^{T2}|\Phi_0\rangle = 0 \qquad (105)$$

expanding the exponential

$$\langle\Phi_{\gamma\delta}^{rs}\, |H\, (1 + T_2 + 1/2\, T_2^2)\, |\Phi_0\rangle + \langle\Phi_{\gamma\delta}^{rs}\, |- T_2\, H\, (1 + T_2)\, |\Phi_0\rangle = 0 \qquad (106)$$

Evaluating the matrix elements in eq.106 explicitly gives the following equation for the cluster amplitudes

$$(\varepsilon_r + \varepsilon_s - \varepsilon_{\gamma} \text{-}\, \varepsilon_\delta)\; t_{\gamma\delta}^{rs} = \langle rs \,||\, \gamma\delta\rangle - \Sigma_{t>u} \langle rs \,||\, tu\rangle\, t_{\gamma\delta}^{tu} - \Sigma_{\mu>\eta} \langle\mu\eta \,||\, \gamma\delta\rangle\, t_{\gamma\delta}^{rs}$$

$$+ \Sigma_\mu \Sigma_t \{\langle\mu b \,||\, jt\rangle\, t_{\gamma\mu}^{rt} - \langle\mu r \,||\, jt\rangle\, t_{\gamma\mu}^{st} - \langle\mu s \,||\, \gamma t\rangle\, t_{j\mu}^{rt} - \langle\mu r \,||\, \gamma t\rangle\, t_{j\mu}^{st}\}$$

$$+ \Sigma_{\mu>\eta}\, \Sigma_{t>u} \langle\mu\eta \,||\, tu\rangle\, [t_{\gamma\delta}^{tu}\, t_{\mu\eta}^{rs} - 2\, (t_{\gamma\delta}^{rt}\, t_{\mu\eta}^{su} + t_{\gamma\delta}^{su}\, t_{\mu\eta}^{rt})$$

$$- 2\, (t_{\gamma\mu}^{rs}\, t_{\delta\eta}^{tu} + t_{\gamma\mu}^{tu}\, t_{\delta\eta}^{rs}) + 4\, (t_{\gamma\mu}^{rt}\, t_{\delta\eta}^{su} + t_{\gamma\mu}^{su}\, t_{\delta\eta}^{rt})] \qquad (107)$$

The perturbative solution of eq.107 shows clearly the connection between CC and MBPT methods and this is now discussed in some detail. First of all, the truncation, $T=T_2$, leads to an equation which is quadratic in the cluster amplitudes. Hence, eq.107 may be solved iteratively by finding an initial guess of the $\{t_{\gamma\delta}{}^{rs}\}$ and then inserting these into the right-hand-side of eq.107 to obtain an improved set of $\{t_{\gamma\delta}{}^{rs}\}$, etc. In practise the initial guess is obtained by assuming all the cluster amplitudes on the right-hand-side of eq.107 are zero. This gives for the amplitudes

$$t_{\gamma\delta}{}^{rs} = \langle\gamma\delta|| rs\rangle / (\varepsilon_r + \varepsilon_s - \varepsilon_\gamma - \varepsilon_\delta) \qquad (108)$$

which inserted into eq 104 yields

$$E = E_{HF} + \Sigma_{\gamma<\delta}\Sigma_{r<s} \langle\gamma\delta || rs\rangle \langle rs || \gamma\delta\rangle / (\varepsilon_r + \varepsilon_s - \varepsilon_\gamma - \varepsilon_\delta) \qquad (109)$$

So our initial guess of the CCD energy is simply the second-order MPPT result of eq.89c. For the second iteration, we substitute the cluster amplitudes obtained in eq.108 into the right-hand-side of eq.107. If we neglect the terms which are quadratic in the cluster amplitudes and substitute again into eq.104 we obtain an energy expression which is equivalent to the third-order energy in MPPT given in eq.89d. Iterating again by substituting in eq.107 and keeping the quadratic terms leads to new amplitudes which give (on inserting back into eq.104) an energy which contains all the contributions to the fourth-order MPPT arising from quadruple excitations. Furthermore, from the form of the cluster products we note that fourth-order MPPT contains quadruple excitations of the form $1/2\, T_2{}^2$ and $-T_2T_2$ only. These terms correspond to pair-pair interactions. The T_4 terms (corresponding to true four-body effects) first enters at fifth-order in MPPT. However, all contributions to the fourth-order MPPT cannot be obtained when T is approximated by T_2. Since inclusion of all fourth-order contributions would require T_1 and T_3 to be included in the cluster expansion. As the iteration process is continued, by substituting each new set of amplitudes into eq.104 convergence is obtained. The converged result corresponds to summing all contributions in MPPT that contain double excitations and products of double excitations to infinite-order.

III Conclusions

Our objective in these notes has been to look at methods for the computation of electron correlation from a very general point of view, concentrating on the details of the standard single reference methods in part II. While single reference CISD, CCD and MPPT are now part of standard software of molecular structure codes, multi-reference methods are still mainly used by specialists. This situation cannot persist much longer since many chemical problems (viz. homolytic bond cleavage) require a multi-reference description (see for example the recent benchmark computations of Cave and Davidson [36] where the argument single vs multi-reference has been explored in a very detailed way).

References

1a. Ab initio methods in Quantum Chemistry I, Ed. H.P. Lawley Adv. Chem.Phys. **67** (1987).
b. Ab-initio methods in Quantum Chemistry II, Ed. K.P. Lawley Adv.Chem.Phys. **69** (1987).
2. O. Sinanoglu, J. Chem. Phys. **36**, 706 (1962)
3. I. Shavitt, in *Methods of Electronic Structure Theory*, H.F. Schaefer III (Ed.), Plenum Press, New York, 1977.
4. C. Bloch, Nucl. Phys., **6**, 329, (1958).
5.Ph. Duand and J. P. Malriu in ref. 1a
6. J. Paldus and B. Jeziorski, Theoret. chim. Acta **73**, 81 (1987)
7) J. Paldus ,in H. Eyring and D.G. Henderson Eds. ,Theoretical Chemistry, Advances and Perspectives, **2**, 131 (1976)
8)A.F. Voter and W.A. Goddard, J. Chem. Phys. **57**, 253 (1981), J. Chem. Phys. **75**, 3638 (1981), J.Amer.Chem Soc. **108**,2830 (1986)
9)H.F. Schaefer,R.A. Klemm and F.E. Harris, J. Chem. Phys. **51**,4643 (1969)
10)J. Hay, T.H.Dunning, W.A.Goddard, J. Chem. Phys. **62**,3912 (1969)
11)J. Hay, T.H.Dunning, J. Chem. Phys. **64**,5077 (1976)
12a)B. H. Brandow, in *Effective Interactions and Operators in Nuclei* (Ed. B. R. Barrett) Lecture Notes in Physics **40** ,1 (1975)
12b)N.H. March, W.H. Young and S. Sampanthar, *The Many-Body Problem in Quantum Mechanics*, University Press, Cambridge, 1967.
13)R.J. Bartlett, S.J. Cole, G.D. Purvis, W.C. Ermler, H.C.Hsieh and I. Shavitt,*J. Chem. Phys.*, **87**, 6579, (1987).
14) N.C. Handy, Faraday Symp. Chem. Soc. **19**, 17 (1984)
15)S.R. Langhoff and E.R. Davidson, *Int. J. Quantum Chem.*, 8, 61, (1974).
16)S. Prime, C. Rees, and M.A.Robb , Mol. Phys. **44**, 173 (1981)
17)P.E.M. Siegbahn, in *Methods in Computational Molecular Physics*, G.H.F. Diercksen and S. Wilson (Eds.), D. Reidel, Dordrecht, 1983.
18a) I. Shavitt, in Ed. D.G. Truhlar *Mathematical Frontiers in Computational Chemical Physics,* Springer IMA Series Vol.15 (in Press)
b) J. Paldus in Ed. D.G. Truhlar *Mathematical Frontiers in Computational Chemical Physics,* Springer IMA Series Vol.15 (in Press)
19) M.A.Robb and U. Niazi, Computer.Phys. Reports **1**, 127 (1984)
20)P. Saxe, D.J. Fox, H.F. Schaefer III, and N.C. Handy, J. Chem. Phys. **77**, 5584 (1982)
21)K.A. Brueckner, *Phys. Rev.*, **97**, 1353, (1955); **100**, 36, (1955).
22)J. Goldstone, *Proc. Royal Soc. (London)*, **A239**, 267, (1957).
23)P. Jørgensen and J. Simons, *Second Quantization Based Methods in Quantum Chemistry*, Academic Press, New York, 1981.
24)A. Szabo and N.S. Ostlund, *Modern Quantum Chemistry: Introduction to Advanced Electronic Structure Theory*, Macmillan, New York, 1982.
25)J. Paldus and J. Cizek, in *Advances in Quantum Chemistry Vol. 9*, P.O. Löwdin (Ed.), Academic Press, New York, 1975.
26) M.A. Robb, in *Computational Techniques in Quantum Chemistry and Molecular Physics*, G.H.F. Diercksen, B.T.Sutcliffe and A. Veillard(ed.) D. Reidel, Dordrecht, 1975

27)R.J. Bartlett, C.E. Dykstra and J. Paldus, in *Advanced Theories and Computational Approaches to the Electronic Structure of Molecules*, C.E. Dykstra (Ed.), D. Reidel, Dordrecht, 1984.
28)B. H. Brandow, Rev. Mod. Phys. **39** 771 (1967)
29)R. J. Bartlet, Ann Rev. Phys. Chem. **32** 359 (1981)
30)I. Lindgren, Phys. Scr. **32** (1985)
31)D. Mukhergee, Int. J. Quantum Chem. **S20** 409 (1986)
32)K. Wolinski, H. Sellers and P. Pulay, Chem. Phys. Lett. **140** 225 (1987)
33)Ph. Durand and J. P. Malrieu , Adv. Chem.Phys. **67** (1987)
34)J.J.W. McDouall, K.Peasley,M.A.Robb, Chem.Phys. Lett **148**,183 (1988)
35) B. Huron, P. Rancurrel and J.P. Malrieu, J. Chem. Phys. **73** 5745 (1973)
36)R.J. Cave and E.R. Davidson, J.Chem. Phys. **88** 5770, 6798 (1988)

MOLECULAR MODELLING. SEMI-EMPIRICAL AND EMPIRICAL METHODS OF THEORETICAL CHEMISTRY

Jean-Louis RIVAIL
Laboratoire de Chimie théorique, Université de Nancy 1
B.P. 239
54506 Vandœuvre lès Nancy Cedex
France

ABSTRACT. The basic equations of the semi-empirical methods of quantum chemistry are developed and the most commonly used approximations are given and their utility in solving chemical problems is dicussed. The principles of molecular mechanics are presented with a special emphasis on the molecular force field and on the possible chemical applications.

1.0 Introduction

Ab initio computations on large molecules are still out of reach for most chemical laboratories and many chemical or biochemical problems can be solved by means of less elaborate approaches.

In particular, chemists are used to some very efficient qualitative reasoning based on analogies or differences between related molecules. This reasoning may well be supported by some quantitative data which do not deserve high accuracy provided that the variations of the quantities of interest are larger than the error bars which affect the results.

This situation justifies the survival and even the development of several semi-empirical quantum chemical or purely classical empirical methods in computational chemistry and molecular modelling.

2.0 Semi-empirical quantum chemical methods [1]

Nowadays, the simplest description of the electronic structure of a molecule is obtained within the framework of the orbital approximation and the molecular orbitals are expressed as a linear combination of atomic orbitals. The coefficients of the combination are obtained by solving the self consistent-field or Hartree-Fock-Roothaan equations [2]. The necessity of using an iterative (self-consitent) procedure together with the enormous number of integrals to compute and to handle during the course of the computation make this procedure long and costly. This is especially true when one uses a large basis of atomic orbitals. The aim of the semi-empirical methods is to reduce the computation time drastically and also the size of the tables to be stored in

C. Ögretir and I. G. Csizmadia (eds.),
Computational Advances in Organic Chemistry: Molecular Structure and Reactivity, 229–259.

order to make the computations much faster and feasible on a small laboratory computer.

The strategy employed consists of :

i using the smallest possible set of atomic orbitals

ii neglecting a large number of integrals in the electron-electron interaction energy

iii setting most of the remaining integrals in a parametric form and adjusting the value of the parameters on atomic or molecular experimental data.

2.1 THE ATOMIC ORBITALS BASIS SET

A first reduction of the number of atomic orbitals can be achieved by assuming that the electrons of the inner shells of the atoms are not greatly affected by the formation of the bonds. This «frozen core approximation» reduces the description of the molecules to the valence electrons only, and the number of the atomic orbitals needed is usually limited to the different atomic orbitals of the valence shell of the atoms.

The next step consists of chosing the proper mathematical form for these atomic orbitals such that the number of functions needed for properly describing the valence electrons should be as low as possible. The Slater orbitals [3] are known to have the proper behaviour at medium and long distance from the nuclei. They may be written in the form :

$$\phi_{n,l,m}(r, \theta, \chi) = N_n r^{n-1} e^{-\zeta r} Y_l^m(\theta, \chi) \tag{1}$$

in which n, l, m are the usual atomic quantum numbers, r, θ and χ the spherical polar coordinates. $Y_l^m(\theta, \chi)$ denotes a spherical harmonic which may be either complex or put in an equivalent real form [4] N_n is a normalization factor.

The mathematical simplicity of the Slater orbitals has another important consequence : the atomic Coulomb or exchange integrals computed with these functions appear as a combination of a limited number of elementary integrals. In particular the atomic multiplet structure which corresponds to a large number of electronic transitions can be rationalized by means of these quantities which can be treated as empirical constants : the Slater-Condon parameters [5]. These quantities are adjusted on few selected electronic transitions and can be used to predict the energy of many other atomic levels. This remark has played an important rôle in the development of the semi-empirical methods.

The main inconvenience of the Slater orbitals is that they give rise to serious difficulties in *ab-initio* molecular computations arising from multicentric two electron integrals which occur in the expression of the exchange or Coulomb terms. Therefore gaussian functions have been introduced in these computations. In the case of semi-empirical methods this difficulty is overcome by neglecting these terms as we shall see below.

2.2 THE ELECTRONIC ENERGY IN THE FROZEN CORE APPROXIMATION

Let us expand the molecular orbital number i as a combination of atomic orbitals φ_μ, in the form :

$$\psi_i = \sum_\mu c_{\mu i} \varphi_\mu \tag{2}$$

in which $c_{\mu i}$ denotes a coefficient. Let us call H^c the core Hamiltonian of the molecule, written in atomic orbitals, as :

$$\mathrm{H}^C = \sum_\ell \left[-\frac{1}{2}\Delta(\ell) - \sum_K \frac{Z_K}{r_{K\ell}} \right] = \sum_\ell \mathrm{H}^C(\ell) \tag{3}$$

where $\Delta(\ell)$ is the Laplace operator acting on the coordinates of electron ℓ, K denotes a nucleus of atomic number Z_K and $r_{K\ell}$ is the distance between this nucleus and the point associated with electron ℓ. Therefore if k and ℓ are two electrons whose representative points are separated by $r_{k\ell}$ and L another nucleus distant from K of r_{KL}, the molecular Hamiltonian, in the Born-Oppenheimer approximation is written, in atomic units :

$$\mathscr{H} = \sum_\ell \mathrm{H}^C(\ell) + \sum_{(k\ell)} \frac{1}{r_{k\ell}} + \sum_{(KL)} \frac{Z_K Z_L}{r_{KL}} \tag{4}$$

in which $\sum_{(ab)}$ stands for a sum over all the couples (ab).

If one assumes that the ground state wavefunction of the molecule is represented by a single determinant of N doubly occupied molecular orbitals expressed as a linear combination of atomic orbitals as in (2), the molecular energy takes the form :

$$E = 2\sum_{i=1}^{N}\sum_\lambda \sum_\mu c_{\lambda i} c_{\mu i} \left\{ H^C_{\lambda\mu} + \sum_\nu \sum_\eta P_{\nu\eta} \left[(\lambda\mu \mid \nu\eta) - \frac{1}{2}(\lambda\eta \mid \lambda\mu) \right] \right\} + \sum_{(KL)} \frac{Z_K Z_L}{r_{KL}} \tag{5}$$

where $H^c_{\lambda\mu}$ is the (λ,μ) element of the matrix associated to H^c in the atomic orbitals basis :

$$H^C_{\lambda\mu} = \langle \varphi_\lambda \mid \mathrm{H}^C \mid \varphi_\mu \rangle = \langle \mu \mid \mathrm{H}^C \mid \nu \rangle \tag{6}$$

$P_{\nu\eta}$ is an element of the one electron density matrix defined as

$$P_{\nu\eta} = \sum_{j=1}^{N} 2c_{\nu j}\, c_{\eta j} \tag{7}$$

and the two electron integrals are written in the usual convention :

$$(\lambda\mu \mid \nu\eta) = \int_{(1)} \int_{(2)} \varphi_\lambda(1)\, \varphi_\mu(1)\, \frac{1}{r_{12}}\, \varphi_\lambda(2)\, \varphi_\eta(2)\, d\tau_1\, d\tau_2 \tag{8}$$

In equation (5) the sums over the atomic orbitals run over all the orbitals of the basis set.

Since we want to reduce the basis set to the n valence orbitals, let us denote the $N-n$ orbitals belonging the inner atomic shells by λ', μ' ... separately, the unprimed indices still denoting the orbitals of the valence shells.

If one assumes that the inner shell orbitals remain unchanged in the molecule *i.e.* that the first $N-n$ molecular orbitals reduce to atomic orbitals, one concludes that

$$P_{\lambda'\mu'} = 2\,\delta_{\lambda'\mu'} \tag{9}$$

where $\delta_{\lambda'\mu'} = 1$ if $\lambda' = \mu'$ and 0 otherwise.

Therefore the molecular energy can be written as :

$$\begin{aligned}
E = & \sum_K \sum_{\lambda' \in K} \left\{ 2 < \lambda' \,|\, -\frac{1}{2}\Delta - \frac{Z_K}{r_K} \,|\, \lambda' > + \sum_{\nu' \in K} [2\,(\lambda'\lambda' \mid \nu'\nu') - (\lambda'\nu' \mid \lambda'\nu')] \right\} \\
& + \sum_\lambda \sum_\mu P_{\lambda\mu} H^c_{\lambda\mu} + \sum_{\eta'} [2(\lambda\mu \mid \eta'\eta') - (\lambda\eta' \mid \mu\eta')] \\
& + \sum_\lambda \sum_\mu P_{\lambda\mu} \sum_\nu \sum_\eta P_{\nu\eta} \left[(\lambda\mu \mid \nu\eta) - \frac{1}{2} (\lambda\eta \mid \nu\mu) \right] \\
& + \sum_{(KL)} \left\{ \frac{Z_K Z_L}{r_{KL}} + \sum_{\lambda' \in K} \sum_{\eta' \in L} [2(\lambda'\lambda' \mid \eta'\eta') - (\lambda'\eta' \mid \lambda'\eta')] \right\} \\
& + \sum_K \sum_{\eta' \notin K} - 2 < \eta' \,|\, \frac{Z_K}{r_K} \,|\, \eta' >
\end{aligned} \tag{10}$$

In this expression, the first line is a sum of atomic terms corresponding to the inner shell electrons only. It is a constant term which does not depend on the molecular electronic structure and can be forgotten. The last two lines represent the *core-core repulsion term*, which is the sum of the interaction between the nuclei, the repulsion between the atomic inner shells and the attraction of each inner shell by the other nuclei. We shall denote the contribution of atoms K and L to this sum :

$$E^c_{KL} = \frac{Z_K Z_L}{r_{KL}} + \sum_{\lambda' \in K} \sum_{\eta' \in L} [2(\lambda'\lambda' \mid \eta'\eta') - (\lambda'\eta' \mid \lambda'\eta')]$$
$$- \sum_{\lambda' \in K} \sum_{\eta' \in L} 2\left[< \lambda' \mid \frac{Z_L}{r_L} \mid \lambda' > + < \eta' \mid \frac{Z_K}{r_K} \mid \eta' > \right] \quad (11)$$

The only contribution to energy which depends on the density matrix, and therefore on the coefficients of the atomic orbitals, is represented by the second and third lines of equation (10). It is the part of the molecular energy which has to be considered in the variational process that minimizes the energy with respect to the coefficients. It is well known that the coefficients of the canonic molecular orbitals obey the Hartree-Fock-Roothaan equations which have to be written in a reduced form in order to limit them to the valence atomic orbitals.

2.3 REDUCED HARTREE-FOCK-ROOTHAAN EQUATIONS

Following the reduced expression of the molecular energy, the molecular orbitals built on the basis set limited to the valence orbitals are solutions of reduced Hartree-Fock-Roothaan equations whose expression is deduced from the density matrix dependent part of the molecular energy.

2.3.1 Diagonal terms of the Hartree-Fock matrix. Owing to the previous assumptions, the diagonal elements are written :

$$F_{\mu\mu} = H^c_{\mu\mu} + \sum_{\nu'} [2(\mu\mu \mid \nu'\nu') - (\mu\nu' \mid \mu\nu')]$$
$$+ \sum_{\eta} \sum_{\sigma} P_{\eta\sigma} \left[(\mu\mu \mid \eta\sigma) - \frac{1}{2} (\mu\eta \mid \mu\sigma) \right] \quad (12)$$

Let us assume that the atomic orbital μ is centered on atom A. From the explicit expression of $H^c_{\mu\mu}$ one can extract, as above :

$$F_{\mu\mu} = < \mu \mid - \frac{1}{2} \Delta - \frac{Z_A}{r_A} \mid \mu > + \sum_{\lambda' \in A} [2(\mu\mu \mid \lambda'\lambda') - (\mu\lambda' \mid \mu\lambda')]$$
$$+ \sum_{B \neq A} \left\{ < \mu \mid - \frac{Z_B}{r_B} \mid \mu > + \sum_{\eta' \in B} [2(\mu\mu \mid \eta'\eta') - (\mu\eta' \mid \mu\eta')] \right\} \quad (13)$$
$$+ \sum_{\eta \neq \mu} \sum_{\sigma \neq \mu} P_{\eta\sigma} \left[(\mu\mu \mid \eta\sigma) - \frac{1}{2} (\mu\eta \mid \mu\sigma) \right]$$

By introducing the following general notation, in which ν stands for an orbital which is centered on A (like μ) and which can be μ :

$$U_{\mu\nu} = <\mu| - \frac{1}{2}\Delta - \frac{Z_A}{r_A}|\nu> + \sum_{\lambda' \in A} [2(\mu\nu \mid \lambda'\lambda') - (\mu\lambda' \mid \nu\lambda')] \tag{14}$$

and

$$V'^{B}_{\mu\nu} = - <\mu|\frac{Z_B}{r_B}|\nu> + \sum_{\eta' \in B} [2(\mu\nu \mid \eta'\eta') - (\mu\eta' \mid \nu\eta')] \tag{15}$$

one obtains

$$F_{\mu\mu} = U_{\mu\mu} + \sum_{B \neq A} V'^{B}_{\mu\mu} + \sum_{\eta \neq \mu}\sum_{\sigma \neq \mu} P_{\eta\sigma}\left[(\mu\mu \mid \eta\sigma) - \frac{1}{2}(\mu\eta \mid \mu\sigma)\right] \tag{16}$$

2.3.2 Monocentric, off-diagonal terms. Let us still assume that orbitals μ and ν are both centered on atom A. By using the previous decomposition and notation one obtains :

$$F_{\mu\nu} = U_{\mu\nu} + \sum_{B \neq A} V'^{B}_{\mu\nu} + \sum_{\eta}\sum_{\sigma} P_{\eta\sigma}\left[(\mu\nu \mid \eta\sigma) - \frac{1}{2}(\mu\eta \mid \nu\sigma)\right] \tag{17}$$

2.3.3 Bicentric terms. These terms corresponding to couples of atomic orbitals μ and λ centered on two different atoms are simply defined by :

$$F_{\mu\lambda} = H'^{c}_{\mu\lambda} + \sum_{\eta}\sum_{\sigma} P_{\eta\sigma}\left[(\mu\lambda \mid \eta\sigma) - \frac{1}{2}(\mu\eta \mid \lambda\sigma)\right] \tag{18}$$

in which $H'^{c}_{\mu\lambda}$ is the matrix element of an «effective core Hamiltonian» which includes the interaction with the inner shell electrons and is defined by :

$$H'^{c}_{\mu\lambda} = H^{c}_{\mu\lambda} + \sum_{\nu'} [2(\mu\lambda \mid \nu'\nu') - (\mu\nu' \mid \lambda\nu')] \tag{19}$$

These equations can readily be transformed into the Unrestricted Hartree-Fock scheme by reducing the Berthier-Pople-Nesbet equations [6] in a similar way as we did above.

2.4 CONSTRUCTION OF THE SEMI-EMPIRICAL HARTREE-FOCK MATRIX

At the end of this first simplification step, there are still many integrals which occur in the matrix elements ; most of them being the two electron integrals like $(\lambda\mu \mid \nu\eta)$ whose number increase like n^4. Further simplifications are needed. They are achieved by :

i A drastic reduction of the number of bi-electronic integrals

ii A semi-empirical evaluation of most of - or even all - the remaining integrals.

2.4.1 Semi-empirical evaluation of one center integrals based on the Slater-Condon parameters. We already mentioned that the one center two electron integrals, which are atomic Coulomb and exchange integrals, can be expressed from the Slater-Condon parameters F^n or G^n. For instance, the one center integrals containing $2s$ and $2p$ Slater orbitals (with the same exponent) are related to the F^0, F^2 and G^1 parameters by means of the following relationships :

$$(ss \mid ss) = (ss \mid xx) = F^0$$

$$(sx \mid sx) = \frac{1}{3} G^1$$

$$(xy \mid xy) = \frac{3}{25} F^2 \tag{20}$$

$$(xx \mid xx) = F^0 + \frac{4}{25} F^2$$

$$(xx \mid yy) = F^0 - \frac{2}{25} F^2$$

...

where s, or x stands for $2s$ or $2p_x$ and so on.

Similarly the ionization potential I_μ or the electron affinity A_μ corresponding to the removal or the addition of one electron to the atomic level corresponding to atomic orbital μ can be expressed, in the orbital approximation and if one assumes that the Koopmans theorem holds, as a combination of $U_{\mu\mu}$ (14) and of two electron integrals. Unfortunately, owing to the approximation on which these equations are based, the values of $U_{\mu\mu}$ deduced from the ionization potential or from the electron affinity are usually different. Pople [7] proposed to evaluate $U_{\mu\mu}$ from the experimental value of $\frac{1}{2}(I_\mu + A_\mu)$ by using the following theoretical expressions :

For hydrogen (orbital 1s) :

$$- \frac{1}{2} (I_s + A_s) = U_{ss} + \frac{1}{2} F^0 \tag{21}$$

For elements from boron to fluorine (orbitals 2s and 2p) :

$$- \frac{1}{2} (I_s + A_s) = U_{ss} + (Z'_A - \frac{1}{2}) F^0 - \frac{1}{6} (Z'_A - \frac{3}{2}) G^1 \tag{22a}$$

$$- \frac{1}{2} (I_p + A_p) = U_{pp} + (Z'_A - \frac{1}{2}) F^0 - \frac{1}{3} G^1 - \frac{2}{25} (Z'_A - \frac{5}{2}) F^2 \tag{22b}$$

2.4.2 Alternative analysis of atomic data. The electronic energy of an atom can be written :

$$E = \sum_{\mu} n_\mu U_{\mu\mu} + \frac{1}{2} \sum_{\mu} \sum_{\nu \neq \mu} n_\mu n_\nu g_{\mu\nu} + \frac{1}{2} \sum_{\mu} n_\mu (n_\mu - 1) g_{\mu\mu} \tag{23}$$

where n_μ is the occupation number of the atomic level μ and $g_{\mu\mu} = (\mu\mu \mid \mu\mu)$

$g_{\mu\nu} = (\mu\mu \mid \nu\nu) - \frac{1}{2} (\mu\nu \mid \mu\nu)$

Let us consider a reference atomic state whose energy is E_0 with all the occupation numbers noted as $n_{0\mu}$. If these occupation numbers vary, the energy may be expressed as [8] :

$$E = E_0 + \sum_{\mu} a_\mu (n_\mu - n_{0\mu}) + \frac{1}{2} \sum_{\mu,\nu} b_{\mu n} (n_\mu - n_{0\mu}) (n_\nu - n_{0\nu}) + \ldots \tag{24}$$

This energy can be written as

$$E = C + \sum_{\mu} U^*_\mu n_\mu + \frac{1}{2} \sum_{\mu,\nu} g^*_{\mu,\nu} (n_\mu - n_{0\mu}) (n_\nu - n_{0\nu}) + \frac{1}{2} \sum_{\mu} g^*_{\mu\mu} n_\mu (n_\mu - 1) \tag{25}$$

Where the quantities C, U^*_μ, $g^*_{\mu\nu}$ are deduced from the constants of equation (24) by identification.

The experimental ionization potential I_μ and the corresponding electron affinity A_μ are expressed by :

$$- I_\mu = E(n_\mu = 1) - E(n_\mu = 0) = U_\mu^* + \sum_{\nu \neq \mu} n_\nu g_{\mu\nu}^* \tag{26a}$$

$$- A_\mu = E(n_\mu = 2) - E(n_\mu = 1) = U_\mu^* + \sum_{\nu \neq \mu} n_\nu g_{\mu\nu}^* + g_{\mu\mu}^* \tag{26b}$$

TABLE 1. Atomic integrals (e.V.) for Carbon atom

Reference state	$V_2\,(s^2,p,p')$	$V_1\,(s^2,p)$	$V_0\,(s^2)$
C	- 24.47	- 9.13	- 1.30
U_s^*	- 52.14	- 58.82	- 64.27
U_p^*	- 40.88	- 48.90	- 54.09
g_{ss}^*	13.23	13.95	17.49
g_{sp}^*	10.26	12.50	15.35
g_{pp}^*	11.08	14.36	16.37
$g_{pp'}^*$	9.53	12.44	14.45

These quantities differ from the theoretical vertical ionization potential and electron affinity which are related to the theoretical $U_{\mu\mu}$, $g_{\mu\nu}$ and $g_{\mu\mu}$ integrals occurring in (23). By comparing several valence states one is able to extract the $g_{\mu\nu}^*$ parameters corresponding to a given reference state, and from the expression of $\frac{1}{2}(I_\mu + A_\mu)$ resulting from (26a) and (26b) one can evaluate the corresponding value of U_μ^*. These integrals vary from one reference valence state to another as shown in table 1 in the case of three different reference states for the carbon atom.

2.5 THE NDDO LEVEL OF APPROXIMATION

The denomination *Differential Overlap* refers to an electronic distribution represented by the product of two different atomic orbitals : $\varphi_\mu(\ell)\,\varphi_\lambda(\ell)$.

The Neglect of Diatomic Differential Overlap consists of setting all such products of two orbitals centered on two different atoms equal to zero.

In addition, the only one center integrals which are considered are of the form $(\mu\mu \mid \nu\nu)$ and $(\mu\nu \mid \mu\nu)$.

2.5.1 The original NDDO method. Owing to the basic assumption of the method [9], equations (16), (17), (18) simplify to :

$$F_{\mu\mu} = U_{\mu\mu} + \sum_{B} V'^{B}_{\mu\mu} + \sum_{\nu\in A} P_{\nu\nu}\left[(\mu\mu \mid \nu\nu) - \frac{1}{2}(\mu\nu \mid \mu\nu)\right] + \sum_{B}\sum_{\eta,\sigma\in B} P_{\eta\sigma}(\mu\mu \mid \eta\sigma) \tag{27}$$

$$F_{\mu\nu} = \sum_{B} V'^{B}_{\mu\nu} + P_{\mu\nu}\left[(\mu\nu \mid \mu\nu) - \frac{1}{2}(\mu\mu \mid \nu\nu)\right] + \sum_{B}\sum_{\eta,\sigma\in B} P_{\eta\sigma}(\mu\nu \mid \eta\sigma) \tag{28}$$

$$F_{\mu\lambda} = H^{c}_{\mu\lambda} - \frac{1}{2}\sum_{\nu\in A}\sum_{\sigma\in B} P_{\nu\sigma}(\mu\lambda \mid \nu\sigma) \tag{29}$$

All the one center integrals are evaluated empirically using the Slater-Condon parameters, the ionization potential and the electron affinity. The two center two electron integrals are computed.

2.5.2 The modified NDDO methods. The original NDDO method has seen little application. In turn, a modified method known under the abbreviation of MNDO [10] (Modified NDDO) is quite popular for organic molecules. The two electron one center integrals are now evaluated by the alternative (Oleari's) method, as well as the first guess for the U's integrals which are further optimized together with the other adjustable parameters.

The evaluation of the two electron two center integrals is performed by an electrostatic approximation : each electronic distribution is replaced by a series of multipoles and the system is scaled in order to ensure that when the interatomic distance goes to zero the limit of the integral is the corresponding one center integral.

The quantities $V'^{B}_{\mu\nu}$ which represent the interaction of the electronic distribution $(\mu\nu)$ on A with the core of atom B are set proportional to the interaction of the distribution $(\mu\nu)$ with a s electronic distribution on B, symbolized by $s^B s^B$:

$$V'^{B}_{\mu\nu} = -Z'_{B}(\mu\nu \mid s^{B}s^{B}) \tag{30}$$

where Z'_B is equal to the number of valence electrons of atom B.

The integral $H^{c}_{\mu\lambda}$ is set proportional to the overlap integral $S_{\mu\lambda}$:

$$H^{c}_{\mu\lambda} = \frac{\beta^{A}_{\mu} + \beta^{B}_{\lambda}}{2} S_{\mu\lambda} \tag{31}$$

Finally the core-core interaction terms (11) are estimated by another empirical formula :

$$E^c_{AB} = Z'_A Z'_B (s^A s^A \mid s^B s^B) [1 + f(R_{AB})] \quad (32)$$

with

$$f(R_{AB}) = e^{-\alpha_A R_{AB}} + e^{-\alpha_B R_{AB}} \quad (33a)$$

where α_A and α_B are two parameters and R_{AB} the interatomic distance or

$$f(R_{XH}) = R_{XH}\, e^{-\alpha_X R_{XH}} + e^{-\alpha_H R_{XH}} \quad (33b)$$

for X = N or O.

This method contains a fairly large number of parameters : the Slater exponents ζ of the atomic orbitals, the β parameters for each kind of atomic orbitals, and one parameter α per atom. In addition the U integrals are optimized. These parameters are adjusted in order to reproduce some selected experimental values of a series of test molecules. The MNDO method has been parametrized on the geometry and the heat of formation of organic compounds and reproduces these data fairly well. In addition, the electronic charge distributions and the force constants are also computed with good accuracy.

The main failure of the method is a very bad description of hydrogen bonding. Several attempts have been made to overcome this inconvenience. The first one consists in using a particular form of the $f(R_{AB})$ function which occur in the empirical formula of E^c_{AB} (32) in the case of a pair of hydrogen bonded atoms [11]. This improvement is known as the MNDO/H method.

Later, Dewar and coworkers [12] reconsidered the core-core repulsion term. They proposed to add to the empirical formula (33a) a second term of the form

$$\frac{Z'_A Z'_B}{R_{AB}} \left[\sum_k \left(a_{kA}\, e^{-b_{kA}(R_{AB} - c_{kA})^2} + a_{kB}\, e^{-b_{kB}(R_{AB} - c_{kB})^2} \right) \right] \quad (34)$$

In the so-called AM1 method, the sum of the k gaussian repulsive functions is extended to 4 terms, so that 12 new parameters per atom are introduced. The improvement is effective, but the hydrogen bonds are still not very well represented as exemplified by figure 1 in the case of the water dimer.

Finally a complete and systematic reparametrization of AM1 has been performed recently by Stewart [13] who used computer function optimization techniques and a huge set of test molecules to determine the best values of the parameters. He limited the gaussian repulsive function (34) of the core-core repulsion to two terms per atom, but treated all the atomic quantities as parameters so that the number of adjustable parameters per element amounts to 18 except for hydrogen, which has 11 parameters.

The results are very convincing, especially in the case of hydrogen bonded systems and hypervalent compounds so that this new method known under the name of PM3 (*parametric method 3*) may well be the best semi-empirical method available for the time being. Nevertheless the table of optimized parameters exhibits some intriguing feactures for those who are familiar with some degree of regularity along a row of the periodic table. Besides the author found two different stable sets of parameters for iodine, so that the PM3 parameters look more like *ad hoc* quantities than semi-empirically evaluated atomic physical data.

It must be pointed out that in all the methods described in this section the overlap integrals are neglected.

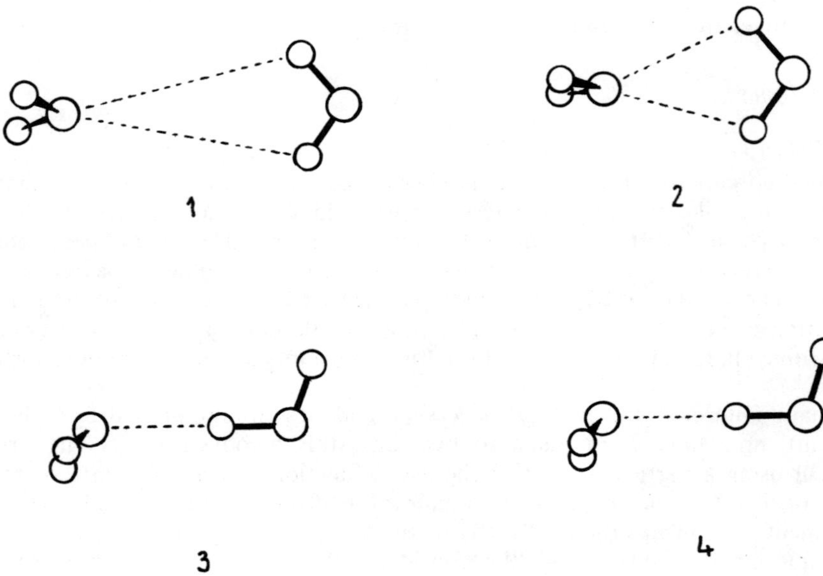

figure 1 The water dimer.
(1) : MNDO, $R_{O...H} = 3.43$ Å, dimerization energy : 0.98 kcal/mol
(2) : AM1, $R_{O...H} = 2.16$ Å, dimerization energy : 5.01 kcal/mol
(3) : PM3, $R_{O...H} = 1.81$ Å, dimerization energy : 3.52 kcal/mol
(4) 6-31G*, $R_{O...H} = 1.79$ Å, (experimental dimerization energy : 5.0 kcal/mol)

2.6 THE CNDO LEVEL OF APPROXIMATION [14,7]

The Complete Neglect of Differential Overlap is located at the lower end of the approximation scale. As indicated in the title, the basic assumption of this level is that all the $(\mu\nu \mid \lambda\eta)$ integrals are assumed to be zero unless $\mu = \nu$ and $\lambda = \eta$.

This drastic simplification introduces spurious non-invariance of the electron-electron repulsion energy in a change of axes. Let us consider, for instance, a bicentric integral of the form $(x^A x^A \mid y^B y^B)$ in which x^A and y^B denote two p orbitals (p_x and p_y respectively) centered on atoms A and B. Let us consider a change in the orientation of the x and y

axes so that in the new frame the p orbitals become x'^A, y'^B related to the previous ones by the following transformations :

$$x' = \frac{1}{\sqrt{2}}(x + y) \quad \text{and} \quad y' = \frac{1}{\sqrt{2}}(x - y)$$

After eliminating the integrals which are zero by symmetry, one obtains

$$(x'^A x'^A \mid y'^B y'^B) = \frac{1}{4}\left[(x^A x^A \mid x^A x^A) + (x^A x^A \mid y^B y^B) + (y^A y^A \mid x^B x^B) + (y^A y^A \mid y^B y^B)\right] - (x^A y^A \mid x^B y^B)$$

At the NDDO level, all the integrals in the left hand side are considered, so that the total interaction energy in invariant. At the CNDO level the last integral is neglected so that the interaction energy is no longer invariant. To overcome this difficulty, Pople [14] had to assume that the two center integrals are orientation independant. Therefore, the two center integrals take the simple form :

$$(\mu\mu \mid \lambda\lambda) = \gamma_{AB} \tag{35}$$

where γ_{AB} only depends on the nature of atoms A and B and on their distance. This quantity is computed with the help of the valence s orbitals of atoms A and B.

In accordance with the basic assumptions, the only non zero Slater-Condon parameter is F^0 which is also denoted by γ_{AA}. It permits the evaluation of the $U_{\mu\mu}$ integrals from the experimental value of ½ $(I_\mu + A_\mu)$.

The integrals $V'^B_{\mu\mu}$ are obtained in the electrostatic approximation by :

$$V'^B_{\mu\mu} = \gamma_{AB} Z_B \tag{36}$$

and the corresponding off-diagonal terms are neglected. The other quantities are defined by the simple empirical equations

$$H^c_{\mu\lambda} = \beta^0_{AB} S_{\mu\lambda} \tag{37}$$

where β^0_{AB} is a parameter and $S_{\mu\lambda}$ the overlap integral and

$$E^c_{AB} = \frac{Z'_A Z'_B}{R_{AB}} \tag{38}$$

Therefore the Hartree-Fock-Roothaan equations (16) reduce to :

$$F_{\mu\mu} = -\frac{1}{2}(I_\mu + A_\mu) + \left[P_{AA} - Z'_A - \frac{1}{2}(P_{\mu\mu} - 1)\right]\gamma_{AA} + \sum_{B \neq A}(P_{BB} - Z'_B)\gamma_{AB} \tag{39}$$

$$F_{\mu\nu} = -\frac{1}{2} P_{\mu\nu}\gamma_{AA} \tag{40}$$

$$F_{\mu\lambda} = \beta^0_{AB} S_{\mu\lambda} - \frac{1}{2} P_{\mu\lambda}\gamma_{AB} \tag{41}$$

in which

$$P_{AA} = \sum_{\mu \in A} P_{\mu\mu}$$

denotes the electronic population of the atom.

In the SCF calculations the overlap integrals are neglected.The few parameters which are needed in the CNDO method (mainly the β^0_{AB}) have been adjusted by reference to *ab initio* computations so that this method usually gives fair electron distributions and the physical quantities which depend on this property (such as the dipole moment) are, in general, well reproduced. In turn the energies cannot be used safely for evaluating enthalpies of formation.

The current CNDO method (also called CNDO/2) is the prototype of the semi-empirical methods. Many molecular systems have been studied by this technique for almost 25 years.

2.6.1 The intermediate level of approximation. It rapidly became obvious that the CNDO assumptions are too drastic to allow the computation of some important physical constants. In particular the one center integrals corresponding to a *monoatomic differential overlap* of the kind $(\mu\nu \mid \mu\nu)$ play a major rôle in the difference of interaction energy of two electrons when they have the same spin and when they have opposite spin. It is clear that this improvement is not sufficient to restore the invariance of electron interaction energy so that the basic features of INDO are similar to CNDO. The original INDO method [15] is still very often used to evaluate some molecular magnetic properties such as NMR parameters.

Besides this original method, several versions of a modified (MINDO) version have been proposed. The latest one : MINDO/3 [16] uses the same principles as MNDO at the INDO level to evaluate the integrals and, in fact, this methodology was first tested at the INDO level. These methods which are quite convenient to compute binding energies, heats of formation and molecular geometries also fail in describing hydrogen bonded systems properly.

A SCF method at the INDO level in which the overlap integrals are not neglected in the solution of the Hartree-Fock-Roothaan equations has been proposed under the name of SINDO [17]. More recently a novel parametrization, SINDO/1 [18] has been tested quite successfully on a large number of compounds containing first row elements, especially a series of molecules containing lithium, beryllium and boron atoms. The results compare favourably with those obtained with MINDO/3, MNDO and even STO-3G.

2.7 SEMI-EMPIRICAL STUDIES OF SPECIAL SYSTEMS

2.7.1 Open shell systems. The UHF computation of open shell systems can be readily performed by using the Berthier-Pople-Nesbet [6] set of equations instead of Hartree-Fock-Roothaan's and the same integrals as in the closed shell systems.

2.7.2 Excited states. Various computational procedures including configuration interaction have been proposed at various levels, to compute transition energies and transition moments. At the CNDO level a modified version for spectroscopy, CNDO/S [19] has been applied successfully to several systems.

2.7.3 Transition metal compounds. A set of parameters for some transition metal atoms is available for INDO [20].

2.8 APPLICATION OF SEMI-EMPIRICAL METHODS TO CHEMICAL PROBLEMS

The gain in computation time and on memory size which the semi-empirical methods allow compared to the *ab-initio* techniques, even with limited basis sets, is quite noticeable. In table 2 this statement is illustrated by the number of two electron integrals needed for the computation of propane.

TABLE 2. Two electrons integrals for propane

Integrals	**ab-initio, minimum basis**	**MNDO**	**INDO**	**CNDO**
1 center	368	173	26	11
2 centers	6652	568	55	55
3 and 4 centers	31206	0	0	0
Total	38226	741	81	66

Therefore, the computation of a rather large molecule (having nearly 100 atoms) which is an enormous task at the *ab-initio* level becomes quite feasible with a semi-empirical method. The problem is the degree of confidence which can be granted to the results. As mentioned briefly during the presentation of the various methods, their qualities and their failings are now well-known and the knowledge is supported by a large number of examples.

Several integrated computer packages offer the possibility of easily applying any of the current semi-empirical methods to a problem. The first example is probably found in GEOMO [21]. Nowadays, MOPAC [22] is also widely used and some fully interactive packages using the recent facilities offered by molecular graphics are appearing. CHIMISTE [23], for instance, belongs to this category.

3.0 The extended Hückel method

By introducing the *net electric charge* of an atom, say A

$$Q_A = Z'_A - P_{AA}$$

expressions (39), (40), (41) of the Hartree-Fock matrix obtained at the CNDO level can be written :

$$F_{\mu\mu} = -\frac{1}{2}(I_\mu + A_\mu) - Q_A \gamma_{AA} - \frac{1}{2}(P_{\mu\mu} - 1)\gamma_{AA} - \sum_{B \neq A} Q_B \gamma_{AB} \quad (42)$$

$$F_{\mu\lambda} = \beta^0_{AB} S_{\mu\lambda} - \frac{1}{2} P_{\mu\lambda} \gamma_{AB} \quad (43)$$

These expressions can be further simplified by neglecting $P_{\mu\mu} - 1$ and γ_{AB} . Then, if one writes:

$$-\frac{1}{2}(I_\mu + A_\mu) = \alpha_\mu$$

and $\gamma_{AA} = \gamma_\mu$; $\beta^0_{AB} = \beta_{\mu\nu}$, which again may vary with the nature of the atomic orbitals, one obtains :

$$F_{\mu\mu} = \alpha_\mu - \gamma_\mu Q_A \quad (44)$$

$$F_{\mu\lambda} = \beta_{\mu\lambda} S_{\mu\lambda} \quad (45)$$

which are the basic expressions of the Iterative Extended Hückel Hamiltonian [24]. It departs from the original extended Hückel method [25] by the dependance of the $F_{\mu\mu}$ terms on the electric charge of the atoms, which makes the computation iterative like in other SCF computations.

Starting from the equation :

$$-\frac{1}{2}(I_\mu + A_\mu) = I_\mu + \frac{1}{2}\gamma_{AA} \quad (46)$$

it is usual to assimilate $-\alpha_\mu$ to the *Valence State Ionization Potential.*

γ_μ is treated as a parameter and $\beta_{\mu\lambda}$ is very often evaluated by the traditional approximation [25] :

$$\beta_{\mu\lambda} = 1.75 \frac{\alpha_\mu + \alpha_\lambda}{2} S_{\mu\lambda} \tag{47}$$

The Extended Hückel Method is very fast and is still appreciated by chemists who want to have an insight into the hierarchy of molecular orbitals at fixed geometry in order to discuss the molecular properties in view of the nature and order of the occupied orbitals [26].

4.0 Empirical methods. Molecular mechanics [27]

4.1 MOLECULAR FORCE FIELDS

The energy of a molecule depends on the relative positions of its atoms. Starting from an equilibrium geometry which is defined as a minimum of the molecular energy with respect to all the distances and angles defining this geometry, the molecular energy V can be expanded in powers of the variations Δx_i of the coordinates. One may therefore write :

$$V = V_o + \frac{1}{2} \sum_i \left(\frac{\partial^2 V}{\partial x_i^2} \right)_e \Delta_x^2 + \frac{1}{2} \sum_{(ij)} \left(\frac{\partial^2 V}{\partial x_i \partial x_j} \right)_e \Delta x_i \Delta x_j + \ldots \tag{48}$$

This equation, in which the subscript e refers to an equilibrium geometry, defines the *molecular force field* which is said to be *harmonic* when it is limited to the terms of degree 2 in Δx_i as in equation (48).

The second derivatives define the *force constants* :

$$f_{ij} = \frac{\partial^2 V}{\partial x_i \partial x_j} \tag{49}$$

4.1.1 The spectroscopic force fields. When all the force constant are known, it is rather easy to calculate the frequencies of vibration [28]. In the absence of degeneracy, a non-linear molecule of n atoms has $3n$ - 6 different modes of vibration *i.e.* $3n$ - 6 different frequencies of vibration which may be observed either in infra-red or in Raman spectroscopy. Nevertheless, some of these modes may be inactive in both spectroscopies or may not be observed and there are more than $3n$ - 6 force constants in equations (48). The consequence of this is the impossibility of determining the whole set of force constants from experimental data. Practically, one simplifies the molecular force field by retaining only the most important force constants.

In a system of *internal coordinates* which are the interatomic distances between bonded atoms r_i , the valence angles θ_{ij} and the dihedral angles φ_{ijk} (see figure 2) the off-diagonal force constants may be neglected. One then obtains the *valence force field* which can be written, if V_o is taken as origin of energies :

$$V = \frac{1}{2}\sum_i f_{r,i}(r_i - r_i^e)^2 + \frac{1}{2}\sum_{(ij)} f_{\theta,ij}(\theta_{ij} - \theta_{ij}^e)^2 + \frac{1}{2}\sum_{(ijk)} f_{\varphi,ijk}(\varphi_{ijk} - \varphi_{ijk}^e)^2 \qquad (50)$$

in which the superscript *e* refers to the equilibrium quantity.

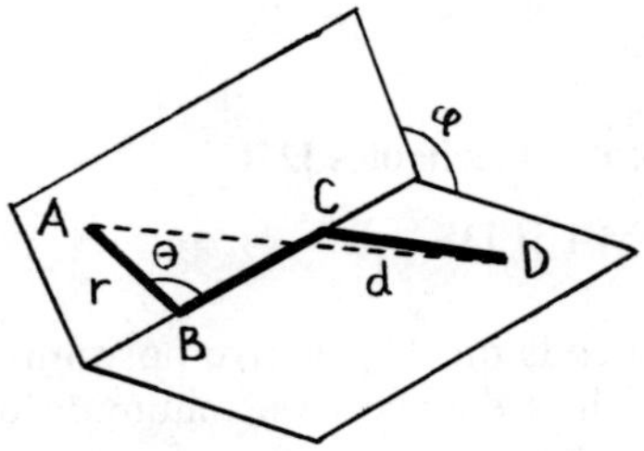

figure 2

The force constants which appear in these equations are characteristic of a given bond, angle or dihedral angle in a given molecule. In general, they cannot be used to describe a different molecule even within the same chemical family. The valence force field is said to be *non-transferable.*

The transferability can be achieved by adding some well chosen off-diagonal terms to equation (50). The *Urey-Bradley* force field [29] consists in adding to the valence force field extra terms which couple the displacements of two atoms bonded to the same center (1-3 interactions). This force field was later extended by Shimanouchi [30] and a slighly different strategy in choosing the off-diagonal terms to be included led to a *generalized valence force field* which also contains transferable quantities.

4.1.2 The force field in molecular mechanics. The basis of molecular mechanics is the use of an *a priori* reliable force field to compute structural and energetic quantities on molecules. Most of the molecular mechanics projects are based on extended valence force fields which are close to Shimanouchi-Urey-Braley's. The general expression may be written as a sum of various contributions :

$$V_{MM} = V(r) + V(\theta) + V(\phi) + V(d) + \ \dots \qquad (51)$$

in which $V(r)$ represents the terms containing the variation of bond lengths, $V(\theta)$ the valence angle deformations, $V(\phi)$ the dihedral angles and $V(d)$ the interactions between «non-bonded» atoms as indicated on figure 2.

4.2 PARAMETRIZATION OF THE MOLECULAR MECHANICS FORCE FIELD

4.2.1 Terms depending on bond lengths and bond angles. In the harmonic approximations these terms may be written

$$V(r) = \frac{1}{2} \sum_i k_{r,i} (r_i - r_i^o)^2 \tag{52}$$

and

$$V(\theta) = \frac{1}{2} \sum_{(ij)} k_{\theta,ij} (\theta_{ij} - \theta_{ij}^o)^2 \tag{53}$$

in which the sums refers to bonds i, j ... . One notices that the equilibrium bond lengths r_i^e and bond angles θ_{ij}^e which appear in equation (50) are replaced here by *reference bond lengths* r_i^o *and bond angles* θ_{ij}^o which are usually different from the equilibrium values since equilibrium corresponds to a minimum of the potential energy with respect to all the contributions.

The harmonic approximation is known to be rather poor, especially for the bond lengths variations which are better represented by a Morse potential. Equation (52) can be greatly improved by adding a cubic term leading to the general expression

$$V(r) = \frac{1}{2} \sum_i \left[k_{i,r} (r_i - r_i^o)^2 + k'_{i,r} (r_i - r_i^o)^3 \right] \tag{54}$$

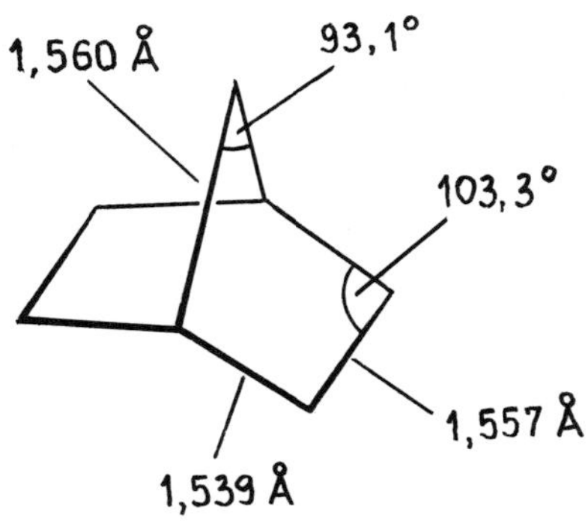

figure 3 The carbon-carbon bonds in adamantane

The necessity of including off-diagonal terms appears clearly in some strained systems in which a bond angle smaller than the usual value induces a lengthening of the corresponding bonds. A typical example is found in adamantane (figure 3). This effect is represented by a crossed term in the potential energy :

$$V(r,\theta) = \frac{1}{2} \sum_{ij} k_{r\theta,ij} (r_i - r_i^o + r_j - r_j^o) (\theta_{ij} - \theta_{ij}^o) \quad (55)$$

Similarly, in planar systems containing π bonds, the expression of $V(\theta)$ has to account for the anisotropy of the motion since the in-plane deformations, which do not greatly modify the energy of the π system are usually easier than the out-of-plane deformations.

The parameters which occur in all these contributions are determined on the basis of experimental data mainly vibrational spectra and molecular geometries given by crystallography.

4.2.2 The torsional potential energy. The torsion of a molecule around a bond may take a large amplitude and the harmonic approximation is no longer valid in many cases. The mathematical expression of $V(\phi)$ is usually expanded in the form of a Fourier series

$$V(\varphi) = \frac{1}{2} \sum_{i} [V_{1i}(1 + \cos \varphi_i) + V_{2i}(1 + \cos 2\varphi_i) + \ldots + V_{ki}(1 + \cos k\varphi_i) + \ldots] \quad (56)$$

The V_{ki} parameters are adjusted in order to fit experimental data on internal rotations. Most of them come from spectroscopic measurements (vibrational and rotational spectra, NMR, ...). These parameters express the hinderance to internal rotation which may be analysed in several ways. Among the possible contributions one finds the interactions between non-bonded atoms which have to be taken into account anyway. Therefore, the parameters of equation (56) cannot be determined independendly of the other contributions.

4.2.3 The special case of conjugated systems. In a conjugated molecule, such as butadiene, no bond can be compared with a standard single or double bond. The bond-length and all the constants characterizing these bonds in the molecular deformation vary according to the nature of the bond which is defined by the π system. A bond order can be defined by simple Hückel calculation and the bond parameters can be related to this quantity by empirical formulae which are calibrated on some selected standard molecules.

4.2.4 Interactions between non-bonded atoms. These interactions cannot be ignored because they are responsible for the steric hinderance which plays a major rôle in the equilibrium geometry of some molecules. The analysis of such interactions may be more or less refined. Usually, it is carried on by assuming that these interactions have the same behaviour as the intermolecular interactions.

The repulsion-dispersion term is assumed to be expanded in atom-atom contributions. The simplest model is the hard sphere model in which the interaction energy is zero for all interatomic distances greater than the sum of the radii of the interacting atoms and becomes infinite when this distance equals this sum. By analogy with intermolecular potentials, some more elaborate atom-atom contributions to the repulsion dispersion term are often introduced. The reference expressions are :

- The «n - 6» Lennard-Jones potential [31]

$$V(d) = \frac{n\varepsilon}{n-6}\left[\frac{6}{n}\left(\frac{d^o}{d}\right)^n - \left(\frac{d^o}{d}\right)^6\right] \tag{57}$$

with $n=12$ or $n=9$

- The «exp - 6» Buckingham potential [32]

$$V(d) = \varepsilon\left[A\,e^{-Bd/d^o} - C\left(\frac{d^o}{d}\right)^6\right] \tag{58}$$

These potentials are isotropic. Some degree of anisotropy can easilybe introduced by a small displacement of the origin which is no longer at the nucleus of the atom of interest, or by adding dummy atoms in the vicinity of the real ones.

The parameters in these potentials are often extracted from crystallographic data on selected molecules since a proper estimate of the intermolecular energies is necessary to reproduce the structure and some properties of a molecular crystal.

The electrostatic term arises from the polarity of the chemical bonds which may be substantial, especially in molecules containing heteroatoms.

In many cases this polarity is represented by partial charges q_i on atoms and the electrostatic energy is expressed in atomic units by :

$$V_{e\ell} = \sum_{(i,j)} \frac{q_i\, q_j}{\varepsilon\, d_{ij}} \tag{59}$$

in which the dielectric constant ε may be taken greater than unity to simulate some environment effects. This picture is rather naive. A better representation of the electrostatic interaction can be achieved by considering local multipole expansions of the charge distribution (net charge + dipole + quadrupole + ...) [33].

These quantities can hardly be extracted from experimental data. They can be determined on the basis of accurate *ab initio* computations on small model molecules, selected in such a way that the assumption of transferability appears reasonable.

The induction term is the consequence of the distortion of the electronic distribution under the influence of the electric field created by the other molecules (or the other parts of the molecule of interest), which generates induced electric moments. It can be included in the force-field if the electronic polarizability of the molecule is decomposed into local contributions. These distributed molecular polarizabilities [33,34] have to be computed theoretically. They are sometimes omitted, and the permanent charge distribution is modified in order to include an average value of the induction term in the electrostatic energy.

4.2.5 Hydrogen bonding. Although the electrostatic term represents a major part of a H bond, an accurate representation of this special kind of interaction may deserve a special treatment. In most cases, extra contributions are added to the corresponding repulsion-dispersion electrostatic and occasionally induction terms. Various analytical expressions have been proposed. Among the most widely used empirical formulae there is a 10-12 potential of the general form [35] :

$$V_{HB}(d) = A\,d^{-12} + B\,d^{-10} \tag{60}$$

but some more elaborate expressions have been proposed [36]. The parameters in these equations have to be fitted to the results of accurate quantum chemical computations.

As mentioned above, the specific V_{HB} contribution is added to the other terms representing the interaction between non-bonded atoms, so that the values of its parameters depends on the detailed expressions and on the parameters of these terms.

4.3 MOLECULAR MECHANICAL COMPUTATIONS

Any force field appears as a more or less sophisticated analytical equation which relates the geometry dependent part of the molecular energy to internal atomic coordinates. It is therefore very easy to compute the first and second derivatives of energy with respect to these coordinates and initialize a very efficient geometry optimization of the molecule.

Concomitantly, it is possible to compute the frequencies of the normal modes of vibration and to simulate infra-red and Raman spectra.

Molecular mechanics provide us with a very direct may to compute the departure from simple additivity rules in the evaluation of heats of formation of molecular compounds. It is therefore possible to assume that the heat of formation ΔH_f^o of a compound obeys the simple equation :

$$\Delta H_f^o = V_{MM} + E_b \tag{61}$$

in which E_b is the bond energy obtained as a sum of bond increments.

In the same way a strain energy can be defined as the difference between ΔH_f^o and a «strainless bond energy» which is evaluated like the bond energy as a sum of increments obtained in such a way that the strain energy is zero in the case of small molecules considered to be strainless.

Molecular computations on large molecules, and in particular biological macromolecules (nucleic acids, proteins) are all based on some kind of molecular mechanics force field. Owing to the enormous number of degrees of freedom which exist is such systems, the force field has to be simplified in order to reduce the number of variables which enter the energy expression. In many cases, only the torsional angles are considered, but the interaction between non-bonded atoms and hydrogen bonds play a very important rôle in these complex molecules. This force field can be used not only to compute the stability of the various molecular configurations but also to compute statistical averages on these configurations leading to an evaluation of the configurational entropy and the corresponding free energies. Similarly molecular dynamical simulations can be performed on the basis of a molecular mechanics force field leading to the representation of time dependant processes.

The prototype of Molecular Mechanics packages is offered by Allinger's MM programmes. MM2 [37] and its modification for treating π bonded systems, MMP2 are very well known. BIGSTRN-2 [38] is an exemple of an integrated system offering the choice between 5 different force fields. Several other packages, intended for treating special problems, are also available such as ECEPP [39] for linear polypeptides.

5.0 Conclusion. Molecular Modelling

All the methods described in this paper contain some parametrized empirical formulae to reduce the time needed to compute the energy of complex systems such as molecules or macromolecules.

This empiricism may appear at the interatomic level. One then deals with molecular mechanics which may be quite appropriate to describe any kind of property bound to the molecular geometry.

It may also be limited to the electronic level, leading to more general methodology based on the semi-empirical quantum chemical methods which are meant to account for the electronic distribution in the molecule. In addition to the molecular geometry and the bonding energy, these methods provide us with far more detailed information on the electronic structure. Besides, they are able to represent any kind of non conventionnally bound molecular structure, such as transition states or reaction intermediates which can hardly be described safely by molecular mechanics. Generally speaking, the analysis of any kind of chemical process involving an electronic displacement requires a more or less realistic description of the system at the electronic level which may be achieved by the proper semi-empirical approach. Conversely, a semi-empirical computation takes a considerably longer time than molecular mechanics would need for the same molecule.

For the time being, to the three classes of computational methods correspond three sizes of molecule :

- small molecules (up to 10 or 20 atoms) are within the reach of elaborate *ab initio* computations
- medium size systems (up to about 100 atoms) can be described reasonably well by semi-empirical methods, which can either be tested or even parametrized on a smaller reference model previously computed at the *ab initio* model
- large systems come within the scope of molecular mechanics. Again, the force field may be adjusted on a model molecule in order to make the method compatible with the previous one.

A continuity then appears from the most fundamental rigorous computations, to the most approximate ones. Obviously, neither the empirical nor the semi-empirical approach can claim to provide an accurate and definite answer to a chemical or a biochemical problem. Nevertheless, they provide us with *molecular modelling* techniques which fill the gap between the purely qualitative picture of molecules used by the majority of chemists, and the highly elaborate description of systems which are still, in the majority of cases, too small to be considered as a realistic model of the chemist's molecule. Molecular modelling is therefore becoming increasingly popular among the experimentalists, and especially the organic chemists.

6.0 Appendix. First and second derivatives of energy in the semi-empirical methods

The simplification introduced in semi-empirical methodology makes the computation of the first derivatives of energy with respect to atomic coordinates considerably easier than in an *ab-initio* computation, as has been recognized for several years [39] and applied to perform fast and efficient geometry optimization [40]. The computation of the second derivatives is also feasible [41] enabling the force constants to be determined and potential energy surfaces to be characterized.

6.1 GENERAL FORMALISM

The formalism presented here was first proposed by Bratos [42]. We shall develop it here in the case of closed shell systems and by using the following notation

C_i denotes an eigenvector of the Hartree-Fock matrix and $C_i^{\dagger}$ is the transposed vector of C_i

e_i is the corresponding eigenvalue.

C is the square matrix formed by the set of eigenvectors

H^c is the matrix corresponding to the core Hamiltonian

G is the electron-electron repulsion operator whose general term is written here

$$G_{\mu\nu\lambda\sigma} = <\mu\nu \mid \lambda\sigma> - \frac{1}{2} <\mu\lambda \mid \nu\sigma> \tag{62}$$

Therefore, the molecular energy is written

$$E = 2\sum_{i}^{occ} C_i^{+} (H^c + \sum_{j}^{occ} C_j^{+} G\, C_j)\, C_i + \sum_{(AB)} E_{AB}^{c} \tag{63}$$

and the Hartree-Fock equation

$$F\, C_i = e_i\, S\, C_i \tag{64}$$

with

$$F = H^c + 2\sum_{j} C_j^{+} G\, C_j \tag{65}$$

Let us consider the first derivative of E with respect to the coordinate q_k and let us use for any quantity the simplified notation

$$\partial_k A = \frac{\partial A}{\partial q_k} \tag{66}$$

Therefore

$$\begin{aligned}\partial_k E = {} & 2\sum_i^{occ} C_i^+ (\partial_k H^c + \sum_j^{occ} C_j^+ \partial_k G\, C_j)\, C_i + 4\sum_i^{occ} \partial_k C_i^+ (H^c + 2\sum_j^{occ} C_j^+ G C_j)\, C_i \\ & + \sum_{(AB)} \partial_k E_{AB}\end{aligned} \tag{67}$$

This expression can be simplified by replacing F C_i which appears in its right hand side by $\varepsilon_i S C_i$ as given by (64) and (65) and by making use of the equation:

$$2\, \partial_k C_i^+ S\, C_i + C_i^+ \partial_k S\, C_i = 0 \tag{68}$$

which is obtained by deriving the normalization condition:

$$C_i^+ S\, C_i = 1 \tag{69}$$

Therefore (67) reduces to :

$$\partial_k E = 2\sum_i^{occ} C_i^+ (\partial_k H^c - e_i \partial_k S + \sum_j^{occ} C_j^+ \partial_k G C_j)\, C_i + \sum_{(AB)} \partial_k E_{AB}^C \tag{70}$$

This expression shows that the first derivative of energy can be computed with the help of the first derivatives of the H, S and G matrices, which are easy to obtain in the semi-empirical formalism.

The second derivatives are obtained by direct derivation of equations (70).

$$\begin{aligned}\partial_{k\ell}^2 E = {} & 2\sum_i^{occ} C_i^+ (\partial_{k\ell}^2 H^c - e_i \partial_{k\ell}^2 S - \partial_\ell e_i \partial_k S + \sum_j^{occ} C_j^+ \partial_{k\ell}^2 G\, C_j)\, C_i \\ & + 4\sum_i^{occ} \partial_\ell C_i^+ (\partial_k H^c - e_i \partial_k S + 2\sum_j C_j^+ \partial_k G\, C_j)\, C_i + \sum_{(AB)} \partial_{k\ell}^2 E_{AB}^c\end{aligned} \tag{71}$$

The first derivatives of the eigenvectors, which did not occur in equations (70) now appear and need to be computed. An efficient algorithm can be devised by introducing a matrix which we shall denote by D^k and whose s,t element is defined by :

$$D^k_{st} = C^+_t \, S \, \partial_k C_s \tag{72}$$

in which s and t denote any eigenvector of the F matrix, regardless of the fact that it corresponds to an occupied or a virtual orbital.

After left multiplication of D by C and taking into account the normalization condition (69) one obtains :

$$\partial_k C = C \, D^k \tag{73}$$

The diagonal elements of D^k are related to the derivative of S by equation (68)

$$D^k_{ss} = - \frac{1}{2} C^+_s \, \partial_k S \, C_s \tag{74}$$

The off-diagonal elements arise from the derivation of the Hartree-Fock equation (64)

$$\partial_k F \, C_s + F \, \partial_k C_s = \partial_k e_s \, S \, C_s + e_s \, \partial_k S \, C_s + e_s \, S \, \partial_k C_s \tag{75}$$

This equation is left multiplied by C^+_t $(t \neq s)$ and is rearranged owing to the hermiticity of F into :

$$C^+_t \, \partial_k F \, C_s = e_s \, C^+_t \, \partial_k S \, C_s + (e_s - e_t) \, C^+_t \, S \, \partial_k C_s$$

Hence

$$D^k_{ts} = C^+_t \, \frac{\partial_k F - e_s \, \partial_k S}{e_s - e_t} \, C_s \tag{76}$$

Similarly, the derivatives of the eigenvalues e_i are obtained by derivation of

$$e_i = C^+_i \, F \, C_i$$

giving :

$$\partial_k e_i = C^+_i \, (\partial_k F - e_i \, \partial_k S) \, C_i \tag{77}$$

This expression has been obtained after taking into account the normalization condition and the hermiticity of F.

One notices that the derivatives of C_i and e_i contain the derivatives of F which again contain the derivatives of the eigenvectors. Therefore, this part of the computation can only be performed iteratively and is expected to take a much longer time than the computation of the first derivatives.

6.2 APPLICATION TO SEMI-EMPIRICAL METHODS

The derivatives of the F matrix elements are directly deduced from the definition of these elements given in equations (16), (17), (18).

$$\begin{aligned}\partial_k F_{\mu\nu} = & \sum_{B \neq A} \partial_k V'^{B}_{\mu\nu} + \sum_{B} \sum_{\lambda,\sigma}^{B} \partial_k P_{\eta\sigma} \left[(\mu\nu \mid \eta\sigma) - \frac{1}{2} (\mu\eta \mid \nu\sigma) \right] \\ & + \sum_{B} \sum_{\lambda,\sigma}^{B} P_{\eta\sigma} \left[\partial_k (\mu\nu \mid \eta\sigma) - \frac{1}{2} \partial_k (\mu\eta \mid \nu\sigma) \right]\end{aligned} \tag{78}$$

for μ and ν on the same atom (A), including $\mu = \nu$.

In this expression we use $\sum\limits_{\lambda,\sigma}^{B}$ for $\sum\limits_{\lambda \in B} \sum\limits_{\sigma \in B}$.

When the two orbitals μ and λ are centred on two different atoms equation (18) gives:

$$\begin{aligned}\partial_k F_{\mu\lambda} = & \partial_k H'^{C}_{\mu\lambda} + \sum_{\eta \neq \mu,\lambda} \sum_{\sigma \neq \mu,\lambda} \partial_k P_{\eta\sigma} \left[(\mu\lambda \mid \eta\sigma) - \frac{1}{2} (\mu\eta \mid \lambda\sigma) \right] \\ & + \sum_{\eta \neq \mu,\lambda} \sum_{\sigma \neq \mu,\lambda} P_{\eta\sigma} \left[\partial_k (\mu\lambda \mid \eta\sigma) - \frac{1}{2} \partial_k (\mu\eta \mid \lambda\sigma) \right]\end{aligned} \tag{79}$$

These expressions are further simplified within one of the various levels of approximation. They all consider only one and two center electron interaction integrals, so that the only non-zero electron interaction integrals are two center integrals when the coordinate q_k modifies the relative position of the centers.

The derivatives of the matrix elements $P_{\eta\sigma}$ are obtained from the definition of the density matrix (7) and the derivative of the C matrix (73) :

$$\partial_k P_{\eta\sigma} = 2 \sum_{i}^{occ} \sum_{j}^{unocc} B_{ij}^{k} (c_{\eta i} c_{\sigma j} + c_{\eta j} c_{\sigma i}) \tag{80}$$

in which $\sum\limits^{occ}$ stands for a sum on all the occupied orbitals, and similarly for all the unoccupied ones, and

$$B_{ij}^{k} = \sum_{\mu} \sum_{\lambda} \frac{c_{\mu i} c_{\nu j}}{e_j - e_i} \partial_k F_{\mu\lambda} \tag{81}$$

These equations, together with the previous one, clearly show the necessity of an iterative computation of the derivatives of F and of P. This computation may be initialised by assuming $\partial_k F_{\mu\lambda} = 0$.

The matrix H^c appear in the expression of the second derivative of E (71). The two center elements of this matrix are $H'^{c}_{\mu\lambda}$ whose first and second derivatives are easy to compute. All the methods assume that these terms are proportional to $S_{\mu\lambda}$ whose first and second derivatives can be obtained analytically. When the S matrix is assumed to be unity in the normalization condition, the derivatives of S disappear in the expression of the second derivatives of E. The first and second derivatives of G can also be obtained analytically by deriving the exact or the approximate formulae used to compute the bicentric two electron integrals. Therefore the problem of computing the first and second derivatives of energy is solved.

6.3 FIRST DERIVATIVES OF THE DIPOLE MOMENT

The intensity of infra-red absorption lines depends on the derivatives of the dipole moment:

$$\mu = \sum_{A} Z_A \mathrm{r}_A - 2\sum_{i}^{occ} \mathrm{C}_i^+ \mathrm{r}\, \mathrm{C}_i$$

The derivatives of this vector are :

$$\partial_k \mu = \sum_{A} Z_A\, \partial_k \mathrm{r}_A - 2\sum_{i}^{occ} (\mathrm{C}_i^+ \partial_k \mathrm{r}\, \mathrm{C}_i + 2\, \partial_k \mathrm{C}_i^+ \mathrm{r}\, \mathrm{C}_i) \tag{82}$$

By using the usual conventions of the semi-empirical methods, this is expanded as

$$\partial_k \mu = \sum_{A} \left[Z'_A\, \partial_k \mathrm{r}_A - \sum_{\nu\in A} P_{\nu\nu}\, \partial_k \mathrm{r}_A - \sum_{\nu,\lambda\in A} \partial_k P_{\nu\lambda}\, r_{\nu\lambda} \right] \tag{83}$$

with :

$$r_{\nu\lambda} = \langle \nu \,|\, \mathrm{r} \,|\, \lambda \rangle$$

If one recalls that

$$\sum_{\nu\in A} P_{\nu\nu} = P_{AA}$$

and that the net electric charge on atom A is defined by :

$$Z'_A - P_{AA} = Q_A \tag{84}$$

one obtains :

$$\partial_k \mu = \sum_A \left[Q_A \, \partial_k \, r_A - \sum_{\nu,\lambda \in A} \partial_k \, P_{\nu\lambda} \, r_{\nu\lambda} \right] \tag{85}$$

The intensities of the Raman lines are related to the first derivatives of the polarizability tensor $\boldsymbol{\alpha}$. These quantities can also be computed by deriving the expression of $\boldsymbol{\alpha}$ given by the variational treatment of the polarizability [43].

7.0 References

[1] Pople J.A. and Beveridge D.L. (1970), *Approximate Molecular Orbital Theory*, J. Wiley, New York

[2] Roothaan C.C.J. (1951), *Rev. Mod. Phys.*, **23**, 69
See also in this book I.G. Csizmadia's paper, equations (38) to (46)

[3] Slater J.C. (1930), *Phys. Rev.*, **36**, 57 ; **42**, 33

[4] See for instance ref [1] p 24

[5] Slater J.C. (1960), *Quantum Theory of Atomic Structure*, Mc Graw-Hill, New York

[6] Berthier G. (1954), *C.R. Acad. Sci. Paris*, **238**, 91 ; *J. Chem. Phys.*, **51**, 363
Pople J.A. and Nesbet R.K. (1954), *J. Chem. Phys.*, **22**, 571

[7] Pople J.A. and Segal G.A. (1966), *J. Chem. Phys.*, **44**, 3298

[8] Oleari L., Di Sipio L. and De Michelis G. (1966), *Mol. Phys.*, **10**, 97

[9] Pople J.A., Santry D.P. and Segal G.A. (1965), *J. Chem. Phys.*, **43**, 5129

[10] Dewar M.J.S. and Thiel W. (1977), *J. Am. Chem. Soc.*, **99**, 4899

[11] Burstein K.Y. and Isaev A.N. (1984), *Theor. Chim. Acta*, **64**, 397

[12] Dewar M.J.S., Zoebisch E.G., Healy E.F. and Stewart J.J.P. (1985), *J. Am. Chem. Soc.*, **107**, 3902

[13] Stewart J.J.P. (1989), *J. Comp. Chem.*, **10**, 209, 221

[14] Pople J.A., Santry D.P. and Segal G.A. (1965), *J. Chem. Phys.*, **43**, 5129
Pople J.A., Segal G.A. (1965), *J. Chem. Phys.*, **43**, 5136

[15] Pople J.A., Beveridge D.L. and Dobosh P.A. (1967), *J. Chem. Phys.*, **47**, 2026

[16] Bingham R.C., Dewar M.J.S. and Lo D.H. (1975), *J. Am. Chem. Soc.*, **97**, 1285

[17] Coffey P. and Jug K. (1973), *J. Am. Chem. Soc.*, **95**, 5231

[18] Nanda D.N. and Jug K. (1980), *Theor. Chim. Acta*, **57**, 95
Jug K. and Nanda D.N. (1980), *Theor. Chim. Acta*, **57**, 107 ; 131

[19] Del Bene J. and Jaffé H.H. (1968), *J. Chem. Phys.*, **48**, 1807

[20] Bacon A.N. and Zerner M.C. (1979), *Theor. Chim. Acta*, **57**, 21

[21] Rinaldi D., *Q.C.P.E.* n° 290. A recent version of GEOMO which includes the computation of force constants and infra-red and Raman spectra and the simulation of solvent effects by the Self Consistent Reaction Field approach is now available under the name of GEOMOS, *Q.C.P.E.* n° 584.

[22] Stewart J.J.P., *Q.C.P.E.*, n° 581

[23] Martins Costa M.T.C., Rinaldi D. and Rivail J.L. (1989), *J. Chim. Phys.*, 1001

[24] Rein R., Fukuda N., Win H., Clarke G.A. and Harris F.E. (1966), *J. Chem. Phys.*, **45**, 4743

[25] Hoffmann R. (1963), *J. Chem. Phys.*, **39**, 1397

[26] Albright T.A., Brudett J.K. and Whangbo M.H. (1985), *Orbital Interaction in Chemistry*, J. Wiley, New York

[27] Burkert V. and Allinger N.L. (1982), *Molecular Mechanics*, American Chemical Society, Washington D.C.
See also Lomas J.S. (1986), *L'Actualité Chimique*, **Mai 1968**, 7

[28] Wilson E.B., Decius J.C. and Cross P.C. (1955), *Molecular Vibrations*, Mc Graw-Hill, New York

[29] Urey H.C. and Bradley C.A. (1931), *Phys. Rev.*, **38**, 1969

[30] Simanouti T. (1949), *J. Chem. Phys.*, **17**, 245

[31] Lennard-Jones J.E. (1924), *Proc. Roy. Soc. London*, **A 106**, 463

[32] Buckingham R.A. (1938), *Proc. Roy. Soc. London*, **A 168**, 264

[33] Claverie P. (1978), *Intermolecular Interactions : from Diatomics to Biopolymers*, p 69, B. Pullman ed., John Wiley, Chichester

[34] Stone A.J. (1981), *Chem. Phys. Letter*, **83**, 233
(1985), *Mol. Phys.*, **56**, 1065

[35] Mc Guire R.F., Momany E.A. and Scheraga H.A. (1972), *J. Phys. Chem.*, **76**, 375

[36] Allinger N.L., Chang S.H.M., Glaser D.H. and Hönig H. (1980), *Isr. J. Chem.*, **20**, 51.

[37] Allinger N.L. and Yuh Y.H., *Q.C.P.E.*, n° 395

[38] Iverson D.J. and Mislow K., *Q.C.P.E.*, n° 410

[39] Browman J. *et al.*, *Q.C.P.E.*, n° 454

[40] Rinaldi D. and Rivail J.L. (1972), *C. R. Acad. Sci. Paris* , **274 C**, 1664

[41] Rinaldi D., Ruiz Lopez M.F. and Rivail J.L. (1984), *J. Chim. Phys.*, **81**, 295

[42] Bratos S. (1958), *Coll. Int. CNRS (Paris)*, **82**, 287

[43] Martins Costa M.T., Rinaldi D. and Rivail J.L., *to be published*

NON-VISUAL MOLECULAR SHAPE ANALYSIS: SHAPE CHANGES IN ELECTRONIC EXCITATIONS AND CHEMICAL REACTIONS

Paul G. Mezey,
Department of Chemistry and
Department of Mathematics,
University of Saskatchewan,
Saskatoon, Canada, S7N 0W0

ABSTRACT. The analysis of molecular shapes is of major importance in the study of molecular structures and chemical reactivity. Computational analyses of shapes of molecular bodies are usually based on computer graphics techniques and on visual inspection. However, for accurate shape comparisons of a large number of molecules, visual inspection is not the best choice, and fully automatic, non-visual, algorithmic shape analysis by computer is more reliable. The shape group method (SGM) and various shape codes, derived from the 3D topological properties of molecular contour surfaces, provide efficient computational tools for non-visual shape analysis. Molecular shapes are subject to change when electronic excitation or a chemical reaction occurs. Several rules are derived, providing constraints on molecular shape changes in electronic excitations. The shape matrices, defined by the curvature properties of molecular contour surfaces, are applicable for a computer-based, non-visual evaluation of molecular shape and molecular similarity. The invariance domains of shape matrices within the dynamic shape space D suggest a general method for a dynamic description of molecular shapes.

C. Ögretir and I. G. Csizmadia (eds.),
Computational Advances in Organic Chemistry: Molecular Structure and Reactivity, 261–288.

1. INTRODUCTION

Computer representations of three-dimensional (3D) molecular shapes are often defined in terms of contour surfaces of electronic charge densities [1-5], electrostatic potentials [6-26], or Van der Waals surfaces generated by fused atomic spheres [27-42]. According to quantum mechanics, a molecule does not have a precisely defined surface, separating the molecule from the rest of the universe, nevertheless, formal molecular surfaces may be defined in an approximate sense. For example, an isodensity contour $G(a)$ for a suitable, constant value a of the electronic density $\rho(\mathbf{r})$ along the surface may serve as an approximate molecular surface. The representation of molecular shapes by fused sphere Van der Waals surfaces is simpler, and their computation and graphical display on a computer screen are tasks easily accomplished even for larger molecules. However, some of the artificial features of fused sphere Van der Waals surfaces, such as the local constancy of the distance of surface points from the nearest nucleus and the seamlines of interpenetrating spheres, are disadvantages.

Formal molecular surfaces play an important role in recent advances of computer graphics techniques applied to molecular problems, for example, in theoretical studies of electron densities, as well as in pharmaceutical drug design, toxicity analysis, and the development of new herbicides and pesticides [43-46]. In particular, the detection of common shape features within families of molecules showing similar biochemical activity is of major importance in modern pharmaceutical research. Molecular surfaces are among the tools used for the analysis of molecular similarity [47-55]. In molecular modeling, three-dimensional representations and the analysis of molecular shapes help the identification of reactive sites of molecules, the elucidation of conformational rearrangements, and an understanding of the interplay of molecular size and volume effects. Shape analysis of molecules provide a new perspective in the development of various aspects of quantitative structure-activity relationships (QSAR) [56-83].

In what follows below we shall place the emphasis on the electronic density and isodensity contour surfaces. Note, however, that the shape analysis techniques are applicable in identical form or with trivial modifications to many other shape descriptors, for example, to electrostatic potential contours and solvent accessible surfaces.

A formal molecular "body" F(a) can be defined by taking the set of all those points **r** where the electronic density $\rho(\mathbf{r})$ generated by the molecule is greater than a threshold value a ,

$$F(a) = \{ \mathbf{r} : \rho(\mathbf{r}) > a \} . \tag{1}$$

The above set theoretical expression reads as " F(a) is the collection of all points **r** such that $\rho(\mathbf{r}) > a$ ". Set F(a) is a *level set,* with respect to the threshold level a . By convention, a large *negative* charge means a large *positive* value for the density function $\rho(\mathbf{r})$; the range for density parameter a is $0 \leq a \leq \infty$.

If molecular bodies are represented by level sets F(a) of electronic charge densities, then their 3D shape properties can be described by their boundary surfaces, also called *isodensity contour surfaces,* denoted by G(a):

$$G(a) = \{ \mathbf{r} : \rho(\mathbf{r}) = a \}, \tag{2}$$

The isodensity surface G(a) is the collection of all those points **r** of the space where the value of the electronic charge density function $\rho(\mathbf{r})$ is equal to the threshold value a. Since the electronic density $\rho(\mathbf{r})$ is a continuous function, G(a) is a continuous surface.

The concept of isodensity contour surface is a physically motivated representation for an approximate molecular surface. By assuming a formal fixed nuclear configuration K , and a small value a for the electronic density, one may define a formal molecular body F(a) and its surface G(a). If an explicit reference to the nuclear configuration K is required, then the F(K,a) and G(K,a) notations may be used. The set G(K,a) may be regarded as the surface of the essential part of the molecule of a formal, fixed configuration K, that is, as an approximate molecular surface.

The assumptions of a fixed nuclear configuration K and a fixed density value a are severe limitations and the above model is much too restricted. Real, quantum mechanical molecules do not possess a fixed, rigid configuration K, neither have a precise boundary at some nonzero density value a. However, some of these restrictions can be released by taking a *family* of nuclear configurations K, as well as an *interval* of small threshold values a. For example, one may take *all configurations* K from some *configurational domain* C,

$$C = \{ K \} , \tag{3}$$

and *all threshold values* a from a *density interval* (a_0,a_1),

$$a_0 > a > a_1. \tag{4}$$

This leads to the generation of a *family* of formal molecular bodies,

$$F(C,a_0,a_1) = \{ F(K,a) : a_0 > a > a_1,\ K \in C \} , \tag{5}$$

and one may investigate the *common* shape features of members of this family. Geometrically, these formal molecular bodies are all different, however, they do have common topological features. By focusing on their common topological properties, one may arrive at a *topological definition of molecular body and molecular shape.* One may define an abstract, topological body $\mathcal{F}(C,a_0,a_1)$ by the following condition: $\mathcal{F}(C,a_0,a_1)$ has no other property than those common to all F(K,a) bodies of the family $F(C,a_0,a_1)$.

Similarly, in the topological model the analysis is not restricted to a single contour surface G(K,a). By taking the entire *family* of contour surfaces,

$$G(C,a_0,a_1) = \{ G(K,a) : a_0 > a > a_1,\ K \in C \} , \tag{6}$$

for all configurations K of family C and all density threshold values a from the interval (a_0,a_1), the shape characterization can be given by the common topological features. One defines an abstract, topological surface $\mathcal{G}(C,a_0,a_1)$ by the following condition: $\mathcal{G}(C,a_0,a_1)$ has no other property than those common to all G(K,a) surfaces of the family $G(C,a_0,a_1)$.

The energy E of formal nuclear arrangements of molecular species may be regarded as a function of the nuclear configuration K. It is natural to use the energy value E(K) assigned to each K as a criterion for deciding which configurations belong to a family C. In particular, one may use some of the topological properties of the energy function E(K) to define those families C of nuclear configurations that may be associated with a given molecule.

The collection of all possible relative nuclear configurations (also called internal configurations) of N nuclei (N>2) can be represented by a (3N-6)-dimensional internal configuration space (also called reduced configuration space) M. Each internal configuration corresponds to one and only one point K of M. It is possible to construct this space with a global distance function d defined on it, where the number $d(K_1,K_2)$ expresses the dissimilarity of any two configurations K_1 and K_2 [84]. Families C of configurations correspond to subsets of M. The energy function E(K) for a specified electronic state may be regarded as a potential energy surface (hypersurface) over this space. With respect to a given energy function E(K) it is meaningful to ask the following question: what is the range of deformations of formal nuclear configurations that preserve chemical identity? A deformation of a nuclear configuration corresponds to a displacement of the corresponding point K within the space M. Small deformations in the vicinity of an equilibrium configuration are usually not expected to convert one molecule into another. On the other hand, extensive deformations, corresponding to large displacements within M, may be sufficient to convert one molecule into another. It is natural to classify deformations by a relaxation criterion. All those deformed nuclear configurations K for which an infinitely slow, vibrationless relaxation along the energy surface E(K) lead to the same equilibrium configuration K_e are regarded to represent the same chemical species. By contrast, if a deformation of an equilibrium configuration K_e is severe enough that a relaxation leads to a different equilibrium configuration $K_{e'}$, then the deformed configuration K is classified as one representing the same chemical species as $K_{e'}$. The above classification scheme generates various families of configurations, called *catchment regions* of the potential surface E(K).

Catchment regions can also be defined by paths of steepest descent, with respect to properly mass weighted coordinates:

Each catchment region $C(\lambda,t)$ is defined as the collection of all those configurations K of M from where a path of steepest descent of the given potential energy surface E(K) leads to a common critical point $K(\lambda,t)$ of serial index t [85]. The critical point index λ is defined as the number of negative eigenvalues of the Hessian matrix of second partial derivatives of the potential energy surface at point $K(\lambda,t)$. For example, the catchment region C(0,t) of a minimum point K(0,t) is a topological representation of a stable chemical species. Note that catchment regions are defined for all critical points, not only for minima; in particular, a catchment region C(1,t) of a saddle point K(1,t) is a topological representation of a transition structure. Each internal nuclear configuration K belongs to one and only one catchment region $C(\lambda,t)$.

Reaction topology is the topological model of the nuclear configuration space M [86]. In reaction topology, chemical species are represented by catchment regions, defined in terms of a given potential energy hypersurface of some specified electronic state [84-88]. The catchment regions $C(\lambda,t)$ generate a decomposition of the nuclear configuration space M:

$$M = \bigcup_{\lambda,t} C(\lambda, t). \tag{7}$$

The (3N-6)-dimensional *metric space* M of all internal configurations of any N-atom system of fixed stoichiometry may be regarded as the union of all the catchment regions of any given potential energy hypersurface.

A configuration domain C of eq. (3) is required in the definitions (eqs.5 and 6) of topological representations of molecular bodies $F(C,a_0,a_1)$ and molecular surfaces $G(C,a_0,a_1)$. For each chemical species this configuration domain C can be taken as the catchment region $C(\lambda,t)$ representing the chemical species. Hence the resulting shape description is not restricted to a single configuration, and reflects the dynamic shape properties of a nonrigid molecule.

If the analysis is carried out by displaying contour surfaces on a computer screen then the recognition of shape features and decisions concerning similarity between molecules are based on visual inspection, that may be affected by the subjectivity of the observer. This is not necessarily the most reliable approach, particularly, if many (for example, several thousand) molecules are involved.

Alternative methods have also been developed, where shape recognition and the evaluation of molecular similarity are based on the mathematical (geometrical and topological) properties of the contour surfaces. This provides a fully automatic, computer-based analysis of large families of molecules, for example, by using the algebraic technique of *symmetry-independent shape groups,* proposed recently [89-93]. In the first step of this method each contour surface G(K,a) is decomposed into domains of various curvature properties, or domains defined by some other physical criteria. Some of the topological relations among these domains are common for an entire family of contour surfaces, where small variations in the nuclear configuration K and density threshold a are allowed. These common topological features define a family of *homology groups,* using the methods of algebraic topology [89-93]. These groups, called *shape groups*, are independent of the symmetry properties of the contour surfaces, and provide a detailed shape characterization, as well as a precise evaluation of molecular similarity. The shape group technique has been applied to molecular problems [94-97], in combination with alternative shape analysis techniques [98-108].

2. SHAPE CHANGES IN ELECTRONIC EXCITATIONS

Electronic excitation may be regarded as a formal, vertical process over the nuclear configuration space M, leading from one potential energy surface, E_1, to another, E_2. It is useful to model actual excitation processes by assuming an initial electronic change that leaves the nuclear configuration K intact, followed by a relaxation of the nuclear configuration along the product energy surface E_2. It is important to realize that the molecular shape may change in both processes. In general, an electronic excitation alters the electron distribution, hence even if the nuclear geometry K remains unchanged, the isodensity contours G(K,a), representing molecular shapes, may undergo significant deformations. In fact, the initial shape change may be greater than the net shape change obtained after completing the relaxation step along the product energy surface E_2. That is, if electronic excitation is involved, then two identical nuclear configurations correspond to molecular shapes that may deviate more from one another than those belonging to two different nuclear configurations of the two relaxed structures on their respective energy surfaces E_1 and E_2.

It is evident, however, that the point symmetry of the nuclear configuration is not affected by the initial electronic change, since the configuration K itself is preserved. It also follows that whatever shape change occurs for the isodensity contour surfaces in the initial process, the resulting isodensity surfaces (as well as the density difference contours) must also have all the symmetry elements of the initial nuclear configuration K. In fact, one has a much stronger result that we shall describe below.

The catchment region point symmetry theorem [ref.84, p367, and refs. 88,109] states the following: among all configurations K within each catchment region $C(\lambda,t)$ the critical point $K(\lambda,t)$ has the highest point symmetry. This theorem is valid for each electronic state. Since electronic excitation of any fixed nuclear geometry K cannot change the point symmetry of K, it follows that the catchment region point symmetry theorem can be applied to the catchment region of the product energy surface, using symmetry information of the reactant K. A relaxation of the configuration K along the product energy surface E_2 leads to a critical point $K(\lambda,t)$ of E_2 that must have all the point symmetry elements of the initial configuration K of the reactant of the first electronic state.

We may summarize this result as follows:

In any one-step electronic excitation or de-excitation process the equilibrium configuration of the product has all the point symmetry elements of the actual (possibly nonequilibrium) reactant configuration K.

For isodensity contours we have an analogous result:

In any one-step electronic excitation or de-excitation process, the isodensity contour $G_2(K(\lambda,t), a')$ of the product electronic state *equilibrium* nuclear configuration has all the symmetry elements of the isodensity contour $G_1(K, a)$ of the initial (possibly nonequilibrium) nuclear configuration K of the reactant electronic state, where the two density thresholds a and a' may be different.

3. SHAPE CHANGES IN CHEMICAL REACTIONS: THE SHAPE MATRIX METHOD AND THE DYNAMIC SHAPE SPACE

Each molecular contour surface G(K, a) may be subdivided into domains, based on local curvature properties of the surface or on some other properties. The fundamental interrelations among these domains and their boundary segments define algebraic-topological relations. These relations can be represented by homology groups of various truncated surfaces derived from G(K,a); these groups are the molecular shape groups (89-93). Alternatively, some of these interrelations can be represented by various matrices, called *shape matrices*. It is possible to introduce these matrices relying only on simple intuitive notions, without direct reference to algebraic topology and shape groups. Here we shall follow this approach, and the reader may find further information on shape groups in recent reviews listed among the references [54,95,96].

Choose an arbitrary point **r** on the contour surface G(K,a). We may regard a small neighborhood of **r** on the surface as a function expressed over the plane $^{2}R(\mathbf{r})$ that is tangent to G(K,a) at this point **r**. We regard this function negative along the normal vector of the tangent plane pointing toward the interior of G(K,a), zero at point **r**, and positive in the opposite direction. Evidently, this function has a *critical point* at point **r**. Since the surface G(K,a) is twice continuously differentiable, the Hessian matrix

$$\mathbb{H}(\mathbf{r}) \tag{8}$$

of second partial derivatives of this function is well defined at point **r**.

The tangent plane $^{2}R(\mathbf{r})$, the local coordinate system defined on it, and the Hessian matrix $\mathbb{H}(\mathbf{r})$ vary with point **r**, hence we refer to $\mathbb{H}(\mathbf{r})$ as the *local Hessian* at point **r**. The local curvature properties of the contour surface G(K,a) are characterized by $\mathbb{H}(\mathbf{r})$ at each point **r** . The *local canonical curvatures* h_1 and h_2 are the *eigenvalues* of the local Hessian matrix $\mathbb{H}(\mathbf{r})$ [110].

The local curvature properties of G(K,a) are compared to a reference curvature parameter b. The number $\mu = \mu(\mathbf{r},b)$ for each point $\mathbf{r}$ of G(K,a) is defined as the number of local canonical curvatures that are less than b. In the special case of b=0, μ is the *index* of point $\mathbf{r}$, that is, the number of negative eigenvalues of $\mathbb{H}(\mathbf{r})$. We recall that $\mathbf{r}$ is a critical point of the locally defined surface.

Each point $\mathbf{r}$ of the contour surface G(K,a) belongs to one of three disjoint subsets A_0, A_1 or A_2 of G(K,a),

$$\mathbf{r} \in A_0 \quad \text{iff} \quad b \le h_1, h_2 , \tag{9}$$

$$\mathbf{r} \in A_1 \quad \text{iff} \quad h_1 < b \le h_2 , \tag{10}$$

$$\mathbf{r} \in A_2 \quad \text{iff} \quad h_1, h_2 < b . \tag{11}$$

The union of these sets is the contour surface,

$$G(K,a) = A_0 \cup A_1 \cup A_2, \tag{12}$$

and these subsets are, indeed, disjoint,

$$A_0 \cap A_1 = A_0 \cap A_2 = A_1 \cap A_2 = \varnothing. \tag{13}$$

The local characterization of of curvature at each point $\mathbf{r}$ depends on whether none, one or both of the local canonical curvatures h_1 and h_2 at point $\mathbf{r}$ are smaller than the reference value b .

Each of these sets A_μ, $\mu = 0, 1, 2,$ may be disconnected. The k-th maximum connected component of a set A_μ is denoted by $D_{\mu,k}$. A typical $D_{\mu,k}$ domain is a two-dimensional subset of the isodensity contour surface G(K,a) for each index $\mu = 0, 1, 2$. Each point $\mathbf{r}$ belongs to one and only one of the $D_{\mu,k}$ domains of some index μ and serial index k, depending on which one of the following conditions is fulfilled:

$$\mathbf{r} \in D_{0,k} \text{ for some } k \quad \text{iff} \quad b \le h_1, h_2 , \tag{14}$$

$$\mathbf{r} \in D_{1,k} \text{ for some } k \qquad \text{iff} \qquad h_1 < b \leq h_2 \ , \tag{15}$$

$$\mathbf{r} \in D_{2,k} \text{ for some } k \qquad \text{iff} \qquad h_1, h_2 < b \ . \tag{16}$$

In the above notation for the $D_{\mu,k}$ domains the nuclear configuration K, the density parameter a, and the reference curvature b are not specified. If explicit reference is required then one may use the $D_{\mu,k}(K,a,b)$ notation. Typically, the $D_{\mu,k}$ domain decompositions for intermediate a and b values provide the most chemically interesting information.

The above decomposition of the contour surface G(K,a) by the $D_{\mu,k}$ domains is the basis of the shape matrix method. The shape matrices $\mathbb{M}(K,a,b)$ are defined by the relative distribution of various $D_{\mu,i}$ domains along the contour surface G(K,a). In addition to shape information, these matrices are used to encode size information for the $D_{\mu,i}$ domains, at least in some relative sense.

Consider the family $\{D_{\mu,i}\}$ of all $D_{\mu,i}$ domains, and assume that these domains are ordered according to *decreasing size,* as defined by their surface area:

$$D_1 , D_2 , \ldots , D_s , \ldots , D_z , \tag{17}$$

where the single index s is the size index, and each

$$D_s = D_{\mu,i} , \tag{18}$$

for some index pair μ, i .

For the $D_{\mu,i}$ domains we define the following N-neighbor relation, analogous to the "symmetric strong neighbor" relation of potential surface catchment regions of reaction topology [84,111]:

$$N(D_s , D_{s'}) = \begin{cases} 1 & \text{if } (D^c{}_s \cap D_{s'}) \cup (D_s \cap D^c{}_{s'}) \neq \emptyset \\ 0 & \text{otherwise,} \end{cases} \tag{19}$$

where $D_s = D_{\mu,i}$ and $D_{s'} = D_{\mu',i'}$. In the above definition the superscript c stands for the set-theoretical closure within the Euclidean metric of the laboratory frame.

The shape matrix $\mathbb{M}(K,a,b)$ is defined as follows:

$$\mathbb{M}(K,a,b)_{s,s'} = \mu_s \delta_{s,s'} + N(D_s , D_{s'})(1- \delta_{s,s'}) , \tag{20}$$

where $\delta_{s,s'}$ is the Kronecker δ symbol.

Shape matrix $\mathbb{M}(K,a,b)$ gives a detailed shape characterization of contour surface G(K,a), with respect to reference curvature b.

The typical nonzero off-diagonal matrix elements, that is, the nonzero N-neighbor relations are those involving the following types of pairs of $D_{\mu,i}$ domains:

$$(D_{0,i} , D_{1,i'}) \tag{21}$$

and

$$(D_{1,i'} , D_{2,i''}) . \tag{22}$$

No nonzero shape matrix element is possible between two D_μ domains of the same μ index, since each $D_{\mu,i}$ domain is a maximum connected component of set A_μ of all points of G(K,a) with index μ. One should note, however, that a nonzero matrix element between a D_0 and a D_2 domain is not impossible. A surface point $\mathbf{r} \in G(K,a)$ may simultaneously belong to a $D_{0,i'}$ domain and to the closure of a $D_{2,i}$ domain,

$$\mathbf{r} \in D^c{}_{2,i} \cap D_{0,i'}, \tag{23}$$

if at point $\mathbf{r}$ the local Hessian matrix $\mathbb{H}(\mathbf{r})$ has degenerate eigenvalues, and *both local canonical curvatures*, $h_1(\mathbf{r})$ and $h_2(\mathbf{r})$ are equal to the reference curvature b. This condition is sufficient and necessary for the existence of a nonzero $(D_{2,i}, D_{0,i'})$ matrix element within shape matrix $\mathbb{M}(K,a,b)$.

Within the *closure* $D^c_{2,i}$ of a domain $D_{2,i}$ there are points $\mathbf{r}$ where for the eigenvalues of matrix $\mathbb{H}(\mathbf{r})$ the relations

$$h_1 \leq b = h_2 \tag{24}$$

hold. If the strict inequality $<$ holds in eq. (24), then the point $\mathbf{r}$ belongs to some $D_{1,i'}$ domain, and these two domains, $D_{2,i}$ and $D_{1,i'}$ are N-neighbors, as follows from definition (19). If the closure $D^c_{1,j}$ of a domain $D_{1,j}$ contains points $\mathbf{r}$ for which the relations

$$b = h_1, \qquad b < h_2 \tag{25}$$

hold, then point $\mathbf{r}$ must belong to some domain $D_{0,j'}$, consequently, domains $D_{1,j}$ and $D_{0,j'}$ are N-neighbors. A point $\mathbf{r}$ within the closure $D^c_{2,i}$ of a $D_{2,i}$ domain can belong to a domain $D_{0,i'}$ *only* if

$$h_1 = h_2 = b \tag{26}$$

holds for the eigenvalues of $\mathbb{H}(\mathbf{r})$. A pair of domains $D_{2,i}$ and $D_{0,i'}$ are N-neighbors if and only if at a boundary point $\mathbf{r}$ the local Hessian matrix has two degenerate eigenvalues, which happen to be equal to the reference curvature value b. As mentioned above, this is a sufficient and necessary condition for an N-neighbor relation between D_0 and D_2 domains. As follows from their definition, each D_0 domain is a *closed* set, hence intersections of the type $D^c_0 \cap D_2$ are necessarily empty and they cannot contribute to a nonzero N-neighbor relation, according to definition (19).

The above properties have an interesting consequence regarding the types of shape matrix elements $\mathbb{M}(K,a,b)_{s,s'}$ that may occur for actual molecules. For a specified reference curvature b the typical boundary points $\mathbf{r}$ of a D_μ domain on the contour surface G(K,a) fulfill either one of the two pairs of relations: relations (24) with the strict inequality $<$, or relations (25). In the typical cases only the (1,2) and the (0,1) μ index combinations occur in the

N-neighbor relations on G(K,a).

These shape matrices $\mathbb{M}(K,a,b)$ provide a concise description of the shape characteristics of molecular charge density. They are invariant within certain intervals of the two parameters a and b . Those parameter values a_k and b_ℓ where the shape matrix $\mathbb{M}(K,a,b)$ changes are characteristic to the given molecular configuration. On the other hand, for a given, fixed parameter pair a, b, the shape matrices are invariant to minor changes of configuration K within certain domains of a nuclear configuration space. Within these configurational domains the essential shape features of G(K,a), with respect to the specified curvature parameter b, are preserved. The corresponding minor changes of the nuclear configurations do not change the shape classification of the molecule with respect to the curvature parameter b and the electronic charge density threshold a , as described by G(K,a). However, for a different curvature parameter b, the invariance domains of shape matrices $\mathbb{M}(K,a,b)$ within the nuclear configuration space are in general different.

In order to provide a more complete description, one should analyse the dependence of shape matrices on parameters a and b, as well as on the nuclear configuration K. A complete description of all shape features along the entire configuration space M , as encoded in the shape matrix $\mathbb{M}(K,a,b)$, can be given by considering the *product space*

$$D = M \otimes P, \tag{27}$$

where M is the metric nuclear configuration space and P is the parameter space of parameters a and b. For molecules of N nuclei (N>2), the dimension of M is 3N-6 , whereas the dimension of P is 2 . Consequently, the dimension of D is 3N-4 . Within space D one may follow the shape variations with changes in nuclear configurations K as well as in parameters a and b , hence D is suitable to study the dynamic shape properties of molecules. Space D is referred to as the *dynamic shape space*.

For a single nuclear configuration K the shape characterization based on shape matrices $\mathbb{M}(K,a,b)$, defined by the D_μ domains, provides a discrete, algebraic shape description that depends on two continuous parameters, a and b.

This combination of the discrete nature of the algebraic model, and the presence of continuous parameters permits a detailed description. By considering the entire, chemically accessible range of the two continuous parameters a and b, one obtains a complete description of the shape of the three-dimensional molecular charge density for a given nuclear configuration K, that can be extended to a family C of nuclear configurations within space M.

First we shall investigate the dependence of shape matrices $\mathbb{M}(K,a,b)$ on the two parameters a and b.

The classification of local curvature at contour points **r** of G(K,a) refers to an orientation of the surface, with respect to the interior and exterior of G(K,a). Consequently, the reference curvature b may take both positive and negative values:

$$-\infty < b < \infty \; . \tag{28}$$

The chemically interesting shape matrices $\mathbb{M}(K,a,b)$ belong to b parameters of absolute values bounded by some constant b_0. If one takes the special choice of b=0 , then the shape matrices $\mathbb{M}(K,a,b)$ describe the interrelations among convex, concave and saddle type domains on G(K,a). In this case the point set of the boundary lines of all D_μ domains consists of all the parabolic points of isodensity surface G(K,a).

For a negative curvature parameter b of a large enough absolute value,

$$b \ll 0 \, , \tag{29}$$

relation (9) holds for eigenvalues h_1 and h_2 at all points $\mathbf{r} \in$ G(K,a). Consequently, the entire isodensity contour surface G(K,a) is a single, simply connected D_0 domain. In this special case the shape matrix $\mathbb{M}(K,a,b)$ is 1×1, and its single element is equal to zero.

For a large enough positive curvature parameter b,

$$0 \ll b \, , \tag{30}$$

relation (11) holds for eigenvalues h_1 and h_2 at each point $\mathbf{r} \in G(K,a)$ of the isodensity surface. The entire contour surface is a single, simply connected D_2 domain. In this case the shape matrix $\mathbb{M}(K,a,b)$ is also 1×1, and its single element is equal to two.

The chemically more interesting intermediate values for curvature parameter b lead to nontrivial shape matrices $\mathbb{M}(K,a,b)$. A continuous change in reference curvature b leaves the shape matrix $\mathbb{M}(K,a,b)$ invariant within certain intervals (b_1, b_2), however, the matrices change at the interval limits. Typically, the change is an increase or decrease of the matrix dimension, accompanied by the addition or elimination of matrix elements, corresponding to the changes in the N-neighbor relations. Simultaneous additions or eliminations of several domains on the surface, and the accompanying changes in the shape matrix $\mathbb{M}(K,a,b)$, are usually indications of a nontrivial symmetry of the contour surface $G(K,a)$.

The shape matrices $\mathbb{M}(K,a,b)$ show a similar dependence on the contour parameter a of $G(K,a)$. For any fixed curvature parameter b, the D_μ domain decomposition, and hence the shape matrix $\mathbb{M}(K,a,b)$ obtained, depend on this density value a. The shape matrices $\mathbb{M}(K,a,b)$ are invariant within certain density intervals (a_1, a_2), however, they change at the limits a_1 and a_2 of such intervals of invariance. Just as for parameter b, the typical change is an increase or decrease of the dimension of the shape matrix $\mathbb{M}(K,a,b)$, accompanied by the addition or elimination of matrix elements, reflecting the changes in the N-neighbor relations. Simultaneous additions or eliminations of several domains on the surface, and the accompanying changes in the shape matrix $\mathbb{M}(K,a,b)$, usually indicate some nontrivial symmetry of $G(K,a)$.

In principle, the electronic charge density contour parameter a may be chosen from the entire range of possible values,

$$0 \leq a < \infty . \tag{31}$$

However, very small contour values a correspond to large, essentially spherical contour surfaces at large distances from the nuclei of the molecule. Very large contour values a correspond to a family of small, nearly spherical surfaces near the nuclei. These extreme choices provide little useful insight, since for all but the extreme values of curvature paramater b they correspond to diagonal shape matrices. The chemically more interesting cases of shape matrices $\mathbb{M}(K,a,b)$ are those of intermediate charge density values a.

The dependence of these shape matrices $\mathbb{M}(K,a,b)$ on the two parameters a and b can be represented by the distribution of matrices along the a,b parameter plane P. A similar description has been introduced earlier for shape groups [90]. A complete description of the shape of the electron density of a fixed nuclear configuration K can be given by specifying the boundaries of those regions of the a,b parameter plane P within which the shape matrices $\mathbb{M}(K,a,b)$ are invariant. These are the P space invariance domains of shape matrices $\mathbb{M}(K,a,b)$ for a fixed nuclear configuration K.

We are also interested in the invariance of shape matrices $\mathbb{M}(K,a,b)$ to variations in the nuclear geometry K. Molecules are dynamic, topological entities and not rigid, geometrical objects [84], hence it is not sufficient to consider a single, formal nuclear configuration K . A complete characterization of molecular shapes requires the study of entire families of formal nuclear configurations. In particular, the domains of invariance of shape matrices $\mathbb{M}(K,a,b)$ within the a,b parameter plane P depend on the formal nuclear configuration K. However, the shape matrices $\mathbb{M}(K,a,b)$ remain invariant also to minor geometry variations within certain domains C of the nuclear configuration space M. Consequently, the shape matrices $\mathbb{M}(K,a,b)$ do reflect the topological nature of molecular species, that can be exploited in a topological shape characterization.

In the dynamic shape space model described below, both types of invariances of shape matrices $\mathbb{M}(K,a,b)$ can be studied. The parameter plane P of points (a,b),

$$P = \{(a, b)\}, \tag{32}$$

can be provided with an Euclidean metric d_p ,

$$d_p\,[(a_1, b_1), (a_2, b_2)] = [(a_1-a_2)^2 + (b_1-b_2)^2]^{1/2}\,. \tag{33}$$

The product space $D = M \otimes P$ of dimensions $3N-6+2 = 3N-4$,

$$\dim(D) = \dim(M \otimes P) = 3N-4, \tag{34}$$

is defined in terms of metric spaces M and P, and is suitable for a complete, dynamic description of molecular shape.

One may define a metric d_D for this space D as

$$d_D = [d_p^2 + d_M^2]^{1/2}, \tag{35}$$

where d_M is the metric in space M. Space D, provided with metric d_D, is the *dynamic shape space*, mentioned above.

For any given family of N nuclei and a specified electronic state of k electrons, the shape matrices $\mathbb{M}(K,a,b)$ are invariant within certain domains of the dynamic shape space D. These invariance domains generate a complete partitioning of space D. The specification of the boundaries of these shape invariance domains within D is equivalent to a complete description of the dynamic shape features of all molecular species of the stoichiometry specified by the N nuclei and the given electronic state of the k electrons.

It is useful to introduce the notation

$$\mathbf{d} = (K,a,b) \tag{36}$$

for points of the dynamic shape space D. Each point $\mathbf{d} \in D$ represents a combination of a *pair* $(a,b) \in P$ of a parameter value a for electronic charge density and reference curvature b, and a given nuclear configuration $K \in M$.

The subsets of invariance of shape matrices

$$\mathbb{M}(\mathbf{d}) = \mathbb{M}(K,a,b) \tag{37}$$

within the dynamic shape space D are denoted by $A_i(\mathbb{M}^i)$:

$$A_i(\mathbb{M}^i) = \{\mathbf{d}:\ \mathbf{d} \in D,\ \mathbb{M}(\mathbf{d}) = \mathbb{M}^i\}, \tag{38}$$

where index i is the serial index of the i-th reference matrix $\mathbb{M}^i$.

In equation (38), $\mathbb{M}(\mathbf{d})$ stands for the actual shape matrix at the specified point $\mathbf{d}$ of the dynamic shape space D. For each point $\mathbf{d}$ within a given set $A_i(\mathbb{M}^i)$ the actual shape matrix $\mathbb{M}(\mathbf{d})$ is identical to a specified reference matrix $\mathbb{M}^i$. These $A_i(\mathbb{M}^i)$ shape invariance sets are pairwise disjoint and generate a partitioning of the dynamic shape space D:

$$A_i(\mathbb{M}^i) \cap A_{i'}(\mathbb{M}^{i'}) = \varnothing \qquad \text{if} \quad i \neq i', \tag{39}$$

$$D = \bigcup_i A_i(\mathbb{M}^i). \tag{40}$$

A shape invariance set $A_i(\mathbb{M}^i)$ is not necessarily connected. Each may be regarded as the union of its maximum connected components $D_{i,j}(\mathbb{M}^i)$

$$A_i(\mathbb{M}^i) = \bigcup_j D_{i,j}(\mathbb{M}^i), \tag{41}$$

$$D_{i,j}(\mathbb{M}^i) \cap D_{i,j'}(\mathbb{M}^i) = \varnothing \qquad \text{if} \quad j \neq j'. \tag{42}$$

A detailed partitioning of the dynamic shape space D into shape regions by reference shape matrices $\mathbb{M}^i$ can be given in terms of the above maximum connected components,

$$D = \bigcup_{i,j} D_{i,j}(\mathbb{M}^i). \tag{43}$$

Any continuous change $\mathbf{d}(j) \to \mathbf{d}(j')$ of parameter values a and b, and nuclear configuration K that leads from one component $D_{i,j}(\mathbb{M}^i)$ to another component $D_{i,j'}(\mathbb{M}^i)$ of $A_i(\mathbb{M}^i)$ must involve points $\mathbf{d}$ characterized by a shape matrix $\mathbb{M}^{i'}$ different from $\mathbb{M}^i$. Consequently, the term "domain of shape matrix invariance", or in short, "shape invariance domain", is used for the components $D_{i,j}(\mathbb{M}^i)$, rather than for the set $A_i(\mathbb{M}^i)$.

The boundaries of shape invariance domains $D_{i,j}(\mathbb{M}^i)$ fully specify the above partitioning (43) of the dynamic shape space D.

4. CATCHMENT REGIONS OF CHEMICAL SPECIES AND SHAPE INVARIANCE DOMAINS OF THE DYNAMIC SHAPE SPACE

The partitioning of the dynamic shape space D given by eq. (43) involves all possible chemical species composed from the specified N nuclei. Within the framework of the shape matrix method, this partitioning describes the shapes of all possible conformations of all species of the given stoichiometry and electronic state. However, the shape invariance domains do not necessarily refer to individual chemical species, and it is meaningful to ask the following questions:

(i) What shape matrices $\mathbb{M}^i$ are available for a given chemical species, represented by a catchment region $C(\lambda,t)$? (That is, by allowing only such variations of the formal nuclear geometry K that preserve the chemical identity of the species, which shape matrices occur?)

(ii) Which are those chemical species of the given stoichiometry that may have formal nuclear configurations of a specified shape matrix ? (That is, which species have nuclear configurations K that for some a and b values belong to a specified shape invariance domain $D_{i,j}(\mathbb{M}^i)$?)

The above two questions (i) and (ii) can be rephrased in terms of relations between the shape invariance domains $D_{i,j}(\mathbb{M}^i)$ of the dynamic shape space D , and the catchment regions $C(\lambda,t)$ of nuclear configuration space M.

For each catchment region $C(\lambda,t)$ a subset $D(C(\lambda,t))$ is defined as

$$D(C(\lambda,t)) = P \otimes C(\lambda,t) . \tag{44}$$

Set $D(C(\lambda,t))$ is the family of all points $\mathbf{d} \in D$ which involve any nuclear configuration $K \in C(\lambda,t)$ of catchment region $C(\lambda,t) \subset M$, in combination with any values of parameters a and b of parameter plane P.

Problem (i) can be approached as follows:

A molecular shape characterized by shape matrix $\mathbb{M}^i$ is available to chemical species $C(\lambda,t)$ if and only if

$$D(C(\lambda,t)) \cap D_{i,j}(\mathbb{M}^i) \neq \varnothing \quad \text{for some index } j . \tag{45}$$

One may restrict the study to the chemically realistic density values a for contour surfaces $G(K,a)$ and to the most relevant intermediate values of curvature parameter b. That is, one may study only a subset $D[C(\lambda,t), (a_1,a_2), (b_1,b_2)]$ of set $D(C(\lambda,t))$, defined as

$$D[C(\lambda,t), (a_1,a_2), (b_1,b_2)] =$$

$$= \{\mathbf{d}: \ \mathbf{d} = a \otimes b \otimes K, \ a \in (a_1,a_2), \ b \in (b_1,b_2), \ K \in C(\lambda,t)\} . \tag{46}$$

A molecular shape characterized by shape matrix $\mathbb{M}^i$ is available to molecular species $C(\lambda,t)$ within some chemically realistic parameter ranges (a_1,a_2) and (b_1,b_2) if and only if

$$D[C(\lambda,t), (a_1,a_2), (b_1,b_2)] \cap D_{i,j}(\mathbb{M}^i) \neq \varnothing \quad \text{for some index } j . \tag{47}$$

Problem (ii) can approached by taking the projection $P_M(D_{i,j}(\mathbb{M}^i))$ of the shape invariance domain $D_{i,j}(\mathbb{M}^i)$ on the nuclear configuration space M:

$$P_M(D_{i,j}(\mathbb{M}^i)) = \{K : \exists (a,b) \in P: \ a \otimes b \otimes K \in D_{i,j}(\mathbb{M}^i)\}. \tag{48}$$

A given chemical species $C(\lambda,t)$ may attain a formal nuclear configuration K characterized by shape matrix $\mathbb{M}^i$ if and only if there exists some index j for which

$$P_M(D_{i,j}(\mathbb{M}^i)) \cap C(\lambda,t) \neq \varnothing . \tag{49}$$

Within the topological context of shape matrices, the conditions (45), (47), and (49) specify all elementary relations between molecular identity and molecular shape.

Note that several different chemical species may attain the same topological shape within the dynamic shape characterization scheme, as defined by shape matrices. This fact is of special concern to drug design, where molecular similarity (within a context that involves dynamic considerations) is of major importance. Considering a given shape invariance domain $D_{i,j}(\mathbb{M}^i)$, relation (49) may hold for several different chemical species, for example, it may hold for species $C(\lambda',t')$ in addition to species $C(\lambda,t)$. That is, the relation $P_M(D_{i,j}(\mathbb{M}^i)) \cap C(\lambda',t') \neq \varnothing$ may also be valid, in addition to (49). Evidently, the simultaneous validity of condition (49) for two catchment regions $C(\lambda,t)$ and $C(\lambda',t')$ is required for the two chemical species to exhibit common shape features, as specified by shape matrix $\mathbb{M}^i$.

The above model provides a framework for the analysis of dynamic shape similarity of molecules. The model is of topological nature, providing a discrete characterization of the shape problem of continua, such as the continuous, 3D electronic density of molecules.

Acknowledgment

Research work leading to this report has been supported by operating and strategic research grants from the Natural Sciences and Engineering Research Council of Canada.

REFERENCES

1. S. Fliszár, Charge Distributions and Chemical Effects, Springer, New York, 1983.
2. P. Coppens and M. B. Hall, Eds., Electron Distribution and the Chemical Bond, Plenum, New York and London, 1982.
3. V.H. Smith, Jr., P.F. Price, and I. Absar, Israel J. Chem., **16**,187 (1977).
4. G.D. Purvis III and C. Culberson, Int. J. Quantum Chem., Quantum Biol. Symp., **13**, 261 (1986).
5. P.A. Kollman, J. Am. Chem. Soc., **100**, 2974 (1978).
6. R. Rein, J. R. Rabinowitz, and T. J. Swissler, J. Theor. Biol., **34**, 215 (1972).
7 E. Scrocco and J. Tomasi, Topics Current Chem. **42**, 95 (1973).
8. D.M. Hayes and P.A. Kollman, J. Am. Chem. Soc., **98**, 3335 (1976) .
9. P.A. Kollman, Acc. Chem. Res.,**10**, 36 (1977).
10. E. Scrocco, and J. Tomasi, Adv. Quantum Chem. **11**, 115 (1978).
11. J. Tomasi, On the Use of Electrostatic Molecular Potentials in Theoretical Investigations on Chemical Reactivity, in Quantum Theory of Chemical Reactions, Vol 1, R. Daudel, A. Pullman, L. Salem, and A. Veillard, Eds., Reidel, Dordrecht, 1979, pp.191-228.
12. G. Náray-Szabó, Quantum Chem. Program Exchange, **13**, 396 (1980).
13. P. Politzer and D. G. Truhlar, Eds., Chemical Applications of Atomic and Molecular Electrostatic Potentials, Plenum, New York, 1981.
14. G. Náray-Szabó, A. Grofcsik, K. Kósa, M. Kubinyi and A. Martin, J. Comput. Chem., **2**, 58 (1981).
15. H. Weinstein, R. Osman, J.P. Green and S. Topiol, Electrostatic Potentials as Descriptors of Molecular Reactivity, in Chemical Applications of Atomic and Molecular Electrostatic Potentials, P. Politzer and D.G.Truhlar, Eds., Plenum, New York, 1981 pp. 309-334.
16. J.J. Kaufman, P.C. Hariharan, F.L. Tobin, and C. Petrongolo, Electrostatic Molecular Potential Contour Maps from *Ab Initio* Calculation, in Chemical Applications of Atomic and Molecular Electrostatic Potentials, P. Politzer and D.G. Truhlar, Eds., Plenum, New York, 1981, pp. 335-380.

17. A. Pullman and B. Pullman, Electrostatic Molecular Potential of the Nucleic Acids, in Chemical Applications of Atomic and Molecular Electrostatic Potentials, P. Politzer and D.G. Truhlar, Eds., Plenum, New York, 1981, pp. 381-405.
18. A. Warshel, Acc. Chem. Res., **14**, 284 (1981). Electrostatic Basis of Structure-Function Correlation in Proteins.
19. P. H. Reggio, H. Weinstein, R. Osman and S. Topiol, Int. J. Quantum Chem., Quantum Biol. Symp. **8**, 373 (1981).
20. S. Cox and D. Williams, J. Comput. Chem., **2**,304 (1981).
21. Z. Gabányi, P. Surján, and G. Náray-Szabó, Eur. J. Med Chem., **17**, 307 (1982).
22. H. Weinstein, R. Osman, S. Topiol and C. A. Venanzi, Pharmacochem. Libr., **6**, 81 (1983).
23. J. Angyán and G. Náray-Szabó, J. Theor. Biol., **103**, 777 (1983).
24. G. Náray-Szabó and P.R. Surján, Computational Methods for Biological Systems, in Theoretical Chemistry of Biological Systems, in Studies in Physical and Theoretical Chemistry Vol. 41, G. Náray-Szabó, Ed., Elsevier, Amsterdam, 1986, pp.1-100.
25. J.R. Rabinowitz and S.B. Little, Int. J. Quantum Chem., Quantum Biol. Symp., **13**, 9 (1986).
26. J.C. Culberson, G.D. Purvis III, M.C. Zerner and B.A. Seiders, Int. J. Quantum Chem., Quantum Biol. Symp., **13**, 267 (1986) .
27. F. M. Richards, Annu. Rev. Biophys. Bioeng., **6**, 151 (1977).
28. T. Porter, Comput. Graphics **12**, 282 (1978).
29. N. L. Max, Comput. Graphics **13**, 165 (1979).
30. R. Langridge, T.E. Ferrin, I.D. Kuntz and M.L. Connolly, Science, **211**, 661 (1981).
31. C. Hansch, R.-L. Li, J.M. Blaney and R. Langridge, J. Med. Chem., **25**, 777 (1982).
32. J.M. Blaney, E.C. Jorgensen, M. L. Connolly, T.E. Ferrin, R. Langridge, S.J. Oatley, J.M. Burridge and C.C.F. Blake, J. Med. Chem.,**25**, 785 (1982).
33. P.K. Weiner, R. Langridge, J.M. Blaney, R. Schaefer and P. Kollman, Proc. Natl. Acad. Sci. U.S.A., **79**,3754 (1982).
34. R.J. Feldman, D.H. Bing, B.C. Furie and D. Furie, Proc. Natl. Acad. Sci. U.S.A., **75**, 5409 (1982).

35. P.A. Basch, N. Pattabiraman, C. Huang, T.E. Ferrin,and R. Langridge, Science, **222**, 1325 (1983).
36. A. Gavezotti, J. Am. Chem. Soc., **105**, 5220 (1983).
37. L.H. Pearl and A. Honegger, J. Mol. Graphics, **1**, 9 (1983).
38. M. L. Connolly, J. Mol. Graphics, **3**, 19 (1985).
39. M.L. Connolly, J. Am. Chem. Soc., **107**, 1118 (1985).
40. K. D. Gibson and H. A. Scheraga, Mol. Phys., **62**, 1247 (1987).
41. K. D. Gibson and H. A. Scheraga, Mol. Phys., **64**, 641 (1988).
42. Y.K. Kang, K. D. Gibson, G. Némethy and H. A. Scheraga, J. Phys. Chem., **92**, 4739 (1988).
43. R. Franke, Theoretical Drug Design Methods, Elsevier, Amsterdam, 1984.
44. W. G. Richards : Quantum Pharmacology, 2nd Ed., Butterworth, London, 1983.
45. T. Oie, G.M. Maggiora, R.E. Christoffersen and D.J. Duchamp, Int. J. Quantum Chem., Quantum Biol. Symp., **8**, 1, (1981).
46. J. Åqvist and O. Tapia, J. Mol. Graphics, **5**, 30 (1987) .
47. E.E. Hodgkin and W.G. Richards, J. Chem. Soc., Chem. Commun.,1980, 1342 (1980).
48. R.Carbó, L.Leyda and M. Arnau, Int. J. Quantum Chem., **17**, 1185 (1980).
49. P.E. Bowen-Jenkins, D.L.Cooper and W.G.Richards, J. Phys. Chem., **89**, 2195 (1985) .
50. M. A. Johnson, J. Math. Chem., **3**, (1989).
51. P.G. Mezey, J. Math. Chem.,**2**, 299 (1988).
52. G. A. Arteca and P. G. Mezey, J. Comput. Chem., **9**, 728 (1988).
53. G. A. Arteca and P. G. Mezey, J. Phys. Chem., **93**, 4746 (1989).
54. P.G. Mezey, Three-dimensional Topological Aspects of Molecular Similarity, in Concepts and Applications of Molecular Similarity, G. M. Maggiora and M.A. Johnson, Eds., Wiley Interscience, 1990.
55. P.G. Mezey, Rev. Comput. Chem., **1**, xxx (1990).
56. Y.C. Martin, Quantitative Drug Design: A Critical Introduction, Deker, New York and Basel, 1978.
57. R. Carbo and M. Arnau, Molecular Engineering: A general Approach to QSAR, in Medicinal Chemistry Advances, F.G. de la Heras and S. Vega, Eds., Pergamon, Oxford, 1981.

58. T.P. Lybrand, P.A. Kollman, V.C. Yu, and W. Sadee, Pharm. Res., **3**, 218 (1986).

59. J.R. Rabinowitz, K. Namboodiri, and H. Weinstein, Int. J. Quantum Chem., **29**, 1697 (1986).

60. W. Ito, H. Nakamura, and Y. Arata, J. Mol. Graphics, **7**, 60 (1989).

61. D. S. Goodsell, I.S. Mian, A.J. Olson, J. Mol. Graphics, **7**, 41 (1989).

62. P.M. Dean, Molecular Foundations of Drug-Receptor Interaction, Cambridge University Press, Cambridge, UK, 1987.

63. G.R. Marshall, C.D. Barry, H.E. Bosshard, R. Dammkoehler, and D.A. Dunn, in Computer Assisted Drug Design, E.C. Olson, R.E.Christoffersen, Eds., ACS. Symp. Ser.,**112**, 205 (1979). The Conformational Parameter in Drug Design: The Active Analog Approach.

64. R.P. Sheridan, R. Nilakantan, J.S. Dixon, and R. Venkataraghavan, J. Med. Chem., **29**, 899 (1986).

65. G.M. Crippen, Distance Geometry and Conformational Calculations, Research Studies Press, Chichester, UK, 1981.

66. G.M. Crippen, Distance Geometry Conformational Calculations, Proc. 3rd Computer Aided Molecular Design Conference, 1986.

67. M.C. Marivet, J.-J. Bourguignon, C. Lugnier, A. Mann, J.-C. Stoclet, and C.-G. Wermuth, J. Med. Chem., **32**, 1450 (1989).

68. H. Shinkai, M. Nishikawa, Y. Sato, K. Toi, I. Kumashiro, Y. Seto, M. Fukuma, K. Dan, and S. Toyoshima, J. Med. Chem., **32**, 1436 (1989).

69. E. Pop, M.E. Brewster, J.J. Kaminski, and N. Bodor, Int. J. Quantum Chem., **35**, 315 (1989).

70. A.J. Hopfinger, J. Amer. Chem. Soc., **102**, 7196 (1980).

71. H.Weinstein, M.N.Liebman, and C.A. Venanzi, in New Methods in Drug Research, A. Makriyannis, Ed., J.R. Prous Science, Barcelona, 1985.

72. C.A. Venanzi, P.M. Canzius, Z. Zhang and J.P. Bunce, J. Comp. Chem., **10**, 1038 (1989).

73. C.J. Coulson, Molecular Mechanisms of Drug Action, Taylor & Francis, London, 1988.

74. S. Namasivayam and P.M. Dean, J. Mol. Graphics, **4**, 46 (1986).

75. C. Hansch and T.E. Klein, Acc. Chem. Res., **19**, 392 (1986).

76. P.-L. Chau and P.M. Dean, J. Mol. Graphics, **5**, 97 (1987).

77. P.M. Dean and P.-L. Chau, J. Mol. Graphics, **5**, 152 (1987).

78. P.M. Dean and P. Callow, J. Mol. Graphics, **5**, 159 (1987).
79. C. Thomson, D. Higgins, and C. Edge, J. Mol. Graphics, **6**, 171 (1988).
80. J. Rebek, Jr., Science, **235**, 1478 (1987).
81. K.B. Lipkowitz and R. Zegarra, J. Comput, Chem.,**10**, 595 (1989).
82. H.R. Karfunkel and V. Eyraud, J. Comput, Chem.,**10**, 628 (1989).
83. L.G. Boulu and G.M. Crippen, J. Comput, Chem.,**10**, 673 (1989).
84. P.G. Mezey, Potential Energy Hypersurfaces, Elsevier, Amsterdam, 1987.
85. P.G. Mezey, Theor. Chim. Acta, **58**,309 (1981).
86. P.G. Mezey, Theor. Chim. Acta, **62**,133 (1982).
87. P.G. Mezey, Can. J. Chem., **61**, 956 (1983).
88. P.G. Mezey, Reaction Topology and Quantum Chemical Molecular Design on Potential Surfaces, in New Theoretical Concepts for Understanding Organic Reactions, J. Bertrán and I.G. Csizmadia, Eds., Kluwer Academic Publishers, Dordrecht, 1989, pp 55-76.
89. P.G. Mezey, Int. J. Quantum Chem., Quantum Biol. Symp., **12**, 113 (1986).
90. P.G. Mezey, J. Comput. Chem., **8**, 462 (1987).
91. P.G. Mezey, Int. J. Quantum Chem., Quantum Biol. Symp., **14**, 127 (1987).
92. P. G. Mezey, J. Math. Chem., **2**, 325 (1988).
93. P.G. Mezey, From Geometrical Molecules to Topological Molecules: A Quantum Mechanical View, in Molecules in Physics, Chemistry and Biology, Vol. II, J. Maruani, Ed., Reidel, Dordrecht, 1988, p 61.
94. G.A. Arteca, V.B. Jammal and P.G. Mezey, J. Comput. Chem., **9**, 608 (1988).
95. P.G. Mezey, Topology of Molecular Shape and Chirality, in New Theoretical Concepts for Understanding Organic Reactions, J. Bertrán and I.G. Csizmadia, Eds., Kluwer Academic Publishers, Dordrecht, 1989, pp 77-99.
96. P. G. Mezey, The Topology of Molecular Surfaces and Shape Graphs, in Computational Chemical Graph Theory and Combinatorics, D.H. Rouvray, Ed., Nova Publications, New York, 1989.
97. G.A. Arteca, V.B. Jammal, P.G. Mezey, J.S. Yadav, M.A. Hermsmeier and T.M. Gund, J. Mol. Graphics, **6**, 45 (1988).

98. G.A. Arteca and P.G. Mezey, Int. J. Quantum Chem., Quantum Biol. Symp., **14**,133 (1987).

99. F. Harary and P.G. Mezey, J. Math. Chem., **2**, 377 (1988).

100. G.A. Arteca and P.G. Mezey, J. Comput. Chem., **9**, 554 (1988).

101. G.A. Arteca and P.G. Mezey, Int. J. Quantum Chem., **34**, 517 (1988).

102. G.A. Arteca and P.G. Mezey, Int. J. Quantum Chem., Quantum Biol. Symp., **15**, 33 (1988).

103. J. Pipek and P.G. Mezey, Int. J. Quantum Chem. Symp., **22**, 1 (1988).

104. J. Pipek and P.G. Mezey, J. Chem. Phys., **90**, 4916 (1989).

105. G.A. Arteca and P.G. Mezey, J. Math. Chem., **3**, 43 (1989).

106. G.A. Arteca and P.G. Mezey, Theor. Chim. Acta, **75**, 333 (1989).

107. G.A. Arteca and P.G. Mezey, Int. J. Quantum Chem. Symp., in press (1989).

108. G.A. Arteca, G.A. Heal, and P.G. Mezey, Theor. Chim. Acta, in press (1990).

109. P.G. Mezey, J. Amer. Chem. Soc., in press (1990).

110. P. G. Mezey, Theor. Chim. Acta, **54**, 95 (1980).

111. P.G. Mezey, Theor. Chim. Acta, **60**, 409 (1982).

SOME NEW THEORETICAL METHODS FOR TREATING REACTION DYNAMICS IN POLYATOMIC MOLECULAR SYSTEMS

WILLIAM H. MILLER, YAN-TYNG CHANG, and NANCY MAKRI
Department of Chemistry, University of California, and Materials and Chemical Sciences Division, Lawrence Berkeley Laboratory, Berkeley, California 94720

ABSTRACT. Many useful descriptions of dynamical processes in polyatomic molecular systems are based on a reaction path approach. I.e., the potential energy surface is approximated as a multidimensional harmonic valley about a (curved) line in the 3N-6 dimensional space of the N atom system along which the motion is thought to be most localized. These lectures briefly review this reaction path methodology, but then focus mainly on some more recent approaches that go beyond the reaction path point of view. In particular, it is shown how empirical potential functions can be combined with selected *ab initio* calculations within an empirical valence bond model in order to construct a *global* potential energy surface for polyatomic reactions. Two other topics that are discussed are a new way of handling zero point energy in a classical trajectory simulation of polyatomic dynamics and a new model for including tunneling effects in a trajectory simulation.

1. INTRODUCTION

One of the fundamental goals of theoretical chemistry is the quantitative description of chemical reactions from first principles ("*ab initio*"). One usually envisions accomplishing this in two steps, (1) determination of the (Born-Oppenheimer) potential energy surface by state-of-the-art *ab initio* quantum chemistry calculations, and (2) solution of the dynamical equations for nuclear motion on this potential surface. For the simplest chemical reactions A+BC → AB+C, where A, B, and C are all single atoms, this is rapidly becoming a reality: quantum chemistry calculations[1] continue to improve in accuracy, and there have recently been quite dramatic advances[2-4] in quantum mechanical reactive scattering theory that provides the rigorous solution to the nuclear dynamics. The goal of the work described in this paper, however, is to extend these capabilities to *polyatomic* molecular systems.

The first part of the task, i.e., determination of the potential energy surface, at first seems almost insurmountable for a polyatomic

C. Ögretir and I. G. Csizmadia (eds.),
Computational Advances in Organic Chemistry: Molecular Structure and Reactivity, 289–302.

molecular system because of the high dimensionality of the problem: for an N atom system the potential (i.e., the Born-Oppenheimer electronic energy) depends on 3N-6 independent coordinates. If one were to try to map out the potential function in a straight-forward way on a grid of coordinate values, then $\sim 10^{3N-6}$ points - i.e., this many quantum chemistry calculations of the electronic energy - would be required, clearly an impossible task for N greater than 3 or 4.

One of the ways of dealing with this situation is to introduce the idea of a reaction path.[5-7] The most common choice of reaction path is the minimum energy path (MEP), i.e., the steepest descent path (in mass-weighted cartesian coordinates) that descends from the saddle point (i.e., the transition state) of the potential energy surface forward to products and backward to reactants. It seems intuitively clear that this path, also called the "intrinsic" reaction path, passes through the most important part of the potential energy surface for the reaction through this transition state. A full-dimensional potential energy surface is obtained by computing the force constant matrix along the MEP, so that the model is that of a (multidimensional) harmonic valley about the reaction path in the many directions orthogonal to it.

Although the minimum energy path is often the most appropriate choice for the reaction path, there are cases for which this is not so. A most important example of this is an H-atom transfer reaction. For such reactions the minimum energy path has many sharp kinks as it passes from the transition state to the reactants and products minima; it is thus not a useful path on which to base the reaction path model. In these cases it has been shown that a straight line, or diabatic reaction path[8] is more useful.

For either (or any) type of reaction path it is possible to construct a reaction path Hamiltonian[5a] from which the dynamical motion can be determined. The system corresponds to a reaction coordinate, i.e., the distance along the reaction path, plus harmonic oscillator motion (for the many degrees of freedom) about it.

A wide variety of dynamical treatments can be carried out using the reaction path Hamiltonian (RPH).[5] Statistical theories, e.g., transition state theory and its generalizations, are very easily implemented using the RPH. Inelastic energy transfer among various degrees of freedom induced by motion along the reaction path and also tunneling through reaction barriers are conveniently treated via the RPH. It is even possible to carry out quite rigorous quantum mechanical calculations using the RPH. Many applications using these approaches have been carried out in recent years.[9,10]

This reaction path methodology has all been extensively reviewed in the paper[11] written from lectures presented at a recent NATO Advanced Study Institute on "New Theoretical Concepts for Understanding Organic Reactions". The reader is thus referred to this paper for this part of the present lectures.

The present paper supplements the previous one[11] by describing some non-reaction path approaches to treating reaction dynamics in polyatomic systems. A reaction path model is good if the dynamics of interest is related to motion that deviates not too far from the reaction path. Though this may often be true, it is certainly not always the case. In some cases, therefore, it will be necessary to have a global potential energy surface.

Section 2 therefore describes a recent approach we have been pursuing for doing this, namely an empirical valence bond (EVB) model. Warshel,[12] in particular, has utilized this idea to construct potential energy surfaces for truly complex molecules, e.g., proteins. Our interest is in smaller polyatomic systems, but we wish a more quantitative potential surface than biological simulations typically are concerned with. Section 2 describes how we utilize the EVB approach in a more quantitative framework.

Section 3 then describes some new advances in classical trajectory methodology for use in simulations of polyatomic dynamics. In particular, a recently developed approach is described for handling the zero point energy in a classical trajectory simulation of polyatomic dynamics, and a new model for including tunneling effects in trajectory simulations is presented.

2. Global Potential Energy Surface

To be specific, we consider an isomerization process, e.g.,

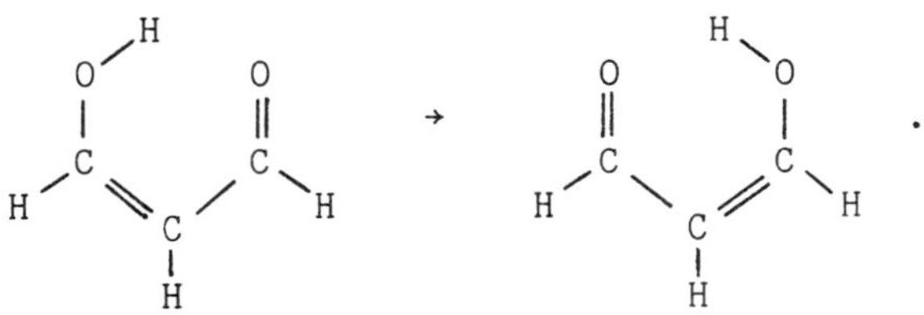

(2.1)

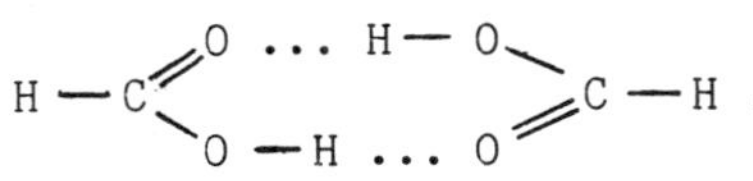

(2.2)

for which the potential surface consists of two local minima separated by a saddle point (in many dimensional space). The idea of the empirical valence bond (EVB) model[12] is to express the potential surface $V(\underset{\sim}{q})$ as the lowest root of a 2x2 secular equation, i.e.,

$$V(\underset{\sim}{q}) = \tfrac{1}{2}[V_{11}(\underset{\sim}{q}) + V_{22}(\underset{\sim}{q})] - \tfrac{1}{2}\{[V_{11}(\underset{\sim}{q}) - V_{22}(\underset{\sim}{q})]^2 + 4V_{12}(\underset{\sim}{q})^2\}^{\frac{1}{2}} . \tag{2.3}$$

In (2.3), $V_{11}(\underset{\sim}{q})$ is a potential energy function that accurately describes the potential in the vicinity of local minimum 1; it has a single local minimum at this geometry. The simplest approximation for $V_{11}(q)$ would be a harmonic approximation about the q_1 (the geometry of local minimum 1), or perhaps more accurately, one might use an empirical potential function, such as the MM2 model,[13] that yields an approximate potential energy surface in the vicinity of q_1. $V_{22}(q)$ is a similar function that describes the potential in the region about q_2, the geometry of the second stable isomer.

One thinks of $V_{11}(q)$ and $V_{22}(q)$ as each describing the potential energy surface for one bonding arrangement (i.e., the left and right hand side of Eqs. (2.1) and (2.2)), or one valence bond configuration; they are sometimes referred to as "diabatic" potential functions. $V_{12}(q)$ is a function which describes the coupling between the two configurations. At any geometry for which

$$V_{12}(q) \ll |V_{11}(q) - V_{22}(q)| , \tag{2.4}$$

Eq. (2.3) gives

$$V(q) \simeq \text{Min}\left(V_{11}(q), V_{22}(q)\right) , \tag{2.5}$$

which will be $V_{11}(q)$ or $V_{22}(q)$ for q near q_1 or q_2, respectively.

As noted above, one has several ways of constructing reasonable approximations to the potential functions V_{11} and V_{22}, each of which corresponds to a single bonding arrangement and thus have only one local minimum. The question is how does one choose the coupling potential $V_{12}(q)$ in a simple, yet accurate fashion. Our answer to this is to choose it so that Eq. (2.3) exactly reproduces an ab initio harmonic approximation about the ab initio transition state geometry. The logic of this prescription is as follows. Empirical approaches (e.g., MM2, etc.) do a reasonably good job of estimating potential functions for stable molecules (and therefore for V_{11} and V_{22}), but they are essentially useless for reactive processes. For medium sized polyatomic systems accurate ab initio calculations are possible for a few selected geometries, even though not at every point in the many dimensional space. The idea is thus to use ab initio input in precisely the region where empirical potential models are unreliable, i.e., the transition state region.

To carry out this idea, note that Eq. (2.3) can be re-written to express $V_{12}(q)$ in terms of $V_{11}(q)$, $V_{22}(q)$, and $V(q)$:

$$V_{12}(q)^2 = [V_{11}(q) - V(q)][V_{22}(q) - V(q)] . \tag{2.6}$$

It is only necessary to know $V_{12}(q)$ accurately near the transition state (because in other regions Eqs. (2.4) and (2.5) pertain so that $V(q)$ is essentially independent of V_{12}), and about the transition state geometry q_0 one makes the following Taylor's series expansions,

$$V(\underset{\sim}{q}) \simeq V_0 + \tfrac{1}{2}(\underset{\sim}{q}-\underset{\sim}{q}_0)\cdot\underset{\approx}{K}_0\cdot(\underset{\sim}{q}-\underset{\sim}{q}_0) \tag{2.7}$$

$$V_{nn}(\underset{\sim}{q}) = V_n + \underset{\sim}{D}_n\cdot(\underset{\sim}{q}-\underset{\sim}{q}_0) + \tfrac{1}{2}(\underset{\sim}{q}-\underset{\sim}{q}_0)\cdot\underset{\approx}{K}_n\cdot(\underset{\sim}{q}-\underset{\sim}{q}_0) \ , \tag{2.8}$$

for n=1, 2, ..., where

$$V_n = V_{nn}(\underset{\sim}{q}_0) \tag{2.9a}$$

$$\underset{\sim}{D}_n = \left(\frac{\partial V_{nn}(\underset{\sim}{q})}{\partial \underset{\sim}{q}}\right)_{\underset{\sim}{q} = \underset{\sim}{q}_0} \ . \tag{2.9b}$$

$$\underset{\approx}{K}_n = \left(\frac{\partial^2 V_{nn}(\underset{\sim}{q})}{\partial \underset{\sim}{q}\partial \underset{\sim}{q}}\right)_{\underset{\sim}{q} = \underset{\sim}{q}_0} \ . \tag{2.9c}$$

The potential and force constant matrix at the transition state, V_0 and $\underset{\approx}{K}_0$, are given by the <u>ab initio</u> calculation, as is the transition state geometry $\underset{\sim}{q}_0$. The values of $V_{nn}(\underset{\sim}{q})$, and its gradient and force constant matrix at the transition state geometry, are evaluated from the empirical functions used for it. Substituting Eqs. (2.7)-(2.9) into (2.6) yields a quadratic approximation for $V_{12}(\underset{\sim}{q})$. We find it useful, furthermore, to exponentiate this expansion, and the result one obtains is

$$V_{12}(\underset{\sim}{q})^2 = A \exp[\underset{\sim}{B}\cdot\Delta\underset{\sim}{q} - \tfrac{1}{2}\Delta\underset{\sim}{q}\cdot\underset{\approx}{C}\cdot\Delta\underset{\sim}{q}] \ , \tag{2.10a}$$

where $\Delta\underset{\sim}{q} = \underset{\sim}{q}-\underset{\sim}{q}_0$, and

$$A = (V_1-V_0)(V_2-V_0) \tag{2.10b}$$

$$\underset{\sim}{B} = \frac{\underset{\sim}{D}_1}{(V_1-V_0)} + \frac{\underset{\sim}{D}_2}{(V_2-V_0)} \ , \tag{2.10c}$$

$$\underset{\approx}{C} = \frac{\cdot\underset{\sim}{D}_2\underset{\sim}{D}_2\cdot}{(V_2-V_0)^2} + \frac{\cdot\underset{\sim}{D}_1\underset{\sim}{D}_1\cdot}{(V_1-V_0)^2} + \frac{\underset{\approx}{K}_0-\underset{\approx}{K}_1}{(V_1-V_0)} + \frac{\underset{\approx}{K}_0-\underset{\approx}{K}_2}{(V_2-V_0)} \ . \tag{2.10d}$$

Eq. (2.10) thus gives the coupling potential as a multidimensional Gaussian function. With Eq. (2.10) for V_{12}, $V(\underset{\sim}{q})$ of Eq. (2.3) has a saddle point at $\underset{\sim}{q}=\underset{\sim}{q}_0$, with value $V(\underset{\sim}{q}_0)=V_0$, and with $\underset{\approx}{K}_0$ as its force constant matrix; i.e., it reproduces the ab initio potential in the vicinity of the ab initio transition state geometry.

It is also clear how one can generalize this procedure to cases where $V(\underset{\sim}{q})$ has more than two local minima: One expresses $V(\underset{\sim}{q})$ as the lowest root of the NxN secular equation (N = number of configurations, i.e., local minima); the diagonal (i.e., diabatic) potential functions $V_{nn}(\underset{\sim}{q})$ are empirical potentials (e.g., MM2) for the various local minima (i.e., bonding arrangements) n=1, ..., N, and the off-diagonal elements $V_{nn'}(\underset{\sim}{q})$ are given by Eq. (2.10) in terms of the ab initio transition state parameters (geometry, barrier height, force constant matrix) for the transition state that connects minimum n to minimum n'.

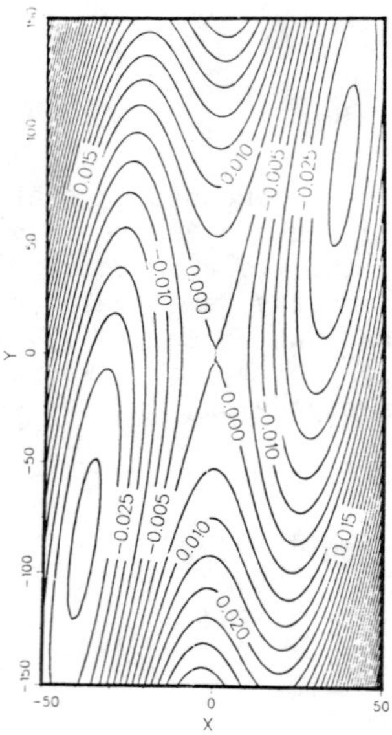

Figure 1. Contour plot of a two dimensional potential surface for formic acid dimer. The energy spacing is 0.005 a.u. The transition state is located at (0,0) with energy set to be 0.0 a.u. The energy of the two minima (locations of the equilibrium state of the two isomers) is about -0.036 a.u. The X axis is the mass-weighted reaction path coordinate which mainly involves the motion of the two tunneling H atoms. The Y axis describes the most strongly coupled normal mode with character of O=C-O motion.

Fig. 1 shows a two-dimensional cut of the EVB potential surface for reaction (2.2), the double H-atom transfer process in formic acid dimer. The normal mode motions for these two degrees of freedom at the transition state are depicted in Figure 2: Figure 2a is mostly motion of the two H atoms that tunnel (the x-coordinate in Fig. 1), and Figure 2b is mostly O=C-O asymmetric stretch and rock (the y-coordinate in Fig 1). This latter mode is the one most strongly coupled to the H atom transfer motion. The diabatic potentials $V_{11}(\underset{\sim}{q})$ and $V_{22}(\underset{\sim}{q})$ in this case are taken simply to be a harmonic normal mode approximation about each local minimum, and $V_{12}(\underset{\sim}{q})$ is given by Eq. (2.10). This EVB potential reproduces the ab initio potential essentially quantitatively over all the relevant region of this two-dimensional configuration space.

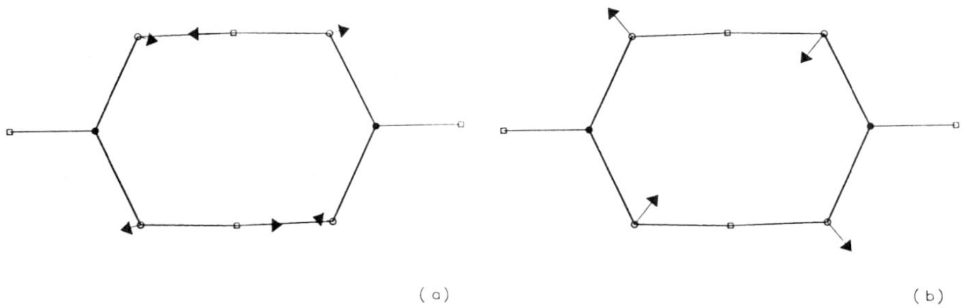

Figure 2. Normal mode representations of the motion of formic acid dimer at the transition state configuration. (a) Motion of the two tunneling H atoms. (b) O=C-O motion: filled circles, open circles and squares stand for C atomc, O atoms, and H atoms, respectively.

3. INCLUDING QUANTUM EFFECTS IN CLASSICAL TRAJECTORY SIMULATIONS

With a global potential surface such as described above, one of the most powerful approaches available to treat dynamics is that of classical trajectory simulations.[14] This is now a widely used technique in many areas of chemistry and biochemistry. There are, however, several serious errors made by these classical methods. In this Section we describe two especially important shortcomings of classical mechanics for polyatomic dynamics and also some recent new work for including the appropriate quantum effects in an approximate framework. The first problem has to do with the zero point vibrational energy and the second with tunneling.

3.1 Simple Model for Handling Zero Point Energy in a Classical Trajectory Simulation

One of the most frustrating shortcomings in using classical mechanics to simulate dynamical processes in polyatomic molecules has to do with a problem involving the zero point energy of vibrational degrees of freedom.[15-17] To describe the problem, recall first the simpler situation of an atom-diatom (gas phase) bimolecular reaction,

$$A + BC(v_i=0) \rightarrow AB(v_f) + C \ , \qquad (3.1)$$

where it is indicated that the reactant diatom BC is initially in its ground vibrational state. It is well-known that a classical trajectory simulation[14] of this process works best if initial conditions for the trajectory are chosen to have the correct zero point vibrational energy in the diatom, with the initial phase of the vibrational motion selected at random (i.e., averaged over), so-called "quasiclassical" initial conditions. Agreement with (the correct) quantum reaction probabilities, or cross sections, would be much worse if the trajectory were begun with no vibrational energy. A problem can arise even for this simple process if the reaction is endoergic and most of the product is produced in $v_f=0$. It is possible classically to obtain reactive trajectories with less than the zero point vibrational energy in the product molecule AB, clearly an unphysical result since this permits reaction below the quantum threshold for the reaction! This problem is usually dealt with[18] by performing the classical simulation always in the exoergic direction and then using microscopic reversibility to obtain probabilities or cross sections in the reverse direction.

One thus believes that a classical simulation of a polyatomic molecular system will mimic nature (i.e., quantum mechanics) more closely if trajectories are begun with (at least) zero point vibrational energy in all vibrational degrees of freedom, with the phases of the vibrational motion selected at random (i.e., averaged over). To simulate vibrational relaxation of CH local mode overtones in benzene,[19,20] for example, it would seem most reasonable to begin trajectories with the appropriate vibrational energy in the CH stretch and zero point vibrational energy in all the other normal modes.

Since the potential energy function for the polyatomic system is in general anharmonic, energy can flow between various degrees of freedom; often, in fact, it is this intramolecular vibrational energy redistribution (IVR) that one is wishing to simulate. The "zero point energy problem"[19,20] mentioned in the first paragraph is that the energy in some vibrational modes may fall below the (quantum) zero point energy ($\frac{1}{2}\hbar\omega_k$, where ω_k is the harmonic frequency for mode k). This may not at first seem like a serious problem, but even in a medium size polyatomic molecule (e.g., benzene) the zero point energy is a sizeable amount of energy (52.2 kcal/mole in benzene). It is a particularly serious problem if the zero point energy flows out of several modes and "pools" into a specific weak bond. For large molecules it may even happen that the classical mechanics is chaotic at its zero point energy. These are all clearly unphysical effects that arise because classical mechanics cannot prevent the energy in each vibrational mode from dipping below its zero point value.

In a recent paper[21,22] a model has been developed for modifying the classical equations of motion in order to remedy thus situation, i.e., to prevent the vibrational energy in each mode from at any time dipping below its zero point value. The algorithm affects the classical trajectory only when the vibrational energy of a mode attempts to decrease below its zero point value; otherwise the trajectory is the ordinary classical one. It conserves the total energy of the polyatomic system, and since it prevents the energy in each mode from decreasing below its zero point value, there can be no unphysical "energy pooling" of the zero point energy from many modes into one bond. The algorithm is actually quite simple: if the energy in any mode k, say, decreases below its zero point value at time t, then at this time the momentum p_k has its sign changed, and the trajectory continues; this is essentially a time reversal for mode k (only!). One can think of the model as supplying impulsive "quantum kicks" to a mode whose energy is trying to fall below its zero point value, i.e., a kind of "Planck demon" analogous to a Brownian-like random force.

No attempt will be made to present the derivation of this approach here, but it is perhaps useful to state it more precisely, at least in its simplest version. Thus let $\underset{\sim}{x}=\{x_k\}$ and $\underset{\sim}{p}=\{p_k\}$, k=1, ..., 3N-6 be the normal coordinates and conjugate momenta of the polyatomic system. The Hamiltonian (for zero total angular momentum) is thus of the simple Cartesian form

$$H(\underset{\sim}{p},\underset{\sim}{x}) = \tfrac{1}{2}\, \underset{\sim}{p}^2 + V(\underset{\sim}{x}) , \qquad (3.2)$$

and the potential energy function consists of a harmonic part plus an anharmonic coupling,

$$V(\underset{\sim}{x}) = V_0(\underset{\sim}{x}) + V_1(\underset{\sim}{x}) , \qquad (3.3a)$$

where

$$V_0(\underset{\sim}{x}) = \sum_k \tfrac{1}{2}\, \omega_k^2\, x_k^2 . \qquad (3.3b)$$

The algorithm may be summarized as follows:

1) Start the trajectory in the appropriate manner (e.g., quasiclassical initial conditions[14]).

2) At the end of each time step in the trajectory, insert the Fortran statement

$$p_k = p_k * \text{Sign}\left(\tfrac{1}{2}\, p_k^2 + \tfrac{1}{2}\omega_k^2 x_k^2 - \tfrac{1}{2}\, \hbar\omega_k\right) , \qquad (3.4)$$

for all k.

3) Keep on computing!

Eq. (3.4) has no effect on the classical trajectory if the energy in mode k ($\tfrac{1}{2}p_k^2 + \tfrac{1}{2}\omega_k^2 x_k^2$) stays above its quantum zero point value ($\tfrac{1}{2}\hbar\omega_k$). If at some time it attempts to dip below this value, though, the sign of p_k is changed, and this has the effect of causing the energy in mode k to increase and thus remain above the zero point value. Since the Hamiltonian involves p_k as p_k^2, this sign change does not change the value of H, i.e., total energy.

The simple version of the model described above pertains to the normal mode vibrational energies. If the dynamics of interest involves motion about a relatively well-defined equilibrium geometry, then this treatment should be adequate. In the more extreme case of a fragmentation process, e.g., unimolecular decomposition, however, it will not be a reasonable description because the physically relevant modes of the system change radically (the frequencies of some vibrations even going to zero as they evolve into rotations). A more general version[21] of the model corresponds to applying essentially the same procedure as above to the vibrational energies of the instantaneous normal modes.

To conclude this description of the model, we note that imposition of the zero point energy constraint within a harmonic approximation should not be a serious limitation. This is because the algorithm affects the classical mechanics of a vibrational mode only when it is near its zero point level, and most such degrees of freedom are reasonably well-described harmonically at this low level of excitation.

Initial applications[21,22] of this approach show that it does indeed prevent the vibrational energy of each degree of freedom from falling below its zero point value. This thus prevents the total zero point energy from "pooling" into any one degree of freedom. Indications are that the model does indeed give a better description of the dynamics than an unconstrained classical simulation.

3.2 A Semiclassical Tunneling Model for Use in Classical Trajectory Simulation

Another serious limitation of classical mechanics, which hinders its application to many interesting chemical problems, is its inability to describe tunneling effects. However, the quantum mechanical phenomenon of tunneling is often quite prominent in chemical reactions that involve significant motion of light atoms. Typical examples include unimolecular dissociation, e.g.,

$$H_2CO \rightarrow H_2 + CO \ , \qquad (3.5)$$

and isomerizations such as Eq. (2.1) and (2.2), as well as bimolecular reactions that involve H atom transfer, e.g.,

$$H_2 + H \rightarrow H + H_2 \ , \qquad (3.6a)$$

$$H_2 + F \rightarrow H + HF \ . \qquad (3.6b)$$

There do exist "rigorous" semiclassical theories that describe how classical trajectories tunnel, e.g., classical S-matrix theory[23] and the "instanton" (periodic orbit in pure imaginary time) model,[24,25] but they are difficult to apply routinely to sizeable (e.g., more than three atom) molecular systems. There also exist a host of simple tunneling corrections to transition state theory[26] expressions for thermal rate constants; these often work well for this purpose, but they are not applicable to more general dynamical phenomena.

What we seek is a semiclassical model, as generally applicable as possible, for including tunneling in a classical trajectory simulation of the full molecular dynamics; such a model has recently been presented.[27] The model is similar in spirit to the Tully-Preston[28] surface hopping model for electronically non-adiabatic processes. In the Tully-Preston model a classical trajectory moving on one potential energy surface (i.e., Born-Oppenheimer electronic state) has a probability of making "hops", i.e., instantaneous transitions, to another potential energy surface at certain times. In the present tunneling model[27] the classical trajectory evolving in one classically allowed region of space has, at specific times, a probability for making an instantaneous (in real time) transition to another classically allowed region of space. The model may also be viewed as the classical version of the semiclassical branching model of Waite and Miller,[29] but generalized to allow for a more general tunneling path. This more general tunneling path is very closely related to that used by Heller and Brown[30] in their semiclassical treatment of radiationless transitions.

Again, the reader is referred to the original paper[27] for the detailed development of the model and its application to a variety of examples, including reactions (2.1) and (2.2), and also to unimolecular decomposition. The examples show that the model provides an excellent description of tunneling in multidimensional systems typical of polyatomic molecules. It is also important to emphasize that the model is implementable for truly multidimensional systems. Applicability was, in fact, the essential criterion kept in mind in constructing the model; there do exist more rigorous semiclassical descriptions of multidimensional tunneling,[23-25] but they are not easily implemented within a standard classical trajectory simulation.

4. CONCLUDING REMARKS

This paper has described a variety of approaches for treating dynamical processes in polyatomic molecules that go beyond the reaction path picture. The first step is devising a tractable procedure for obtaining a global potential energy surface, and the empirical valence bond model described in Section 2 shows how this can be done while still incorporating some ab initio input. Section 3 then noted two recent advances in classical trajectory simulation methodology, namely a new procedure for dealing more realistically with the quantum zero point energy of vibrational degrees of freedom, and a new quite general semiclassical procedure for including tunneling in a trajectory simulation. Both of these developments should help extend trajectory methods to a wider class of phenomena in the dynamics of polyatomic molecules.

ACKNOWLEDGMENT

This research has been supported by the Director, Office of Energy Research, Office of Basic Energy Sciences, Chemical Sciences Division of the U.S. Department of Energy under Contract No. DE-AC03-76SF00098.

REFERENCES

1. See, for example,
   (a) W. J. Hehre, L. Radom, P. v.R. Schleyer, and J. A. Pople, Ab Initio Molecular Orbital Theory, Wiley, N.Y., 1986.
   (b) H. F. Schaefer, The Electronic Structure of Atoms and Molecules, Addison-Wesley, Reading, MA, 1972.
2. (a) J. Z. H. Zhang and W. H. Miller, Chem. Phys. Lett. **140**, 329 (1987).
   (b) J. Z. H. Zhang and W. H. Miller, J. Chem. Phys. **88**, 4549 (1988).
   (c) J. Z. H. Zhang, S. I. Chu, and W. H. Miller, J. Chem. Phys. **88**, 6233 (1988).
   (d) J. Z. H. Zhang and W. H. Miller, Chem. Phys. Lett. **153**, 465 (1988).
   (e) J. Z. H. Zhang and W. H. Miller, J. Chem. Phys. **91**, 317 (1989).
3. (a) K. Haug, D. W. Schwenke, Y. Shima, D. G. Truhlar, J. Z. H. Zhang, and D. J. Kouri, J. Phys. Chem. **90**, 6757 (1986).
   (b) D. W. Schwenke, K. Haug, D. G. Truhlar, Y. Sun, J. Z. H. Zhang, and D. J. Kouri, J. Phys. Chem. **91**, 6080 (1987).
4. For a review, see G. C. Schatz, Ann. Rev. Phys. Chem. **39**, 317 (1988).
5. (a) W. H. Miller, N. C. Handy, and J. E. Adams, J. Chem. Phys. **72**, 99 (1980).
   (b) W. H. Miller, in Potential Energy Surfaces and Dynamics Calculations, edited by D. G. Truhlar, Plenum, New York, 1981, p. 265.
   (c) C. J. Cerjan, S.-h. Shi, and W. H. Miller, J. Phys. Chem. **86**, 2244 (1982).
   (d) W. H. Miller, J. Phys. Chem. **87**, 3811 (1983).
6. For early work on reaction paths and reaction coordinates, see
   (a) S. Glasstone, K. J. Laidler, and H. Eyring, The Theory of Rate Processes, McGraw-Hill, New York, 1941.
   (b) R. A. Marcus, J. Chem. Phys. **45**, 4493, 4500 (1966); **49**, 2610 (1968).
   (c) G. L. Hofacker, Z. Naturforsch. Teil A **18**, 607 (1963); J. Chem. Phys. **43**, 208 (1965).
   (d) S. F. Fischer, G. L. Hofacker, and R. Seiler, J. Chem. Phys. **51**, 3941 (1969).
7. Some of the recent papers by other workers on reaction path models are
   (a) S. F. Fischer and M. A. Ratner, J. Chem. Phys. **57**, 2769 (1972).
   (b) P. Russegger and J. Brickman, ibid. **62**, 1086 (1975); **60**, 1 (1977).
   (c) M. V. Basilevsky, Chem. Phys. **24**, 81 (1977); **67**, 337 (1982); M. V. Basilevsky and A. G. Shamov, ibid. **60**, 349 (1981).
   (d) K. Fukui, S. Kato, and H. Fujimoto, J. Am. Chem. Soc. **97**, 1 (1975); K. Yamashita, T. Yamabe, and K. Fukui, Chem. Phys. Lett. **84**, 123 (1981); A. K. Fukui, Acc. Chem. Res. **14**, 363

(1981).
(e) K. Ishida, K. Morokuma, and A. Komornicki, J. Chem. Phys. **66**, 2153 (1977).
(f) A. Nauts and X. Chapuisat, Chem. Phys. Lett. **85**, 212 (1982); X. Chapuisat, A. Nauts, and G. Durrand, Chem. Phys. **56**, 91 (1981).
(g) J. Pancir, Collect. Czech. Commun. **40**, 1112 (1975); **42**, 16 (1977).
(h) G. A. Natanson, Mol. Phys. **46**, 481 (1982).

8. (a) B. A. Ruf and W. H. Miller, J. Chem. Soc. Faraday Trans. 2 **84**, 1523 (1988).
(b) W. H. Miller, B. A. Ruf, and Y. T. Chang, J. Chem. Phys. **89**, 6298 (1988).
9. (a) S. K. Gray, W. H. Miller, Y. Yamaguchi, and H. F. Schaefer, J. Chem. Phys. **73**, 2733 (1980).
(b) S. K. Gray, W. H. Miller, Y. Yamaguchi, and H. F. Schaefer, J. Am. Chem. Soc. **103**, 1900 (1981).
(c) Y. Osamura, H. F. Schaefer, S. K. Gray, and W. H. Miller, J. Am. Chem. Soc. **103**, 1904 (1981).
(d) B. A. Waite, S. K. Gray, and W. H. Miller, J. Chem. Phys. **78**, 259 (1983).
10. See also,
(a) R. T. Skodje, D. G. Truhlar, and B. C. Garrett, J. Phys. Chem. **85**, 3019 (1981).
(b) R. T. Skodje, D. G. Truhlar, and B. C. Garrett, J. Chem. Phys. **77**, 5955 (1982).
(c) A. D. Isaacson and D. G. Truhlar, ibid. **76**, 1380 (1982).
(d) D. G. Truhlar, N. J. Kilpatrick, and B. C. Garrett, ibid. **78**, 2438 (1983).
(e) R. T. Skodje and D. G. Truhlar, ibid. **79**, 4882 (1983).
(f) R. T. Skodje, D. W. Schwenke, D. G. Truhlar, and B. C. Garrett, J. Phys. Chem. **88**, 628 (1984).
(g) B. C. Garrett and D. G. Truhlar, J. Chem. Phys. **81**, 309 (1984).
11. W. H. Miller, in New Theoretical Concepts for Understanding Organic Reactions, eds. J. Bertran and I. G. Csizmadia, ASI Series, Vol. 267, Reidel, 1989, pp. 347-372.
12. A. Warshel and R. M. Weiss, J. Am. Chem. Soc. **102**, 6218 (1980); A. Warshel, Biochemistry **20**, 3167 (1981).
13. N. L. Allinger, J. Am. Chem. Soc. **99**, 8127 (1977).
14. For reviews, see
(a) R. N. Porter and L. M. Raff, in Dynamics of Molecular Collisions, B, ed. W. H. Miller, Plenum, NY, 1976, p. 1.
(b) L. M. Raff and D. L. Thompson, in Theory of Chemical Reaction Dynamics, Vol III, ed. M. Baer, CRC Press, Boca Raton, FL, 1985, p. 1.
15. R. A. Marcus, Ber. Bunsenges. Phys. Chem. **81**, 190 (1977).
16. W. L. Hase and D. G. Buckowski, J. Comput. Chem. **3**, 335 (1982).
17. G. C. Schatz, J. Chem. Phys. **79**, 5386 (1983).
18. See, for example, J. M. Bowman, G. C. Schatz, and A. Kuppermann, Chem. Phys. Lett., **24**, 378 (1974).

19. D.-H. Lu and W. L. Hase, J. Chem. Phys. **89**, 6723 (1988).
20. D.-H. Lu and W. L. Hase, J. Chem. Phys., submitted.
21. W. H. Miller, W. L. Hase, and C. L. Darling, A Simple Model for Correcting the Zero Point Energy Problem in Classical Trajectory Simulations of Polyatomic Molecules, J. Chem. Phys. **91**, 0000 (1989), in press.
22. See also, J. M. Bowman, B. Gazdy, and Q. Sun, J. Chem. Phys. **91**, 000 (1989), in press.
23. (a) W. H. Miller, Adv. Chem. Phys. **25**, 69 (1974).
    (b) W. H. Miller, Adv. Chem. Phys. **30**, 74 (1975).
    (c) W. H. Miller, Science **233**, 171 (1986).
24. W. H. Miller, J. Chem. Phys. **62**, 1899 (1975).
25. (a) S. Coleman, in The Whys of Subnuclear Physics, edited by A. Zichichi, Plenum, N.Y., 1979, pp. 805-916.
    (b) A. O. Caldiera and A. J. Leggett, Ann. Phys. **149**, 374 (1983).
26. See, for example,
    (a) D. G. Truhlar and B. C. Garrett, Ann. Rev. Phys. Chem. **35**, 159 (1984).
    (b) G. C. Lynch, D. G. Truhlar, and B. C. Garrett, J. Chem. Phys. **90**, 3102 (1989).
    (c) R. A. Marcus and M. E. Coltrin, J. Chem. Phys. **67**, 2609 (1977).
    (d) C. J. Cerjan, S.-H. Shi, and W. H. Miller, J. Phys. Chem. **86**, 2244 (1982).
27. N. Makri and W. H. Miller, "A Semiclassical Tunneling Model for Use in Classical Trajectory Simulations", J. Chem. Phys. **91**, 0000 (1989), in press.
28. J. C. Tully and R. K. Preston, J. Chem. Phys. **55**, 562 (1971).
29. B. A. Waite and W. H. Miller, J. Chem. Phys. **76**, 2412 (1982).
30. E. J. Heller and R. C. Brown, J. Chem. Phys. **79**, 3336 (1983).

THEORETICAL STUDY OF HOMOGENEOUS CATALYSIS

J. Bertrán
Departament de Química
Universitat Autònoma de Barcelona
08193 Bellaterra (Barcelona)
Spain

ABSTRACT. According to the degree of participation of the catalyst parameters in the internal reaction coordinate, one can characterize different catalytic processes. The acid-catalyzed hydrogenation of ethylene, the Friedel-Crafts reaction and the charge transfer processes catalyzed by external electric fields have been selected as examples ranging from a large to a small intervention of the catalyst in the internal reaction coordinate. In the extremum case of an external perturbation, such as that introduced by an electric field, the perturbation belongs also to a correct representation of the reaction coordinate, which cannot be defined exclusively in terms of internal parameters of the chemical system. This new concept opens the possibility of catalyzing a reaction without any catalyst.

1. Introduction

There has been lately an increase in importance and a fast development of homogeneous catalysis. This development is connected to the progress in organometallic chemistry. By means of an adequate selection of metal atoms and ligands, very specific catalysts can be found.

Owing to its simplicity, homogeneous catalysis is especially suited to analyze the factors determining catalytic phenomena. However, since the catalyst modifies the reaction mechanism, thus acting in a very specific way, it is difficult to propose general models. According to the degree of participation of the catalyst parameters in the internal reaction coordinate, one can characterize different processes. On one hand, one can consider bifunctional catalysis, whereas on the other hand one can consider the catalysis originated by ions. Non-intervention of the catalyst in the internal reaction coordinate is even clearer when an external perturbation like an electric field turns out to be the catalyst.

To get an insight in homogeneous catalysis we will select some examples. As an example of bifunctional catalysis the acid-catalyzed hydrogenation of ethylene will be presented and compared with its catalysis by organometallic compounds and with heterogeneous catalysis. A second example will deal with a Friedel-Crafts reaction where both a bifunctional catalysis and a catalysis by an external ion may be found. As a third example, charge transfer processes will be taken into account; proton transfer will be analyzed regarding the catalysis by external ions, as the electron transfer will be regarding the effect of a uniform electric field. Finally, the possibility of carryng out a catalysis without the presence of any catalyst but rather by the effect of external perturbations, like electric and magnetic fields or pressure, will be discussed.

C. Ögretir and I. G. Csizmadia (eds.),
Computational Advances in Organic Chemistry: Molecular Structure and Reactivity, 303–321.

2. Hydrogenation of Ethylene

The molecular addition of the hydrogen molecule to ethylene is a prototype of a symmetry-forbiden reaction. Moreover, recent theoretical calculations have shown that the barrier for a reduced symmetry path is also very high [1]. It is thus not surprising that catalysis of this reaction has been a matter of great interest.

Heterogeneous catalysis of this reaction has been interpreted through a change in symmetry of the H_2 HOMO when it is adsorbed [2]. Nevertheless, the process is far more complex. Recently, state-of-the-art techniques applied do Pt and Rh single crystals, which act as catalysts for the hydrogenation of ethylene, have shown that the reaction occurs on an active overlayer rather than on a metal surface [3-5]. At the temperature of 300-400 K, ethylene is adsorbed as ethylidyne, which is present on the metal surface during ethylene hydrogenation reaction. The adsorbed ethylidyne overlayer facilitates the transfer of hydrogen, dissociated by the metal surface, to the ethylene that is weakly adsorbed to the ethylidyne overlayer. The role of the metal in this case is to dissociate the hydrogen molecule, and to act as a support for the active ethylidyne species. This mechanism changes dramatically going from gas-solid to liquid-solid interface when electrochemical hydrogenation of ethylene at the platinum surface takes place. In the hydrogenation at the liquid-solid interface, ethylene is reduced on the Pt surface by adsorbed hydrogen atoms. The interfering hydrocarbon layer is reduced in the liquid-solid reaction by electrochemical reduction of most of the chemisorbed ethylene. Owing to the absence of a blocking hydrocarbon layer, ethylene molecules can react directly with adsorbed atoms in the liquid-solid interface.

In 1965, Wilkinson [6], using the $Rh(PPh_3)_3Cl$ complex as a catalyst, found the first efficient way to catalyze homogeneously the hydrogenation of olefins. The most commonly accepted mechanism of this catalytic process is due to Halpern [7-9]. It consits of oxydative addition of H_2, coordination of ethylene, ethylene insertion, isomerization, and reductive elimination of ethane. Thus, in this homogeneous catalysis the first step turns out to be also the cleavage of H_2, which originated the large barrier in the uncatalyzed process.

A third way to catalyze olefins hydrogenation involves acids [10,11]. The main steps of the proposed mechanism consist of an initial protonation of the olefin to yield a carbocation, and the attack of a carbocation to molecular hydrogen. In this last step, the hydrogen molecule is broken, the proton is regenerated, and the hydrogenated olefin is obtained. In this mechanism, the first step is protonation.

Many theoretical studies on the main step of the heterogeneous catalysis, the absorption of molecular hydrogen onto the metal surface, have appeared in the last decade [12,13]. The different steps of Halpern's mechanism for the homogeneous catalysis by the Wilkinson catalyst have been investigated as well [14-18]. However, no theoretical studies have been accomplished on the acid-catalyzed hydrogenation of olefins. To explore theoretically this last type of catalytic hydrogenation, our group has compared three processes:

$$C_2H_4 + H_2 \rightarrow C_2H_6$$

$$C_2H_4 + H_2 + HF \rightarrow C_2H_6 + HF$$

$$C_2H_4 + H_2 + H_3O^+ \rightarrow C_2H_6 + H_3O^+$$

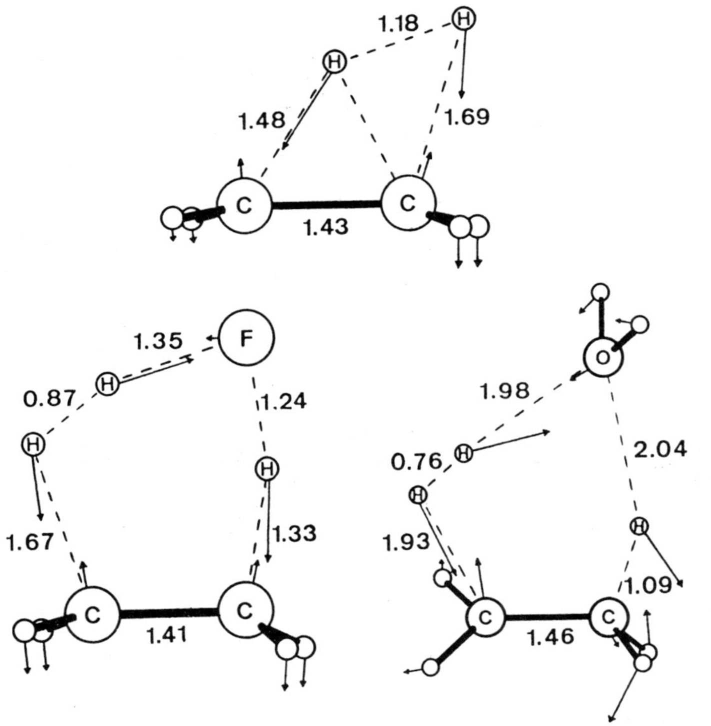

Figure 1: 3-21G optimized structures for the transition states of the three reactions. Distances are given in Å. Arrows indicate the main components of the transition vector.

where HF and H_3O^+ have been chosen as examples for weak and strong acids respectively [19].

2.1. ACID-CATALYZED HYDROGENATION OF OLEFINS

The H_2-, HF- and H_3O^+-C_2H_4 complexes exhibit a C_{2v} symmetry. The distances between the hydrogen atom and the midpoint ot the C-C bond decrease in the series of mentioned complexes. In the trimolecular complexes, the hydrogen molecule is loosely bonded to the acid fragment, and represents just a small perturbation of the ethylene-acid system.

The structures of transition states are depicted in Figure 1. Arrows indicate the main components of the transition vector. Comparison between reactants and transition states in catalyzed processes shows that in transition states the hydrogen molecule is placed in a suitable position to react with ethylene, while in the reactants the hydrogen molecule is far away from ethylene. If the three transition states are compared, one can see that the hydrogen molecule is more broken in the uncatalyzed process than in one catalyzed by FH, whereas in a process catalyzed by H_3O^+ the transition state structure exhibits an almost undistorted hydrogen molecule. Looking to the main components of the transition vector in the uncatalyzed process, the motion proceeds toward formation of ethane, while in the catalyzed processes the main motions correspond to a proton transfer from the acid to ethylene, and a simultaneous transfer of one hydrogen to fluorine or oxygen atoms, together with another hydrogen to ethylene. This transfer produces the rupture of the hydrogen molecule. The proton transfer is far more advanced in the third reaction than in the second one.

Using the optimized 3-21G structures described above, the scheme of energetic levels presented in Figure 2 is obtained. The activation energy from separated reactants or from trimolecular complexes decreases noticeably when the process is catalyzed by a weak acid, and it decreases dramatically when catalyzed by a strong acid, in such a way that the last mechanism appears as an efficient way of catalysis. Quantitative changes, if zero-point and correlation energies are taken into account, or a larger basis set like 6-31G** is used, are always small when compared to barrier height differences. The energetic results can be strongly influenced when free energy is considered, since catalyzed trimolecular systems are highly structured, and are thus unfavoured from an entropic point of view. In fact, in the scheme of free energies (fig. 2), all intermediates dissappear with the exception of $C_2H_4 \cdot H_3O^+ \cdot H_2$ which is stabilized with respect to separated reactants by -7.9 Kcal/mol. The entropic term appears as more unfavourable in the catalyzed than in the uncatalyzed reaction when free energy is taken into account. Nevertheless, the catalyzed processes appear to be more favorable than the uncatalyzed ones, and energy barrier heights are heavily dependent on the acid strength.

When analyzing geometries and transition vectors of the transition states, a concerted process arises where the acid acts as a bifunctional catalyst. One can see that the acid gives a proton to ethylene, and accepts a proton from H_2, so it collaborates in the formation of the nucleophile (hydride ion), which is neutralizing the emerging carbocation. The big difference in potential barrier in the catalyzed reaction comes from the smaller breakage of H_2 in the transition state, which translates into a smaller distortion energy, which is the main component of the barrier.

The acid catalysis of ethylene hydrogenation proceeds via the formation of a six-membered ring which involves both carbons of ethylene, the hydrogen molecule, the oxygen or fluorine atom, and the acidic proton. However, differences between a normal acid and a strong acid arise in reactions two and three. In the reaction catalyzed by HF, the bifunctional catalysis is very concerted. The cleavage and formation of bonds is halfway. The acid acts as a catalyst that simultaneously releases and accepts a proton. In this way, the acid contributes to the breakage of the hydrogen molecule. On the contrary in the reaction catalyzed by a strong acid such as H_3O^+, the first phase of the process consists almost of the full proton transfer from the acid to ethylene to form the carbocation. In a second phase the hydrogen molecule is broken owing to the action of the carbocation helped by the conjugated base of the acid. This mechanism is almost coincident with that proposed starting from experimental data, where a carbocation is formed in a first step, and a hydride ion attacks later the carbocation to yield the alkane. The theoretically proposed mechanism does not proceed through two steps but via a unique step with two phases. A two step process, as that which might be obtained if a polar medium had been taken into account, would not change the bifunctional character of the process.

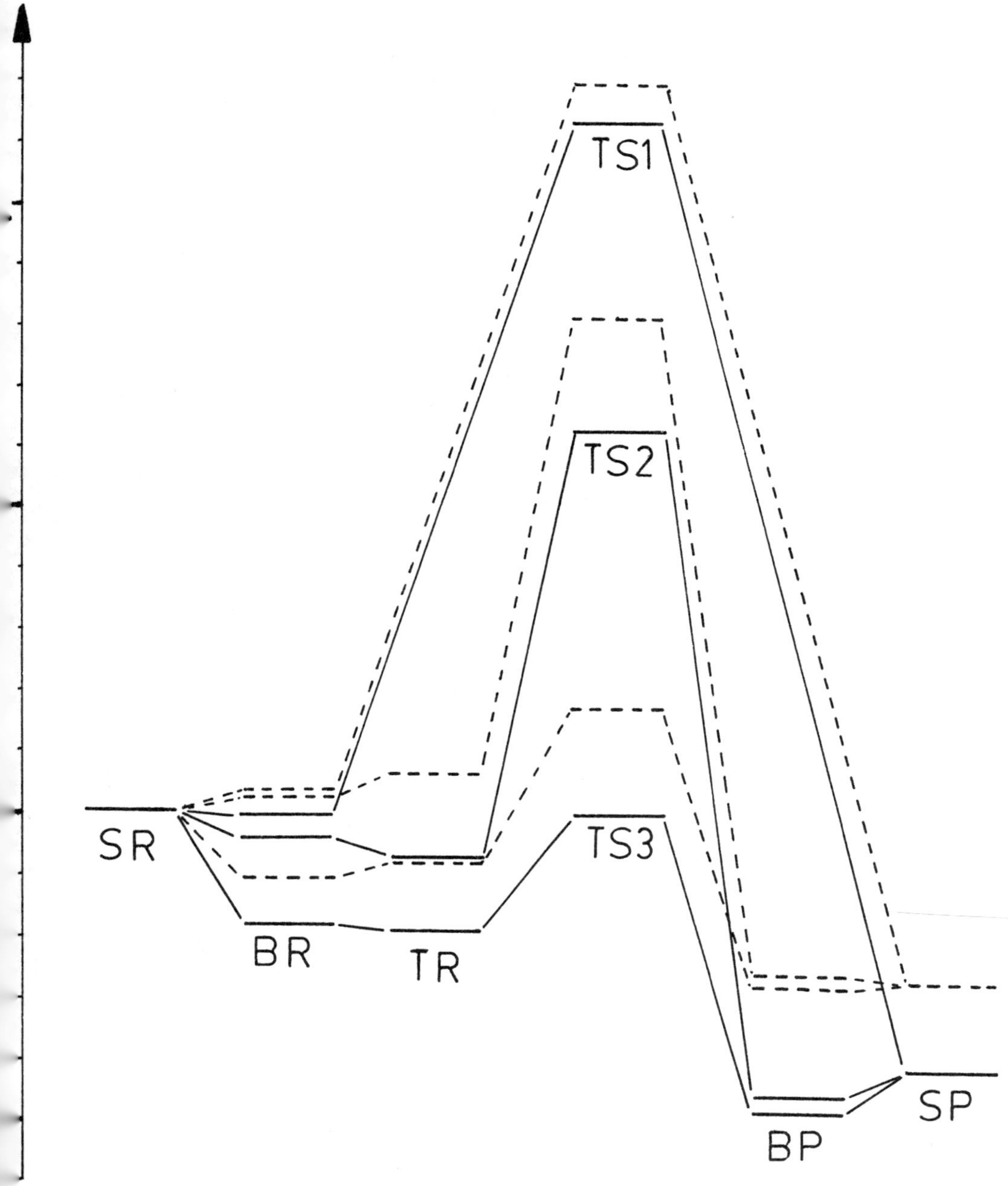

Figure 2: Relative 3-21G potential energies (—–) and Gibbs freen energies (– – –) for the uncatalyzed reaction 1, and for the HF- and H_3O^+-catalyzed reactions 2 and 3, respectively. Energies are given in kcal/mol. SR and SP stand for separated reactants and products, respectively. BR and BP stand for bimolecular reactant complexes and product complexes, respectivley. TR stands for trimolecular reactant complexes. Finally, TS stands for transition states.

If we compare the acid catalysis with the hererogeneous or homogeneous catalysis by metals, a remarkable difference exists. Whereas metal catalysis implies the breakage of the hydrogen molecules as a first step, the catalysis by a strong acid requires the formation of the carbocation in the first phase. The catalysis by normal acids such as HF exhibits an intermediate behavior, because the hydrogen molecule rupture and the carbocation formation proceed simultaneously, as seen in the concertedness of the reaction.

2.2. THE WILKINSON CATALYST.

Let us now get an insight on organometallic catalysis. Morokuma [17,18] has studied the full catalytic cycle of the mechanism proposed by Halpern, where the active species is $RhCl(PPh_3)_2$. The geometries of the intermediates and the transition states have been determined at the Hartree-Fock level of theory through gradient methods. In this way the potential energy profile for the full catalytic cycle has been obtained. This profile is remarkably smooth, no barrier being too high. No intermediate is too stable so the chain of reactions may be broken. Any good catalytic system, of course, must satisfy these criteria, and quite certainly the Wilkinson catalyst system does.

The first two reactions, oxydative addition of H_2 and coordination of olefin, are found to be exothermic. In the oxydative addition of H_2, the formation of a H_2 complex is followed by a transition state with a very small barrier. Thus, the first two steps of Halpern's cycle should take place easily. The olefin migratory insertion has a barrier of 18.4 Kcal/mol. The resulting trans hydride alkyl complex isomerizes with a small barrier very exothermically to yield a cis complex. The olefin insertion combined with the first portion of isomerization constitutes the rate-determining step. The nearly thermoneutral reductive elimination of C_2H_6 takes place with a modest barrier.

The main interest of this study is that the potential energy profile may be compared with the profile obtained changing the metal atom and the ligands in order to sketch the best catalyst for this reaction. For instance, one can compare the potential energy profile between the Rh and the Pt catalytic systems. For the Rh complex, the β-elimination from a stable cis-ethyl hydride complex requires a higher activation energy than that for the final catalytic step of reductive elimination. On the contrary for the Pt complex, the β-elimination exhibits a smaller barrier [20] than the activation barrier for alkane reductive elimination. Thus Halpern's cycle does not take place and the Pt system is not a good catalyst for ethylene hydrogenation.

According to our initial classification of homogeneous catalysis the acid-catalyzed hydrogenation of ethylene was a bifunctional catalysis, because the acid intervenes in the internal reaction coordinate, releasing and accepting a proton. In some way the Wilkinson catalyst can also be classified as bifunctional, because the organometallic complex intervenes in the internal reaction coordinate, giving and accepting electrons. The first step was an oxydative addition and the last step a reductive elimination. This role is even clearer if one pays attention to the rate-determining step, olefin insertion. Morokuma [18] has shown that the reaction takes place via hydride migration to the coordinated olefin and not via olefin insertion into the M-H bound. Hence, it is a nucleophile attack on an ethylene activated by metal complexation, where the metal acts as an electron acceptor. In one hand the metal releases electrons to generate the nucleophile agent, hydride ion; and on the other hand it receives electrons from ethylene with a deficiency in electrons initially in the double bond and finally in the carbon atom bond to the metal.

3. Friedel-Crafts Reaction

The Friedel-Crafts reaction is one of the most important examples of a catalytic process in Organic Chemistry. Among these reactions, the alkylation of alifatic and aromatic compounds by alkyl halides in the presence of acids has been subject of a great interest [21-23].

Few theoretical studies have been devoted to the Friedel-Crafts reaction [24-28]. Most of them have been restricted to consider the formation of the donor-acceptor complexes between alkyl halides and either BF_3 [24,27] or H^+ [26]. Only two works [25,28] have studied the electrophilic attack step. In particular, our group has studied the catalytic action of H^+ and BF_3 in the reaction between CH_4 and HF, which is taken as a model of a Friedel-Crafts process [28,29]. The potential barrier of the electrophilic attack decreases dramatically upon catalysis by H^+ . When the reaction is catalyzed by BF_3, two different mechanisms have been investigated. While one of them corresponds to a normal acid catalysis, the other involves the BF_3 molecule acting as bifunctional catalyst. It turns out that the second mechanism is clearly more efficient. The difference between both types of catalysis may be easily understood if attention is focused on transition state structures. In the uncatalyzed process the internal reaction coordinate is defined by the two C-H and the two F-H distances, so a four-centered cyclyc structure is formed. The internal reaction coordinate does not change in the normal acid catalysis, but in the bifunctional catalysis a six-membered ring is formed and the geometrical parameters of the catalyst intervene in the internal reaction coordinate. In this second mechanism the BF_3 molecule acts as a bifunctional catalyst because in the process it receives a fluorine atom from de chemical system and on the other hand it releases another fluorine atom.

Paying attention to the catalysis by a proton, some important differences appear between the transition state structures of the uncatalyzed and catalyzed processes. The C-H and F-H distances change form 1.37 and 1.20 Å respectively to 1.20 and 1.80 Å at the 3-21G level. The electron density plots show that, whereas in the uncatalyzed process there is a ring critical point, none is found in the catalyzed process, so the transition state is not a real cyclic structure. It turns out to be an intimate ion-dipole pair, as shown from the positive charges of the CH_5 fragment and HF (0.91 and 0.09 respectively) [28,29].

The fact that catalyst parameters do not belong to the internal reaction coordinate does not mean at all that they do not participate in the reaction coordinate. As a matter of fact the F-H distance changes from 0.98 Å in HFH^+ to 0.94 Å in the transition state, whereas it is 0.97 Å in the intermediate. The HFH angle also changes from 123 to 165 degrees, whereas its value is 125 degrees in the intermediate. An inspection of the main components of the transition vector shows that the components corresponding to the proton are very important. Likewise, there is an increase in electron transfer from the chemical system to the proton between from reactants to the transition state. The positive charge of the hydrogen atom, is 0.65 au in HFH^+, 0.62 au in the intermediate, and 0.54 au in the transition state. There is an electron transfer from the chemical system to the catalyst in the first phase of the process, and from the catalyst to the chemical system in the second phase, so this catalysis exhibits some bifunctional character.

If the proton is replaced by a positive charge, it will be possible to separate the electrostatic effect from the charge transfer effect. In figure 3 we present the potentical energy profiles of the uncatalyzed process, the process catalyzed by a proton, and the process catalyzed by a positive charge placed at the same distance of the proton. The

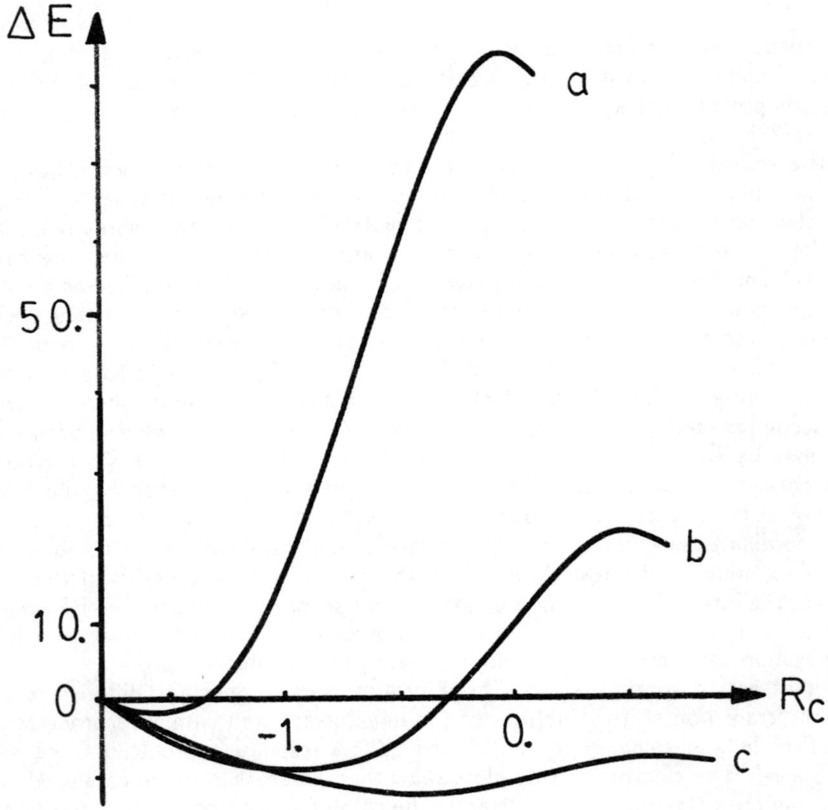

Figure 3: Potential energy profiles of the CH_4 + HF process (a) uncatalyzed (b) catalyzed by a proton and (c) catalyzed by a positive charge. Energies are given in kcal/mol.

reaction coordinate, Rc , is defined as the difference between the H-F and H-C distances. One can see that the position of the stationary points and the height of the potential barriers change dramatically on going from the uncatalyzed to the catalyzed processes. There are differences indeed between the processes catalyzed by a proton and by a positive charge, which must be assigned to the charge transfer term. However the electrostatic term alone furnishes a large amount of the proton catalytic effect. In both transition state structures both C-H distances are 1.20 Å. Further the positive charge on the CH_5 fragment is 0.87 au when the catalyst is a positive charge, and 0.91 au when the proton is the catalyst. Obviously the electric field created by a point charge is only an external perturbation to the chemical system.

When a positive charge is the catalyst, the fluorine atom supports a large negative charge, since no electron transfer is allowed. The use of diffuse functions will help in describing more properly these polar structures. Hence, the 3-21+G basis set [30] has been used. It is interesting to compare the tansition state structures of the uncatalyzed process obtained by means of the 3-21G and 3-21+G basis sets [31]. The most important change appears in the different lenghts of the C-H and H-F bonds in the ring. Whereas with the 3-21G basis set these two distances are 1.37 and 1.26 Å respectively, with the 3-21+G basis set they are 1.26 and 1.42 Å. Thus, with the larger basis set the hydrogens are closer to the carbon atom than they are to fluorine. The positive charge on the CH_5 fragment increases from 0.52 au to 0.65 going from the 3-21G to the 3-21+G basis set. Finally, the difference between both basis sets can be seen more clearly by looking at the electron density plots. Whereas the 3-21G basis set yields a ring point, with the 3-21+G basis set this ring point disappears and two different ionic fragments with an electrostatic interaction are seen.

3.1. EFFECT OF ELECTRIC FIELDS

Let us now examine the effect of a non-uniform electric field, as created by a positive charge placed at several distances from the fluorine atom, and the effect of a uniform electric field with different intensities on the reaction between CH_4 and HF [31]. The electric field is introduced through addition of the electron-field interaction term to the one-electron hamiltonian. It must be mentioned that in all cases the complete set of geometrical parameters of the perturbed systems have been optimized to obtain a stationary point, so the systems have aligned freely with respect to the electric field.

In tables 1 and 2 we present the values of Rc, defined as the difference between the C-H and H-F distances, and the charges on the fluorine atom, Q_F, for the intermediates and transition states of the processes catalyzed by a positive charge (table 1) or by a uniform electric field (table 2). In addition, the potential barriers are also presented. d is the distance between the positive charge and the fluorine atom, whereas F stands for the intensity of the uniform electric field.

When the strength of the perturbation is increased, the intermediate is found later along the reaction coordinate. It means that the transfer of the hydrogen to the alkyl group is more advanced the stronger the perturbation. The ion-pair character of the complex also increases with the strength of the perturbation. Paying attention to the transition states, one can see that upon increase of the field strength they are found also later along the reaction coordinate. In parallel an increase in the ion-pair character of the transition state is found. Finally, the potential barriers relative to the intermediates

Table 1: Values of R_C and Q_F for the stationary points and potential barrier of the reaction between CH_4 and HF catalyzed by a positive charge placed at a distance **d** from the fluorine atom.

	Intermediate		Transition State		
d	R_C	Q_F	R_C	Q_F	$\Delta E^{\#}$
∞	-1.56	-0.50	0.16	-0.65	84.45
5	-1.45	-0.53	0.25	-0.70	78.01
3	-1.34	-0.57	0.35	-0.81	66.29
2	-1.16	-0.65	0.43	-0.90	47.87
0.94	-0.63	-0.80	0.55	-0.97	11.50

Table 2: Values of R_C and Q_F for the stationary points and potential barrier for the reaction between CH_4 and HF catalyzed by a uniform electric field of intensity **F** (in au, $1au=5.14 \cdot 10^{11} V/m$).

	Intermediate		Transition State		
F	R_C	Q_F	R_C	Q_F	$\Delta E^{\#}$
0	-1.56	-0.50	0.16	-0.65	84.45
0.01	-1.43	-0.53	0.30	-0.75	76.24
0.02	-1.32	-0.57	0.43	-0.84	65.19
0.03	-1.22	-0.60	0.73	-0.94	50.60

decrease when the positive charge-fluorine distance is shortened or when the intensity of the uniform electric field is augmented. Given that the transition states exhibit a higher charge separation that the intermediates do, the transition states are more stabilized by the electric fields than the intermediates. It must be remarked that the stabilization energy at the intermediates by the effect of the electric fields used is far higher than the thermal agitation energy, so the chemical system will align spontaneously along the direction of the electric field.

The positions of the stationary points along the reaction coordinate and the barrier heights are dramatically altered by an increase in the intensity of a uniform electric field or by a shortering of the fluorine-positive charge distance. Therefore, the external perturbation of an electric field belongs to a correct representation of the reaction coordinate, which cannot be defined exclusively in terms of the internal parameters of the chemical system. To get a deeper insight into this participation, we have applied the uniform electric field with various intensities, and placed the positive charge at several distances from the fluorine atom at the field-free optimized intermediate. Further, geometries have not been allowed to relax.

Let us discuss, as an example, the results obtained when a uniform electric field of 0.03 au is applied, paying attention first to the charge on the fluorine atom. The negative charge increases from -0.50 au to -0.58 au when the electric field is applied. Thus the ion-pair character of the complex is augmented, which represents an advance in the reaction from an electronic point of view. If attention is paid to the values of the electron density at the C-H and H-F bond critical points, its value increases from 0.0073 to 0.0079 at the C-H bond and decreases from 0.3165 to 0.3047 at the H-F bond upon introduction of the electric field. According to the relationship between electron densities in bond critical points and bond strengths [32], the C-H bond is strengthened and the H-F is weakened as effect of the electric field. A similar modification in the electron density could have been produced by a shortening of the C-H distance and a lengthering of the H-F distance. These two distances obviously belong to the reaction coordinate. Since the electric field causes an effect identical to that obtained by those variations, one can say that the electric field belongs to the reaction coordinate, which must be understood in a less restrictive sense than usual.

In order to understand the reasons of the evolution of the intermediate along the reaction coordinate, we must look at the forces acting on the nuclei of the intermediate. Obviously the forces acting on the nuclei of the field-free intermediate are zero in absence of perturbation. However, upon introduction of the mentioned electric field, a force (-0.0079 au) appears trying to shorten the C-H distance and another one (0.0166 au) trying to lengthen the H-F distance. When an electric field is present, nuclei are subject to a couple of forces of opposite sign. First of all, one appears opposing the field owing to the direct effect of the electric field on positive charges of nuclei. Second, an indirect force appears in the direction of the field, arising from electronic cloud polarization. As a result, depending on the net charge on each atom, a nucleus will be subjected to a force in the direction of the field, if its overall charge is negative, and opposite to the field, if it is positive. In the present system, the total charges on fluorine and carbon are negative, whereas the charges on hydrogen are positive. Thus it can be understood that the induced forces will tend to transfer the hydrogen atom from fluorine to carbon, thus making the intermediate to advance in the reaction coordinate, and leading to a decrease in energy barrier.

In conclusion, the strong polar character of the transition state of the reaction between CH_4 and HF shows that any electric field will align the chemical system in a suitable way for the process to take place. Likewise, the applied field will dramatically lower the energy barrier of the process. Thus, the effect of an electric field can be invoked as one of the reasons of the efficiency of the acid catalysis in Friedel-Crafts reactions. Furthermore, it leads us to better understand the reason why a Friedel-Crafts reaction is largely accelerated in solution. In this case, polar character of the intermediate and the transition state creates a reaction field. Given that the polar character is higher at the transition state than the intermediate, the former will be more stabilized than the latter because the field will be more intense and the interaction between the electric field and the chemical system will be larger. Finally it is worth mentioning that this reaction is especially suited to be catalyzed by strong uniform electric fields, because the chemical system is aligned in a suitable way. Such fields are nowadays accessible by means of different experimental techniques, and may provide a new way to carry out an efficient Friedel-Crafts reaction.

4. Charge Transfer Processes

Table 3: Geometrical parameters and charge on the metal of the intermediate of the $M(H_2O)_2^{+2}$ and $H_5O_2^+$ systems.

M	H-O1	O1-H	H-O2	Q_H
H^+	0.97	1.19	1.19	0.50
Be^{+2}	1.42	1.71	1.01	1.37
Mg^{+2}	1.83	1.15	1.26	1.70
Zn^{+2}	1.77	1.28	1.14	1.59

The charge transfer processes may be especially sensitive to the effect of external electric fields. As a first example, the proton transfer will be analyzed regarding the catalysis by external ions and, as a second example, electron transfer will be analyzed as to the effect of a uniform electric field.

4.1. PROTON TRANSFER

The proton transfer process in the water dimer complex catalyzed by three divalent metal ions (Be^{2+}, Mg^{2+} and Zn^{2+}) has been studied using the 3-21G basis set. Metal ions have been shown to be essential in enzymatic catalysis. In particular, the zinc dication acts very efficiently in hydration and hydrolisis reactions. In the processes catalyzed by carbonic anhydrase, the active species has been suggested to be $ZnOH^+$ [33]. Our aim is to compare the ionizing power of Zn^{2+} in front of other divalent metals so its specifity can be understood. For this purpose, the reaction:

$$MOH_2^{+2} + H_2O \rightarrow MOH^+ + H_3O^+$$

with M = Be, Mg and Zn will be studied [34]

The complex formed by two water molecules that interact with a cation leads to ressemblance with the $H_5O_2^+$ system, where a hydrogen atom has been substituted by the metal. For the $H_5O_2^+$ system, the proton transfer has been shown theoretically to exhibit a U-shaped symmetric energy profile [35,36]. Since presence of the metal in place of the hydrogen indeed modifies the energy profile, the changes made by the presence of the metal ion will be studied. As in Friedel-Crafts reactions, we will try again to separate the effect of the electric field from the effect of charge transfer.

For each dication metal, one intermediate and one transition state are found. In table 3 the main geometrical parameters of the intermediates obtained using the 3-21G basis set are presented. Values obtained for the $H_5O_2^+$ system are also presented for comparison purposes. In addition, the charge on the metal atom is also given. The optimal 3-21G geometry of the $H_5O_2^+$ species exhibits a linear symmetric hydrogen bond where the proton is located midway between both oxygen atoms. Substitution of hydrogen by dications changes the geometry of the minimum and destroys symmetry, as one can see from the values of the distances between the two oxygens and the transferring hydrogen. The most remakable effect is that in all minima the proton is already transferred to some

Table 4: Geometrical parameters and charge on the metal of the transition states of the $M(H_2O)_2^{+2}$ systems.

M	M-O1	O1-M	H-O2	Q_H
Be^{+2}	1.40	2.28	0.98	1.32
Mg^{+2}	1.74	3.07	0.98	1.53
Zn^{+2}	1.72	2.98	0.98	1.42

extent. For Be^{2+}, the proton is completely transferred, for Zn^{2+} the proton is more than halfway transferred; finally, for Mg^{2+} it is just beginning to be transferred. The values of the charge on the metal indicate that the more electron-drawing the metal is, the more advanced the proton transfer will be.

In table 4 the geometrical parameters of the three transition states are presented. In all cases the H_3O^+ ion is completely formed and placed far away from the MOH^+ species. The energy barrier relative to the hydrogen-bonded complexes are 0.8 kcal/mol for $Be(H_2O)^{+2}$, 21.8 kcal/mol for $Mg(H_2O)^{+2}$, and 16.6 kcal/mol for the $Zn(H_2O)^{+2}$ species. The barrier for the Be^{2+} ion is smaller because in the intermediate complex the proton is already transferred, whereas in the other two cases the proton transfer is not so advanced. The eigenvectors associated to the only negative eigenvalues of the Hessian matrices indicate that the transition states do not correspond actualy to the proton transfer process, but rather to the separation between the charged species MOH^+ and H_3O^+.

The energy profiles for the proton transfer reactions are depicted in Figure 4. The reaction coordinate, Rc, is defined as the difference between the O1-H and O2-H distances. It must be remarked that the metal-oxygen distance decreases along the reaction coordinate, so it must be optimized; otherwise the electron transfer would be smaller and the potential barrier would rise. The figure shows clearly that the symmetric shape which characterizes the $H_5O_2^+$ system dissapears when the hydrogen atom is substituted by the metal. The curves for Mg^{2+} and Zn^{2+} follow a trend similar to the $H_5O_2^+$ curve from reactants to transition states. In contrast, the curve for Be^{2+} is fairly different. Further, the positions of the transition states are more advanced in the reaction coordinate the less exothermic the reaction is, in full agreeement with Hammond's principle.

The similarity between the Mg^{2+} and Zn^{2+} profiles suggests that the $3d^{10}$ filled orbitals of Zn^{2+} do not intervene in the proton transfer process. This is confirmed by the 3d Mulliken population values at different points along the reaction coordinate. That population remains almost constant throughout the reaction. The charge transfer to the metal along the reaction goes to the empty s and p orbitals of Zinc.

The proton transfer in the isolated water dimer is very difficult owing to two main reasons: first, the formation of the OH^- and H_3O^+ions; second, their separation. In contrast, the proton transfer process is favoured if a monovalent ion, e.g., a proton, catalyzes the reaction [37]. In this case, there is no longer a charge separation, but rather there is a migration of the positive charge, which is more favourable energetically. The final separation between the created dipole and the positive charge is not so difficult as the

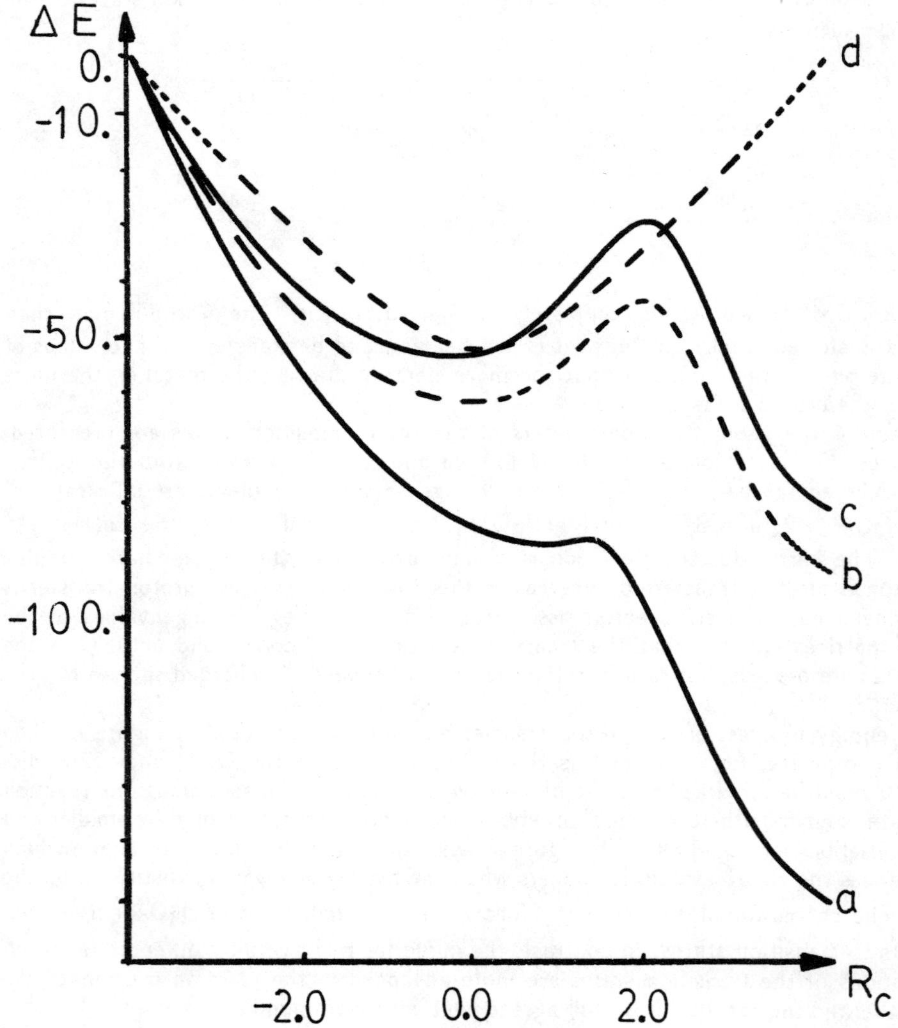

Figure 4: Potential energy profiles for the proton transfer in the $M(H_2O)_2^{+2}$ (M=Be, Mg, Zn) and $H_5O_2^+$ systems. (a) M=Be; (b) M=Zn; (c) M=Mg; (d) $H_5O_2^+$ system.

separation in the non-catalyzed process where two opposed charges must be pulled apart. If the proton transfer is catalyzed by a divalent cation, the process is further favoured, which can be attributed to two reasons. First, the positive charge is spread over two fragments, and second, the appearance of a positive charge in each fragment helps the separation towards the final products. This fact translates into the exothermicity of the proton transfer catalyzed by the three cations considered in the present work.

The catalytic effect of dications on the proton transfer in the water dimer may be analyzed again looking to the electrostatic and charge transfer terms. In order to separate these two factors, we have simulated the cations by a dipositive charge placed at different distances from the oxygen atom (1.9, 1.6 and 1.4 Å). The presence of the dipositive charge causes an important polarization of the electronic cloud of the water dimer. As a first consequence the separation into a positive and a negative charge is more advanced; for instance, the charge on the final hydroxyl group increases from -0.46 to -0.68 au when the charge is placed at 1.4 Å from the oxygen atom. As a second consequence, a weakening of the O-H bond and a strengthening of the hydrogen bond are produced. The respective Mulliken indexes change from 0.25 to 0.16 and from 0.05 to 0.10 for the aforementioned placement of the dipositive charge. Of course, at the optimized isolated water dimer forces acting on the nuclei are zero. On the contrary, when the positive charge is placed close to the oxygen, forces appear trying to lengthen the O-H bond and to shorten the hydrogen bond. The origin of such forces can be found in the attraction between the negative oxygens and the positive charge, and the repulsion between the positive charge and the transferring proton. These forces origin a relaxation of the water dimer, and also modify the energy profiles of the proton transfer process.

The shapes of the three energy profiles are qualitatively similar to those for the systems with the three metal dications. The differences in behaviour between Be^{2+}, Mg^{2+} and Zn^{2+} can already be explained by inspection of the metal-oxygen distances, which are initially 1.52, 1.89 and 1.83 Å respectively. This qualitative agreement is not out at all quantitative. For instance, a dipositive charge must be placed at 1.4 Å so that the energy profile be similar to that of the Zn dication, where the Zn-O distance is about 0.4 Å longer. This fact shows that to interpret the proton transfer process the electrostatic effect is not the only cause to be considered, and thus the charge transfer effect must also be taken into account. The divalent metal cations do not only act like a dipositive charge but like a Lewis acid as well. This behaviour is produced by accepting electrons into the empty orbitals of the metal. In all three metals the s and p orbitals are the only ones which are ready to accept charge. The observed difference between the three metals can also be interpreted by inspection of the energies of the s and p metal dication orbitals. For Zn^{2+}, the energies of these orbitals are found to be between these of Be^{2+} and Mg^{2+}, yet closer to those of Mg^{2+}. Thus, the different behaviour as Lewis acids helps in explaining the different results found in the present study for the three metals. Their behaviour is originated both by an electrostatic effect and by their Lewis acid character.

It has been shown that Zn^{2+} exhibits intermediate characteristics between Be^{2+} and Mg^{2+}, yet closer to Mg^{2+}. This suggests that the catalytic specificity of Zn^{2+} arises from an intermediate behaviour, rather than an extreme one. The filled d shell of Zn^{2+}, which does not participate directly into bonding, places the empty 4s and 4p orbitals in the adequate level for the optimal catalysis.

In this work we have considered only the effect of a naked ion on the proton transfer in the water dimer. Actually, the ion will be surrounded by a first shell of solvation, or in the carbonic anhydrase by three imidazole groups. These solvent molecules or coordination groups will transfer electron charge to the central ion, so the s and p metal orbitals will be partially occupied and destabilized, and the positive charge on the cation will be disminished. In addition, in this first shell a partial negative charge will appear around the ion due to the orientation and polarization of ligands. Thus, both the electrostatic and charge transfer terms are overemphasized in our model. When the dication is surrounded by ligands, it is far more difficult to study the electrostatic component alone. Many works have dealt with the effects of macromolecular environnments on proton transfer processes. Usually the environment is represented by a collection of point charges, dipoles and atomic polarizabilities [38-43]. When we have studied the process catalyzed by a naked ion, the polarization on the electronic cloud of the ion by the chemical system has not been introduced in the electrostatic term. In this case, the approximation looks right, but when the ion is situated in a more complex system it is necessary to introduce the reaction field created by the charge distribution of the reaction system on the cation surrrounding.

4.2. ELECTRON TRANSFER

In the proton transfer a non-uniform electric field has been considered. Let us now apply a uniform electric field to the most typical charge transfer process: the electron transfer. Following Morokuma's work [44] the electron transfer between two oxygen molecules in the system $O_2 \cdot O_2^-$ will be studied. By means of a broken space-symmetry UHF wave function, two diabatic surfaces with the lone electron located in each oxygen molecule can be found. The transition state is defined as the most stable point along the seam of crossing between both diabatic potential energy surfaces, corresponding to the initial and the final state of the electron transfer, respectively. The transition state can be located by means of the Lagrange-Newton method. If the electron transfer is an adiabatic process instead of a diabatic one, this structure will be only a good approximation to the transition state, since the coupling between the initial and final electronic states remains still small.

Three symmetric transition state geometries have been located by using the broken space-symmetry UHF wave function under the constraint of D_{2h}, C_{2v} and C_{2h}, respectively [44]. Here the D_{2h} geometry will be the only one considered, and a C_{2v} symmetry along the reaction coordinate will be kept. Under this restriction three geometrical parameters define the full geometry of the system, D1 and D2, the bond lengths of both oxygen molecules, and DM, the distance between the midpoints of both O-O bonds. When a uniform electric field in the direction of the main C_{2v} axis is applied, the $O_2 \cdot O_2^-$ diabatic surface is stabilized and the $O_2^- \cdot O_2$ one is destabilized. As a consequence, the electron transfer from the first to the second oxygen molecule is favoured. In table 5 [45] the geometrical parameters of the transition state and the potential barrier for three intensities of the uniform electric field are collected.

It is worth mentioning that with the help of the electric field the electron transfer takes place before both oxygen bond lengths (D1 and D2) are the same. The potential energy barrier decreases as a function of the electric field intensity. As a matter of fact, a straight line is obtained as the ratio between both magnitudes. Therefore, the electric

Table 5: Geometrical parameters of the transition states and potential barriers of electron transfer in the $O_2^- \cdot O_2$ system catalyzed by a uniform electric field of intensity $\mathbf{F}$ (in au, 1 au=$5.14 \cdot 10^{11}$ V/m).

F	D1	D2	DH	$\Delta E^{\#}$
0.000	1.31	1.31	3.12	7.25
0.002	1.33	1.29	3.08	4.07
0.004	1.35	1.28	2.99	1.91

field intensity which promotes the spontaneous electron transfer from the first to the second oxygen molecule can be determined. For weak intensities, a catalytic effect on the electron transfer by means of an external uniform electric field will be obtained. Finallly, the electron transfer rate according to Fermi's golden rule must also increase. Although the electronic part of the electron-transfer matrix element for the diabatic process has not been computed yet, one can justify the mentioned increase. The reason can be found in the increase in overlap between both oxygen molecules upon increase of field strength, in parallel with the shortening of the DM distance.

5. Other External Perturbations

The last remark allows us to discuss magnetic field effects in chemical kinetics. A quantitative estimation of magnetic contributions to the Gibbs free enthalpy of activation shows that no major effects are to be expected by magnetic fields in the most favourable examples going from diamagnetic to paramagnetic states or vice-versa. For adiabatic reactions, with strong coupling between initial and final electronic states, the transmission coefficient approaches a value of one. Therefore, for this class of reactions we may exclude any significant magnetic field effect on the chemical reaction rate. In the case of diabatic reactions, however, when the electronic coupling between initial and final states does not provide for an efficient anticrossing of the respective potential surfaces, the transmission coefficient may drop to rather small values, leaving in principle a large dynamic range for magnetic field effects if they have any influence on the transmission factor. A rather discontinuous change of the electronic wave function around the barrier is a requirement for magnetically sensitive reactions. Reactions where a spin change takes place will be a good example. In the radical pair mechanism, the spin evolution may lead to a recombination or to another competitive process. The spin evolution depends upon an external magnetic field. A very good review of the experimental and theoretical work on kinetic effects of magnetic fields has recently appeared [46].

Pressure is another external perturbation, which may increase dramaticllly reaction rates. In a pioneering work [47] Williams showed that the reaction rate of the methyl halyde exchange reaction is increased if two helium atoms are placed in both extrema of the trigonal symmetric axis of the system, because the X-X' distance is shorter at the transition state than it is at the initial complex.

In conclusion, we have shown in this work that external perturbations belong to the reaction coordinate, in such a way that one can play with chemical reactions. The role of the chemist is not only to imitate nature, but also to create new materials through control of chemical processes. To master the matter turns out to be our job.

6. Acknowledgments

The research presented in this chapter has been carried out in the Theoretical Chemistry Group of the Department of Chemistry of the Universitat Autònoma de Barcelona. I must thank all members of the Group for their collaborative effort. This work has been supported by the Spanish Direccion General de Investigacion Cientifica y Tecnica under Project No. PB86-0529.

7. References

[1] M. S. Gordon, T.N. Truong and J.A. Pople *Chem. Phys Lett.* **130**, 245 (1986)
[2] R.G. Pearson *Acc. Chem. Res.* **4**, 152 (1971)
[3] F. Zaera and G. A. Somorjai *J. Am. Chem. Soc.* **106**, 2288 (1984)
[4] F. Zaera, A.J. Gellman and G.A. Somorjai *Acc. Chem. Res* **19**, 24 (1986)
[5] A. Wieckowski, S.D. Rosasco, G.N. Salaita, A. Hubbard, B.E. Bent, F. Zaera, D. Godbey and G.A. Somorjai *J. Am. Chem. Soc.* **107**, 5910 (1985)
[6] F.H. Jardine, J.A. Osborn, G. Wilkinson and J.F. Young *Chem. Ind. (London)* 560 (1965)
[7] J. Halpern and C.S. Wong *J. Chem. Soc. Chem. Commun.* 629 (1973)
[8] J. Halpern, in *Organo-transition Metal Chemistry*, Y. Ishii and M. Tsutsui, Eds. Plenum Press, New York, 1975, pag 109.
[9] J. Halpern, T. Okamoto and A. Zakhariev *J. Mol. Catal.* **2**, 65 (1976)
[10] J. Wristers *J. Am. Chem. Soc.* **97**, 4312 (1975)
[11] J. Wristers *J. Am. Chem. Soc.* **99**, 5051 (1977)
[12] J.Y. Saillard and R. Hoffmann *J. Am. Chem. Soc.* **106**, 2006 (1984) and references therein.
[13] J. Garcia-Prieto, M.E. Ruiz and O. Novaro *J. Am. Chem. Soc.* **107**, 5635 (1985)
[14] A. Dedieu *Inorg. Chem.* **19**, 375 (1980)
[15] A. Dedieu and A. Strich *Inorg. Chem.* **18**, 2940 (1979)
[16] A. Dedieu *Inorg. Chem.* **20**, 2803 (1981)
[17] N. Koga, C. Daniel J. Han, X.Y. Fu and K. Morokuma *J. Am. Chem. Soc.* **109**, 3455 (1987)
[18] C. Daniel, N. Koga, J. Han, X.Y. Fu and K. Morokuma *J. Am. Chem. Soc.* **110**, 3773 (1988)
[19] J.C. Siria, M. Duran, A. Lledós and J. Bertrán *J. Am. Chem. Soc.* **109**, 7623 (1987)
[20] N. Koga, S. Obara, K. Kitaura and K. Morokuma *J. Am. Chem. Soc.* **107**, 7109 (1985)
[21] G.A. Olah *Friedel-Crafts Chemistry*, Wiley, New York, 1975.
[22] G.A. Olah, G. Klopman and H . Schosberg *J. Am. Chem. Soc.* **91**, 3261 (1969)
[23] H. Mayr and W. Strieppe *J. Org. Chem.* **48**, 1159 (1983)
[24] E. Silla, E. Scrocco and J. Tomasi *Theoret. Chim. Acta* **40**, 343 (1975)

[25] G. Alagona, E. Scrocco, E. Silla and J. Tomasi *Theoret. Chim. Acta* **45**, 127 (1977)
[26] W.L. Jorgensen and M.E. Cournoyer *J. Am. Chem. Soc.* **100**, 5278 (1978)
[27] J. Bertrán, F. Mora and E. Silla *J. Chem. Soc. Perkin 2* 647 (1982)
[28] V. Branchadell, A. Oliva and J. Bertrán *J. Mol. Catal.* **44**, 285 (1988)
[29] J. Bertrán, in *New Theoretical Concepts for Understanding Organic Reactions*, J. Bertrán and I.G. Csizmadia (Eds.), Kluwer Academic Publishers, Dordrecht, 1989, pag. 231.
[30] T. Clark, J. Chandrasekhar, G.W. Spitznagel and P. von R. Schleyer *J. Comput. Chem.* **4**, 294 (1983)
[31] E. Carbonell, M. Duran, A. Lledós and J. Bertrán, to be published elsewhere.
[32] R.F.W. Bader, Y. Tal, S.G. Anderson and T. Nguyen-Dang *Isr. J. Chem.* **19**, 8 (1980)
[33] D.N. Silverman and S. Lindskog *Acc. Chem. Res.* **21**, 30 (1988)
[34] M. Solà, A. Lledós, M. Duran and J. Bertrán, to be published elsewhere.
[35] J.E. Del Bene, M.J. Frisch and J.A. Pople *J. Phys. Chem.* **89**, 3669 (1985)
[36] J.E. Del Bene *J. Phys. Chem.* **92**, 2874 (1988)
[37] S. Scheiner, P. Redfern, M.M. Szczesniak and L.D. Bigham *J. Phys. Chem.* **89**, 262 (1985)
[38] O.Tapia and G. Johannin *J. Chem. Phys* **75**, 3624 (1981)
[39] B.T. Thole and P.T. van Duijnen *Theoret. Chim, Acta* **63**, 209 (1983)
[40] A. Warshel and M. Levitt *J. Mol. Biol.* **103**, 227 (1976)
[41] A. Warshel *Biochem.* **20**, 3167 (1981)
[42] H. Weinstein, J.E. Eilers and S. Chang *Chem. Phys. Lett.* **51**, 534 (1977)
[43] A. Mercier, Jr., J.P. Dijkman, R. Osman and H. Weinstein, in *Quantum Chemistry: Basic Aspects, Actual Trends*, R. Carbo (Ed.), Elsevier, Amsterdam, 1989, pag 577.
[44] K. Ohta and K. Morokuma *J. Phys. Chem.* **91**, 401 (1987)
[45] J.L. Andrés, M. Duran, A. Lledós and J. Bertrán, to be published elsewhere.
[46] U.E. Steiner and T. Ulrich *Chem. Rev.* **89**, 51 (1989)
[47] I.W. Williams *J. Am. Chem. Soc.* **106**, 7206 (1984)

CALCULATION OF PHOTOCHEMICAL REACTIVITY. OLIGOSILANES AS AN ILLUSTRATION

Josef Michl and V. Balaji
Center for Structure and Reactivity
Department of Chemistry
The University of Texas at Austin
Austin, Texas 78712-1167, U.S.A.

ABSTRACT. The fundamental photophysical and photochemical processes in the reactions of organic molecules are summarized with particular attention to the role of barriers, minima and funnels in the S_1 and T_1 surfaces. Non-spectroscopic minima often occur at biradicaloid geometries and elementary photochemical reaction steps can be classified according to the nature of the biradicaloid minimum through which the return to the S_0 state occurs. In simple cases, one active orbital per atom is involved. These steps are subdivided into two-center, cyclic multicenter, and linear multicenter processes. In complex cases, more than one active orbital participates on at least one atom. Minimal calculational requirements for a theoretical treatment are outlined.

Only the simple two-center case is considered in some detail. It corresponds to the dissociation of a single non-polar, weakly polar, charged, or dative bond by stretching. The shapes of the S_0, S_1 and T_1 surfaces are discussed and the photochemical implications considered.

Finally, an application of the simple principles to the three known photoreactions of oligosilanes, homolytic chain cleavage, chain cleavage by 1,1-elimination, and chain abridgement by 1,1-elimination, is described, using approximate potential energy curves derived from *ab initio* CI calculations.

1. Scope

The purpose of this brief introduction to theoretical organic photochemistry is to acquaint the reader with the elementary considerations that precede the calculation of a photochemical reaction path and to illustrate such calculations on a fairly simple example, polysilanes. We first list the elementary photophysical and photochemical processes that need to be considered, identify the critical features of the lowest singlet (S_0), lowest excited singlet (S_1) and lowest triplet (T_1) potential energy surfaces, and recognize that crucial parts of the calculation will involve biradicaloid geometries. We then organize organic photochemical paths according to the detailed nature of electronic

C. Ögretir and I. G. Csizmadia (eds.),
Computational Advances in Organic Chemistry: Molecular Structure and Reactivity, 323–354.

interactions at their geometries, and indicate how this defines the minimum acceptable level of complexity in the calculation. We illustrate the results in a qualitative way on the simplest case of the photochemical breaking of a sigma bond between atoms that do not carry lone pairs or empty orbitals, and indicate very briefly the generalization to pi bonds and to multicenter systems. All this material is handled in much more detail in a recent book[1] and the reader is referred there for further information and original literature sources.

Finally, we provide an example of a qualitative theoretical treatment of the photochemistry of polysilanes, a class of compounds that has received much recent attention experimentally[2] but whose distinctive photochemical reaction paths have not been a subject of prior calculations to our knowledge.

2. Fundamentals of photophysics and photochemistry - the elementary processes

A priori theoretical treatment of photochemical reactivity is incomparably more difficult than that of ground-state reactivity, and the goals are normally set appropriately lower. In an investigation of a ground-state reaction, one first generates the ground state potential energy surface, or more typically, merely identifies the structure, energy and normal mode frequencies of the starting material and of a transition state for each primary product, and second, uses transition state theory to obtain reaction rates as a function of temperature in gas phase or a non-polar solvent. Although sometimes a larger section of the ground state surface is calculated and used in an investigation of dynamics, usually by computing classical trajectories, this is often not considered necessary or even feasible.

Compare this now with an ideal *a priori* treatment of a typical photochemical process, again using the standard Born-Oppenheimer approximation with its concept of potential energy surfaces. In principle, the first step still consists of a computation of potential energy surfaces, now a minimum of two, and the second step, of an evaluation of the rate and quantum yield of formation of each primary product as a function of temperature and excitation wavelength. The first point to note is that the calculation of excited state energies and wavefunctions is much more difficult than a similar calculation for the ground state. Indeed, it is at times a major task just to obtain the correct sequence of vertical excited states, which is an absolute prerequisite for any further progress. Yet, this is only the first of many obstacles.

The search for minima in the ground-state surface is facilitated by the availability of the usual structural theory, which provides us with a good initial guess of bond lengths, valence angles, and often even dihedral angles, and permits good initial guesses for efficient use of gradient search techniques. Although the search for the transition structures is much harder, it, too, benefits from a great deal of accumulated experience and "chemical intuition."

These advantages are almost entirely missing when minima and saddle points in barriers are sought on excited state surfaces. True, for many molecules there will be a minimum in the lowest excited singlet (S_1) or triplet (T_1) surface at a geometry fairly close to equilibrium geometry of the ground electronic state S_0, amenable in principle to a gradient-based search starting from the latter geometry. These are the so-called vertical, or spectroscopic, excited states. Although their electronic wavefunctions are frequently intrinsically more complicated than those of the ground states in that they cannot be described, even approximately, by a single-determinantal wave function, these minima in T_1 and S_1 are still relatively easy to find.

Immediate radiationless or radiative return to the ground state S_0 simply regenerates the starting species. In bimolecular photochemical processes, the spectroscopic minima serve as holding areas for excited molecules, permitting them to wait for diffusion to bring in a reaction partner. However, in themselves, they do not directly dictate the nature of the product. Instead, this is determined by the geometry at which the return to S_0 occurs.

With a few exceptions such as proton transfer reactions which often lead to no lasting chemical change, such return takes place at one of the other numerous minima in the S_1 or T_1 surface. These have no close counterpart in the S_0 surface and are called non-vertical or non-spectroscopic, since their existence is difficult to detect in the ordinary absorption spectra of the ground-state species. Many of these minima in S_1 and T_1 surfaces occur at geometries where the ground state energies are very unfavorable and S_0, S_1 and T_1 states quite close in energy. Some occur at geometries where the S_0 and S_1 surfaces actually touch (conical intersections), or nearly touch, and are frequently referred to as funnels.

The search for the non-spectroscopic minima and for the saddle points in the barriers leading to them is very difficult. Much less "chemical intuition" is available to guide the search, and the barriers often result from a crossing or avoided crossing or two or more closely spaced states of equal multiplicity, all of which therefore need to be computed with comparable accuracy. Accurate computation of energies and wavefunctions in the vicinity of a conical intersection is notoriously difficult, particularly when the touching states differ significantly in the degree of "ionic" character.

Assuming now that the "static" part of the theoretical treatment, the computation of potential energy surfaces and identification of minima, funnels, barriers and saddle points has been accomplished successfully, attention can be turned to the second part of the task, derivation of product formation rates and yields. Again, this is far more complicated than one is used to from ground-state chemistry.

In fluid or rigid solution or in a gas at relatively high pressures, an initially excited organic molecule usually relaxes rapidly into one of the minima on the S_1 or T_1 surface, the former in the case of direct excitation, the latter in the case of triplet sensitization. Alternatively, it may find a funnel in the S_1 surface through which it relaxes all the way to a minimum in the S_0 surface. These initial relaxation processes are usually completed in a few ps or tens of ps. Only under low-pressure conditions do they take longer and chemical transformations are then very likely to occur in vibrationally hot species.

An accurate description of the initial relaxation phase would require a treatment of the radiationless transitions from the higher excited electronic states, often reached in the initial excitation, to the lowest state of the same multiplicity, S_1 or T_1, as well as a description of vibrational relaxation by collisions with molecules of solvent or buffer gas, and determination of the probabilities with which the molecule ends up in one or another of the minima in the S_1, T_1 or S_0 surface at the end of the initial relaxation period. Although one could imagine doing this in some approximate way, the complexity of the task is discouraging and it has rarely been seriously attempted for a molecule of a size interesting to an organic photochemist.

Fortunately, in dense media the relaxation to a vertical (spectroscopic) minimum in S_1 (or T_1) is frequently quantitative and the photochemical observables do not depend on the wavelength of the exciting light throughout the ordinary UV-visible region. This is the situation assumed in the following. It is still quite complicated relative to that familiar from ground-state chemistry. In addition to escaping from the initial vertical minimum in S_1 or T_1 over one or another saddle point in the surrounding barrier, the molecule can also relax by a radiative or radiationless jump to a lower surface. Ordinary transition state theory may be used to describe the rate of escape over the barrier. The computation of the rates of the competing processes requires the evaluation of electric dipole transition moments for the former case, and of the matrix elements of the coupling operators as well as vibrational frequencies and normal modes of the two states in the latter. For internal conversion, which preserves state multiplicity, the non-adiabatic coupling elements are needed. These are related to the rate of change of the electronic wavefunction with the change in nuclear coordinates. For intersystem crossing, which changes state multiplicity, the elements of the spin-orbit coupling operator and in certain circumstances, also of the hyperfine interaction operator, are needed.

That part of the molecules which remained on an excited surface reached another minimum or funnel. All the competing rates have to be reevaluated for this new minimum. This sequence continues until the integrated probability of return to S_0 by one or another route reaches unity.

The final difficult task is to determine the fate of the molecule once it reached the S_0 surface from a given excited state minimum or funnel by a radiative or radiationless process. This requires a good knowledge of the S_0 surface in regions not normally visited by ground state molecules, a detailed knowledge of the radiative or non-radiative transition and in particular, of the vibrational modes into which excess electronic energy is being deposited, and a fair understanding of the process of vibrational relaxation in the solvent used.

Little wonder, then, that *a priori* calculations of organic photochemical reactivity lag far behind similar calculations of thermal reactivity. Instead of attempting quantitative predictions or even rationalizations of the rates of photochemical events, current theoretical photochemistry struggles with the far simpler task of describing reaction paths in a qualitative fashion. Much of this can be accomplished by identifying the minima and barriers in the S_1 and T_1 surfaces without attempting any calculations of the dynamics, and we shall turn our attention to this task next.

3. Minima and barriers in the T_1 and S_1 surfaces. Implications for the classification of primary organic photochemical processes and for the choice of computational methods.

The computational search for S_1 and T_1 minima, conical intersections, and saddle points in the multidimensional nuclear configuration space is greatly facilitated by a general qualitative understanding of their nature, often provided by simple correlation diagrams. In the past, one often had to be satisfied with such simple tools but improving theoretical methods and computer facilities open new possibilities for the near future.

The structures of spectroscopic minima are usually appreciated without much trouble, since overall bonding in these is reminiscent of that in the corresponding S_0 minimum, and the effects of a missing bonding or non-bonding electron and an extra antibonding electron are fairly readily visualized.

It is perhaps less generally appreciated that most if not all of the non-spectroscopic minima important in organic photochemical reactions occur at biradicaloid geometries, i.e., those at which a first-order description of the electronic wavefunction of low-lying electronic states is characterized by the presence of two approximately or exactly degenerate "non-bonding" orbitals occupied by a total of only two electrons. An orthogonally twisted double bond, a severely stretched single bond, a divalent carbon, a cyclic array of overlapping orbitals with no phase mismatch containing 4N electrons, and a cyclic array of overlapping orbitals with one phase mismatch containing 4N+2 electrons are the most important examples of such geometries.

Molecules at biradicaloid geometries are usually very energy-rich in their S_0 state. Their two non-bonding electrons are wasted as far as bonding is concerned, and distortion towards the more normal geometries, at which they form at least a partial bond instead, is clearly favorable. A distortion that introduces an overlap and interaction between the two non-bonding orbitals will cause them to split into a bonding and a more strongly antibonding combination. This is energetically favorable in the S_0 state since both electrons are located in the resulting bonding orbital. It is, however, unfavorable in low-lying excited states, in which one electron goes into each of the combinations. The rather high energy of S_0 and rather low energy of S_1 and T_1 at biradicaloid geometries agrees with the general notion that biradicals have low S_0-S_1 and S_0-T_1 energy gaps.

It thus becomes clear that a typical organic photoreaction proceeds from an "ordinary" initial geometry to a biradicaloid geometry, and after electronic deexcitation, to another "ordinary" geometry. It follows that an understanding of the electronic states of biradicals and biradicaloids is a prerequisite for a qualitative appreciation of the course of organic photochemical reaction paths. Clearly, only those methods of calculation that are capable of describing the electronic states of these species properly are likely to be useful for tracking the course of the paths of photochemical reactions. We shall return to this subject in more detail below.

A brief survey of the electronic states of biradicals and biradicaloids is therefore called for. This can be profitably set into the more general context of the interaction of two electrons in two orbitals, whether it takes the form of a non-polar or a polar covalent bond, of a charged bond, a dative bond, a biradicaloid, or of a perfect biradical. It can also be viewed as a discussion of the low-energy states of a general two-electron bond in all of its forms and at all of its geometries.

For most organic photochemical processes, the crucial part of the action cannot be confined to the behavior of two electrons in two orbitals, and we need to consider also more complex situations in which the breaking of a bond, as a biradicaloid geometry is reached, is assisted by other electrons in other orbitals. Identification and conceptual separation of that region of the reacting molecule (or two molecules) in which the primary bonding changes take place permits a formulation of a classification scheme for elementary steps in photochemical reactions. Many such schemes are possible; one that we have found particularly useful distinguishes two-center, cyclic multicenter and linear multicenter processes. Although it clearly offers the possibility of further subdivision according to the number of active centers and their relative electronegativities, illustrated on the case of two-center processes below, a more fundamental differentiation is based on the number of active orbitals. In simple processes, only a single orbital participates on each participating atom. In complex processes, two or more participating orbitals are present on at least one participating atom. It is useful to subdivide the two-center processes of this kind by the total number of active orbitals into bitopic, tritopic, tetratopic, etc. The concept of topicity is less clearly defined for multicenter processes.

The qualitative identification of the orbitals actively involved in the reaction process, and of the number of electrons present in them in the S_0, S_1 and T_1 states also permits an identification of the simplest method of calculation that has a reasonable chance of producing anything better than purely qualitative results. This is the "complete-active-space multiconfigurational SCF" method or its equivalent, preferably followed by multireference CI. The minimal active space is dictated by the region active in the photochemical process in the sense used above, i.e., 3×3 for the singlets and 1×1 for the triplets in a two-electron two-center process with one active orbital on each center, 20×20 for the singlets and 15×15 for the triplets in a cyclic or linear four-electron four-center four-orbital process, etc. The actually required active space may be larger since the whole reaction path and not only the vicinity of the biradicaloid minimum needs to be investigated. When state touchings or avoided touchings are present along the way, additional orbitals and electrons may need to be included. In addition, more than one reaction path and thus more than a simple biradicaloid minimum may need to be considered. In general, the active space is smaller for triplets and the investigation of the T_1 surface is simpler than that of the S_1 surface. However, it is important to attain comparable accuracy for the S_0, S_1 and T_1 surfaces, and if avoided touchings with higher states are involved, also for their surfaces, and a balanced treatment in both the CAS-SCF and multireference CI steps is essential.

In the simplest case of two-center two-electron processes the CAS-SCF procedure simplifies to a one-pair GVB method for singlets and a ROHF method for triplets.

In the past, many calculations were performed by simpler methods and provided much useful qualitative insight, but they can hardly be viewed as much more than glorified correlation diagrams. The use of ROHF molecular orbitals calculated for the triplet state as the basis for large-scale CI calculations of singlet states has been particularly popular. Nowadays, there appears to be little justification for such qualitative calculations except in those cases for which even a qualitative understanding does not yet exist, where they can be of heuristic value. Still, we would strongly urge all those contemplating a calculation of photochemical paths to perform them at the CAS-SCF level advocated here, even if they are to serve only qualitative purposes. An example of such calculations is given below.

Even when the choice of the form of the many-electron wavefunction is made sufficiently flexible, considerable pitfalls remain. The difficulties with choosing relevant reaction paths in the multidimensional nuclear configuration space and with identifying the proper saddle points and minima have been mentioned at the outset. Here, we add a brief mention of the difficulties associated with the choice of an adequate basis set. A double-zeta set appears to be the minimum acceptable, and at least one set of polarization functions would in general be needed for any quantitative work. Diffuse functions present particular problems. They are essential for the description of Rydberg states, and these are the lowest states at the ordinary geometries of small molecules, and of many that are not so small. The diffuse orbitals enter heavily into the description of "zwitterionic" excited states, to a degree that is considerably exaggerated when electron correlation is underestimated. When diffuse functions are absent, the excitation energies are often seriously underestimated. In the regions of biradicaloid geometries, the Rydberg states normally do not drop in energy in the dramatic way that valence states do since they roughly follow the sum of S_0 energies and ionization potentials, and the latter are usually not sufficiently lower at biradicaloid than at ordinary geometries to compensate for the increased S_0 energy. Therefore, at biradicaloid geometries, the presence of diffuse orbitals is not particularly critical to the calculation. When the total reaction path is to be calculated properly, diffuse functions must be included, but for qualitative considerations, they are frequently dispensable.

The situation is further complicated by the circumstance that most organic photochemistry is performed in solution, where Rydberg orbitals are pushed up in energy relative to the gas phase and the low-energy state wavefunctions acquire much less diffuse (Rydberg) character. It thus appears possible that diffuse functions may be omitted from the basis set in many cases even if the whole reaction path is to be followed in a semiquantitative fashion.

In summary, whichever method is used, the following requirements must be met for the results to be useful: (i) the correct order of the few lowest pure spin states needs to be reproduced, with at least roughly correct energy differences, (ii) ordinary and biradicaloid geometries must be handled with comparable accuracy and in a balanced way.

Such an approach will permit the calculation to locate barriers, minima, and funnels in the S_0, S_1 and T_1 surfaces and thus to rationalize the observed elementary photochemical processes, to provide a degree of qualitative understanding that will permit extrapolation to related substrates, such as those derived by substitution, and hopefully also to suggest new experiments.

4. Two-center two-orbital processes

In order to provide a specific illustration of the qualitative conclusions that emerge from the consideration of the active space spanned by the reaction region, we describe the results for the dissociation of a sigma bond between atoms that do not carry lone pairs or empty valence orbitals. We shall initially ignore the presence of Rydberg orbitals, for the reasons given above. At their ordinary geometries, such bonds can be covalent and uncharged, either non-polar, as in $H_3Si\text{-}SiH_3$, or weakly polar, as in $(CH_3)_3Si\text{-}SiH_3$, charged, as in $H_3C\text{-}N^+H_3$ and $H_3C\text{-}B^-H_3$, or dative and uncharged, as in $H_3B^-\text{-}N^+H_3$, where the usual formal charges have been indicated.

The biradicaloid geometries of these bonds are reached by stretching (Figure 1). In the dissociation limit, the S_0 and T_1 states of covalent uncharged bonds correspond to a combination of two radicals, as in $H_3Si\bullet$ + $\bullet SiH_3$ or $(CH_3)_3Si\bullet$ + $\bullet SiH_3$ (dot-dot valence-bond structures) and are degenerate, while the S_1 state corresponds to a combination of two ions (a hole-pair structure), as in $(CH_3)_3Si^+$ + $^-SiH_3$, and is therefore much higher in energy.

For a charged bond, the dissociation limit is unambiguous for the T_1 state and again corresponds to a combination of two radicals (a dot-dot structure), as in $CH_3\bullet$ + $\overset{+}{NH_3\bullet}$. The singlet dot-dot structure, as in $CH_3\bullet$ + $\overset{+}{NH_3\bullet}$, and hole-pair structure, as in CH_3^+ + $:NH_3$, are roughly comparable in energy. In the case chosen for illustration here, the hole-pair structure lies about 1 eV below the dot-dot structure, since the ionization potential of the methyl radical is about 1 eV lower than that of ammonia, but in other cases the energetic order of the two structures can easily be the opposite. In this case, then, the S_0 and S_1 states are approximately degenerate at the biradicaloid geometry, one being of the dot-dot and one of the hole-pair type.

For a dative bond, the situation is again clear-cut. In the dissociation limit, the S_0 state corresponds to the combination of uncharged fragments, the hole-pair structure, as in BH_3 + NH_3. The excited singlet and triplet states of the sigma bond are degenerate (A′ in Figure 1) and much higher in energy since they correspond to the combination of two radical ions, as in $BH_3\overset{-}{\bullet}$ + $\overset{+}{NH_3\bullet}$. As a result, the lowest excited singlet and triplet states calculated for H_3BNH_3 actually are of locally excited character (E in Figure 1).

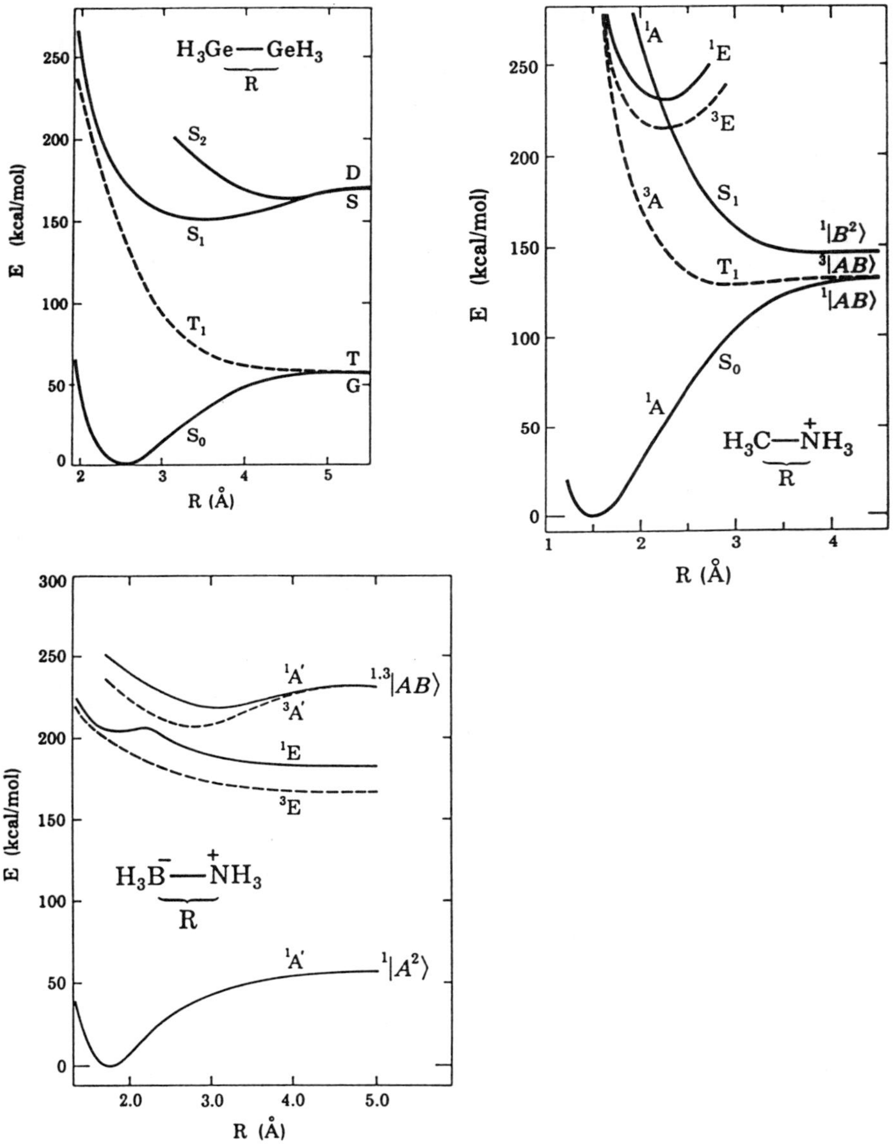

Figure 1. Typical potential energy curves for homolytic dissociation in non-polar, charged and dative σ-bonded systems. Reproduced with permission from ref. 1.

The complete active space of a two-orbital two-electron system contains three singlet states, but we have only discussed the lower two, S_0 and S_1. In the dissociation limit, the third singlet (S_2) corresponds to the other, less favorable hole-pair structure, such as CH_3^- + NH_3^{2+} and is much higher in energy. Only in the case of uncharged covalent bonds is its energy comparable with that of S_1, and can even be degenerate with it, as in H_3Si^+ $^-SiH_3$ and H_3Si^- $^+SiH_3$. Formally, the S_1 and S_2 states then correspond to linear combinations of the two hole-pair structures, but since at infinite distance their interaction vanishes, even an infinitesimal perturbation of the symmetry is sufficient to remove this mixing (this is the sigma analogue of the "sudden polarization" effect familiar from discussions of pi bonds broken by twisting to orthogonality).

At their ordinary geometries, the three kinds of sigma bonds can still be described by mixtures of dot-dot and hole-pair structures, as in valence-bond theory, or more commonly, in terms of the mixtures of the σ^2, $^{1,3}\sigma\sigma^*$ and σ^{*2} configurations built from the σ and σ^* molecular orbitals. In the symmetrical case of a homonuclear bond, as in H_3Si-SiH_3, the antisymmetric $^1\sigma\sigma^*$ configuration cannot mix with the two symmetric singlet configurations, σ^2 and σ^{*2}, and represents the "singly excited" state S_1. In valence-bond language, this wavefunction corresponds to the out-of-phase mixture of the two hole-pair configurations. The S_0 wavefunction is dominated by σ^2, with some out-of-phase admixture of σ^{*2}, and can be described alternatively by the dot-dot structure with some admixture of the in-phase combination of the hole-pair structures. Since these are not orthogonal, a straightforward visualization is difficult. The T_1 wavefunction is described by $^3\sigma\sigma^*$ or alternatively by the triplet dot-dot structure.

In all other cases, the symmetry is lower and all three singlet configurations can mix in the MO-CI description, while all three singlet structures can mix in the valence-bond description. Now, the σ and σ^* orbitals also lose their symmetry. The former is concentrated on the more electronegative center, the latter on the less electronegative one, so that the σ^2 configuration continues to dominate the S_0 wavefunction, and $^{1,3}\sigma\sigma^*$ the S_1 and T_1 wavefunctions. The relative weight of the valence-bond structures in these states is dictated by their energies, but since they are not mutually orthogonal, it is difficult to provide a simple unique description.

There clearly are two important degrees of freedom in the description of the electronic states of the active space of the two-orbital two-electron system. One is the degree of the sigma interaction between the two orbitals (the "resonance" integral of simple semiempirical theories), the other is their electronegativity difference (the difference of the "Coulomb" integrals of simple semiempirical theories). The former is determined by the distance and mutual orientation of the two orbitals, the latter by the nature of the atoms on which they reside, their molecular environment (hybridization, substituents) and their solvent environment. A schematic representation of the effect of the two independent variables on the state energies is given in Figure 2. Motion along the R coordinate corresponds to a variation in the distance of the two sigma orbitals, and the resulting curves to the dissociation curves in the S_0, S_1 and T_1 states. Motion along the δ coordinate corresponds to a variation in the relative electronegativity of the two sigma orbitals, say by a realignment of the molecules of surrounding solvent, and the resulting curves to the state energy diagrams often drawn in discussions of electron transfer between separated species. Explicit numerical calculations on the dissociation of species such as H_3GeGeH_3, $CH_3NH_3^+$ and H_3BNH_3 (Figure 1) agree fully with the qualitative conclusions immediately apparent upon inspection of Figure 2: while the S_0 state is fairly strongly bound in all cases (generally less so for dative bonds), the S_1 state is weakly bound in the cases of an uncharged covalent bond and of a dative bond, and dissociative in the case of a charged covalent bond (since in this case dissociation in S_1 does not involve charge separation). The T_1 state is dissociative.

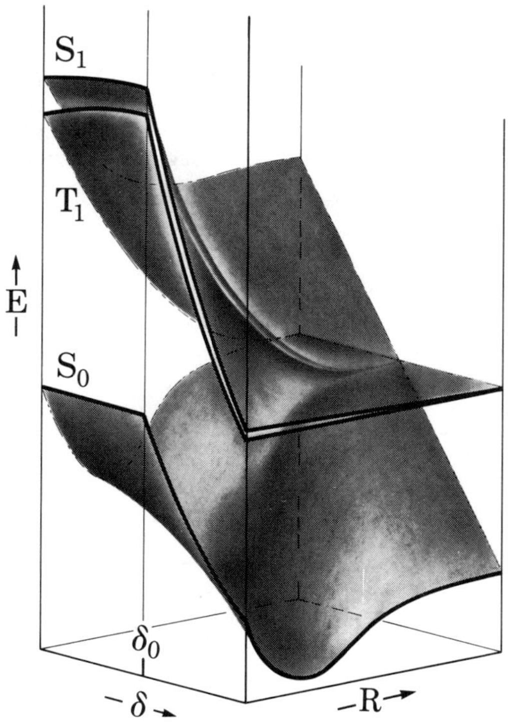

Figure 2. Potential energies of the S_0, S_1 and T_1 electronic states of a two-orbital two-electron σ-bond as a function of its length R and electronegativity difference δ of the two separating centers. Reproduced with permission from ref. 1.

These results have obvious photochemical connotations for molecules whose vertical excitation at ordinary geometries produces S_1 or T_1 states describable in terms of the same active space as those of the dissociation products just discussed, i.e., molecules whose S_1 and T_1 states are of $^1\sigma\sigma^*$ and $^3\sigma\sigma^*$ nature, respectively, where σ and σ^* are the bond orbitals of the bond to be stretched. This is a rare situation. It would be encountered in species such as H_3Ge-GeH_3 or H_3C-N^+H_3 if it were not for Rydberg states, and in species such as $C_6H_5CH_2$-NH_3^+ if it were not for the locally excited states of the benzene chromophore. Such locally excited states of chromophores that have little to do with the actual photochemical path are usually present in the substrates of organic photochemical reactions and indeed in their absence we would have to be using much shorter irradiation wavelengths than normally necessary, since at ordinary geometries $\sigma\sigma^*$ states are relatively high in energy. Often, the question is asked: how does the excitation energy flow from the initially excited chromophore or Rydberg state into the dissociating bond? In terms of potential energy surfaces, the answer is illustrated for T_1 in Figure 3. In the absence of any interaction between the chromophore and the dissociating bond, the purely dissociative curve for the $^3\sigma\sigma^*$ configuration is that familiar from Figures 1A and 2 ($\delta = 0$), and the locally excited states are simply displaced upwards from the S_0 curve by a constant amount given by the excitation energy. When the interaction between the chromophore and the dissociating bond is introduced, the crossing is avoided.

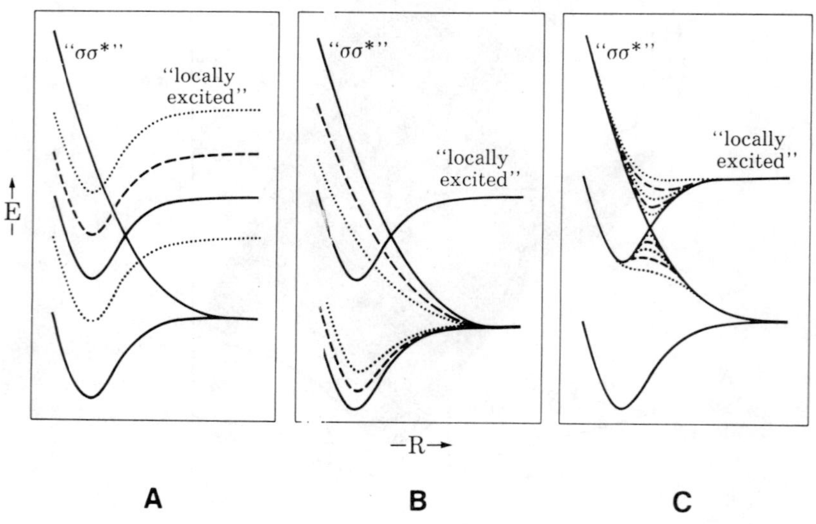

Figure 3. Potential energy curves for dissociation of a σ bond upon excitation of an adjacent chromophore, as a function of A: the energy of local excitation, B: the strength of the sigma bond, and C: the degree of chromophore-bond interaction. Reproduced with permission from ref. 1.

The relative positions of the two crossing curves and the degree to which their crossing is avoided dictate the size of the resulting barrier in the T_1 state. Parts A, B, and C of Figure 3 illustrate the effects of the energy of the local excitation, of the strength of the sigma bond, and of the degree to which the crossing is avoided, respectively. Clearly, a high energy of the local excitation at the ordinary geometry will help reduce the barrier, as will a small bond strength in the S_0 state and the associated low $^3\sigma\sigma^*$ energy, as well as a large degree of interaction between the locally excited state and the $^3\sigma\sigma^*$ state of the dissociating bond. The latter is optimized when the sigma orbitals of the dissociating bond have good overlap with orbitals used in describing the local excitation. This interaction is essentially always small for Rydberg states because of the diffuse nature of the Rydberg orbitals, so that the crossing is avoided only weakly and the "deRydbergization" rather sudden as the bond is stretched. It can be quite strong from suitably aligned $\pi\pi^*$ chromophores, e.g., a double bond located such that the dissociating bond is allylic and nearly parallel to the axes of the p components of the π bond. In extreme cases, the first-order barrier in the dashed curves may be wiped out altogether, but usually, a small barrier remains as a memory of the avoided crossing. Such a barrier separating the "spectroscopic" minimum in S_1 or T_1 from the "non-spectroscopic" biradicaloid minimum occurring further along in the reaction path is very commonly encountered in organic photochemical reactions and it accounts for two remarkable features of these processes.

First, it produces a minimum in S_1 or T_1 near the starting geometry where the excited molecules collect and vibrationally relax before starting their journey back down to S_0 and it thus provides a qualitative rationale for the lack of dependence of the results on the exciting wavelength as long as the initial excitation is into one or another of the locally excited states. At much shorter wavelengths, not ordinarily used in organic photochemistry, directly dissociative states could be excited, the minimum in S_1 circumvented, and wavelength dependence would then set in. This is indeed well known in the photochemistry of transition metal complexes, where the directly dissociative ligand-metal charge transfer states are relatively low in energy and often excited even with the commonly available irradiation sources while the d-d excited states that often play the role of locally excited states are at even lower energies. Wavelength dependence then is the rule rather than the exception. Here, however, we focus on the photochemistry of compounds of main-group elements.

A second remarkable feature is the great sensitivity of the rates and yields of many organic photochemical reactions to what would appear to be minor substituent effects. After all, the excited reactant frequently has well over 100 kcal/mol of excess energy relative to the ground state reactant, and it is not uncommon that several dozen kcal/mol of this are stored as chemical energy in the final relaxed product. With all this energy available, why should it matter if a few kcal/mol are gained or lost by interaction with an additional substituent? The answer, of course, is that the excitation energy is not really available for motion towards the geometry at which the return to S_0 will occur, since the molecule must move on the S_1 (or T_1) surface. Little time is available for such motion because of the competing radiative and radiationless surface jumps to lower surfaces, and it can be effectively hindered already by a barrier of a few kcal/mol, such as those discussed presently. Whether a substituent will increase or reduce such a barrier by a few kcal/mol may well be of critical importance.

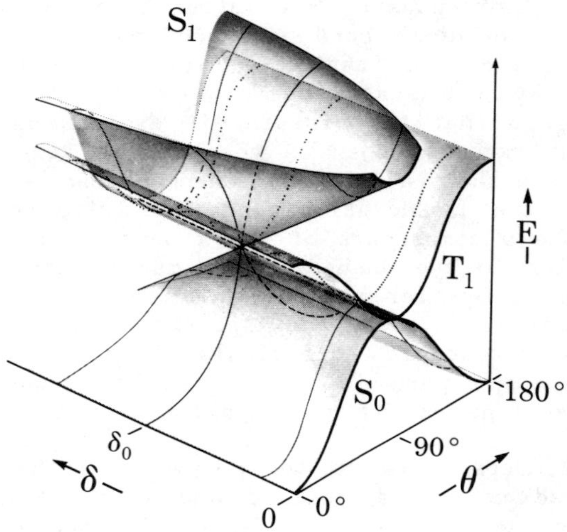

Figure 4. Schematic representation of the energies of the S_0, S_1 and T_1 electronic states of a π bond as a function of the twist angle θ and the electronegativity difference δ of the two π centers. Reproduced with permission from ref. 1.

The general picture of the states of a two-center two-electron system with sigma interaction, complemented by a discussion of the role of locally excited and Rydberg states, can be generalized in a straightforward fashion to pi interactions, in which the dissociation limit is reached not by stretching, but by a 90° twist. We shall not describe the results here in detail, but it is worth noting that the separation into three cases again applies. Orthogonally twisted covalent (e.g., C_2H_4), charged (e.g., $CH_2NH_2^+$) and dative (e.g., BH_2NH_2) pi bonds are again different when it comes to the description of their electronic states (Figure 4). They all have barriers to rotation in the S_0 state (which can be quite small for dative π bonds). The pi cases differ from the sigma cases discussed above in that in all three there is a minimum at orthogonally twisted geometry in the S_1 and T_1 states. The difference is ultimately due to the fixed distance between the centers in the pi case, which allows the biradicaloid geometry to be reached without much electrostatic charge separation in all three cases. As in the sigma case, S_0-S_1 degeneracy or near degeneracy is reached in the biradicaloid limit in the charged pi bond case, providing for a very effective return to the ground state, and it is surely not a coincidence that the natural visual pigment, rhodopsin, is a protonated Schiff base.

The nearly degenerate S_1 and T_1 states of twisted aminoborane, H_2BNH_2, which are most simply described as charge-transfer biradicaloid states of a twisted dative π bond ($H_2\overset{-}{B}\cdot{-}\overset{+}{\cdot}NH_2$), are the simplest models for the TICT (twisted internal charge transfer) states, of considerable interest in photophysics.

Also multicenter processes, both linear and cyclic, are amenable to a similar description. The complete active space is now considerably larger but it is possible to describe the results by a minor modification of those described above for the 3×3 + 1×1 case of two electrons in two orbitals.

In this fashion, photocycloaddition, excimer formation, radical ion recombination and triplet-triplet annihilation can all be fitted into the same state correlation diagram, head-to-head versus tail-to-tail orientation in photocycloaddition can be discussed, etc. A similar qualitative analysis and many qualitative computational results are also available for processes in which one or more atoms participate through more than one orbital.

5. A quantitative 3×3 + 1×1 model for the states of two electrons in two orbitals

The results described in the preceding section can be formalized in the form of a model applicable to any two-electron two-orbital system, and approximately, to larger biradicals and biradicaloids as well.[3] It is based on the recognition of the two degrees of freedom. The first is the interaction between the most localized orbitals A and B, described by γ_{AB}, and the second is the electronegativity difference between A and B, described by δ_{AB}. The orbitals A and B are chosen orthogonal, and all energies are expressed relative to that of the triplet state, E(T).

The following definitions are adopted:

$$E(T) = h_{AA} + h_{BB} + J_{AB} - K_{AB}$$
$$K'_{AB} = [(J_{AA} + J_{BB})/2 - J_{AB}]/2 = K_{(A + B)/\sqrt{2},\ (A - B)/\sqrt{2}}$$
$$\delta_{AB} = h_{AA} - h_{BB} + (J_{AA} - J_{BB})/2$$
$$\gamma_{AB} = 2h_{AB} + (AA|AB)^* + (BB|BA)$$
$$\bar{\gamma}_{AB} = (AA|AB)^* - (BB|BA)$$

where h_{AA}, h_{AB} and h_{BB} are matrix elements of the core Hamiltonian, J is the Coulomb, K the exchange, and (AA|AB) and (BB|BA) the hybrid electron repulsion integrals. It can be shown that $\bar{\gamma}_{AB}$ vanishes when A and B are chosen to be the most localized orbitals, as we have done. In that case, it also follows that $K_{AB} \leq K'_{AB}$.

The Hamiltonian matrix is

$$
\begin{array}{l} {}^1|A^2\text{-}B^2\rangle \\ {}^1|A^2\text{+}B^2\rangle \\ {}^1|AB\rangle \end{array}
\begin{pmatrix}
E(T) + 2K'_{AB} & \delta_{AB} & \gamma^-_{AB} \\
\delta_{BA} & E(T) + 2(K'_{AB} + K_{AB}) & \gamma_{AB} \\
\gamma^-_{BA} & \gamma_{BA} & E(T) + 2K_{AB}
\end{pmatrix}
$$

in the basis of two-electron functions $|A^2\text{-}B^2\rangle$, ${}^1|A^2\text{+}B^2\rangle$, and ${}^1|AB\rangle$. These are the in-phase and the out-of-phase combination of the hole-pair structures $|A^2\rangle$ and $|B^2\rangle$, and the dot-dot structure ${}^1|AB\rangle$.

A convenient starting point for a discussion is a perfect biradical, for which $\delta_{AB} = \gamma_{AB} = 0$. The energies of orbitals A and B are identical, and they do not interact. This situation is normally enforced by symmetry. Since $\gamma_{\bar{A}B} = 0$ as well, the matrix is diagonal in the basis chosen.

Two limiting cases of perfect biradicals are *pair biradicals*, for which $K_{AB} = 0$, and *axial biradicals*, $K_{AB} = K'_{AB}$. Many intermediate cases with $0 < K_{AB} < K'_{AB}$ exist as well.

The energies of the states of a pair biradical are $E(S_0) = E(T)$ and $E(S_2) = E(S_1) = E(T) + 2K'_{AB}$ (Figure 5). The wavefunction of S_0 is ${}^1|AB\rangle$, those of S_1 and S_2 are ${}^1|A^2 \pm B^2\rangle$. Since a degeneracy of S_0 and T_1 is expected in the simple model, and since interactions with other electrons usually stabilize S_0 relative to T_1 somewhat, pair biradicals normally do not obey Hund's rule and have singlet ground states in spite of having exactly degenerate non-bonding orbitals.

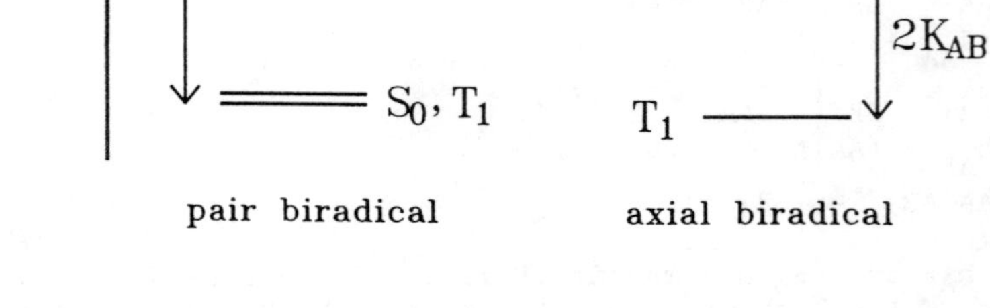

Figure 5. Energies of the low-lying electronic states of a pair biradical and an axial biradical in the 2-electron 2-orbital model.

The previously discussed case of a dissociated sigma bond between two identical atoms that do not carry lone pairs or empty valence orbitals provides a perfect example, and we note that $2K'_{AB} = J_{AA} - J_{AB}$ is just the energy needed to generate an ion pair from the radical pair. Since this can amount to hundreds of kcal/mol, it provides a nice illustration of the fact that the S_0-S_1 gap is not necessarily small in a perfect biradical, although its non-bonding orbitals are exactly degenerate. The case of an orthogonally twisted double bond between two identical atoms that do not carry lone pairs or empty valence orbitals is a nearly perfect example of a pair biradical, since K_{AB}, while nonvanishing, is very small. Here, the relations $E(S_0) \simeq E(T)$ and $E(S_2) \simeq E(S_1)$ hold quite well. Yet another example of nearly perfect pair biradicals are the "antiaromatic" uncharged perimeters, regular cyclic arrays of n overlapping orbitals occupied by n electrons, with $n = 4N$ for Hückel arrays and $n = 4N+2$ for Moebius arrays.

The pattern of the state energies of an axial biradical is quite different (Figure 5): $E(S_0) = E(S_1) = E(T) + 2K_{AB}$, $E(S_2) = E(T) + 4K_{AB}$. Now, S_0 and S_1 are degenerate and have the wavefunctions $^1|AB\rangle$ and $^1|A^2-B^2\rangle$, S_2 is higher in energy and has the wavefunction $^1|A^2+B^2\rangle$. Since T_1 is predicted to be considerably more stable than S_0 in the simple model, it usually is the ground state even in the real system and axial radicals normally obey Hund's rule.

Examples of axial biradicals are molecular oxygen, nitrenes, and the "antiaromatic" charged perimeters, regular cyclic arrays of m overlapping orbitals occupied by n electrons, where $n \neq m$ and n satisfies the conditions given above. An example is perfectly triangular planar $C_3H_3^-$. The difference between charged and uncharged perimeters is readily understood from group theory and has been invoked to account for differences in the pericyclic ring closure reactions of heteroatom-containing triplets of the diphenylamine type and the corresponding reactions of hydrocarbon triplets[1], where the former appear to proceed in a concerted fashion and the latter do not.

A perfect biradical can be perturbed into a biradicaloid in two linearly independent ways, by introducing $\delta_{AB} \neq 0$ or $\gamma_{AB} \neq 0$, and also by any linear combination of these. In any case, the perfect degeneracy of the non-bonding orbitals will be lost. However, the usual expectations, that this will result in an increase in the S_0-S_1 gap and a stabilization of S_0 relative to T_1, are not necessarily valid.

They are fulfilled in the case $\gamma_{AB} \neq 0$, $\delta_{AB} = 0$, in the so-called homosymmetric biradicaloids. Examples are $H_3Si\bullet + \bullet SiH_3$ and orthogonally twisted ethylene. In such cases, the singlet Hamiltonian matrix factors in a 2×2 + 1×1 fashion. The out-of-phase combination of the two hole-pair structures, $^1|A^2-B^2\rangle$, remains as an eigenstate and represents a "zwitterionic" S_1 state. It can also be written as $^1\sigma\sigma^*$ and is referred to as the singly excited state. The S_0 state is no longer described by the pure dot-dot structure $^1|AB\rangle$ but acquires some hole-pair character by mixing in a small amplitude of the $^1|A^2+B^2\rangle$ wave function. As γ_{AB} deviates increasingly from zero, S_0 is stabilized below T_1 and becomes the ground state of an ordinary molecule, say disilane or planar ethylene.

A more intriguing behavior is displayed by the so-called heterosymmetric biradicaloids, for which $\delta_{AB} \neq 0$, $\gamma_{AB} = 0$. Examples are CH_3^+ + NH_3 and orthogonally twisted formaldiminium $CH_2{=}NH_2^+$ or aminoborane $H_2B{-}NH_2$. The Hamiltonian matrix for the singlets still factors in a 2×2 + 1×1 fashion, but now, it is the dot-dot structure $^1|AB\rangle$ that remains as an eigenstate. It continues to represent the S_0 state and the S_0-T_1 gap is not affected. The in-phase and the out-of-phase combination of the hole-pair structures interact. In the lower energy combination, S_1, the more stable of the two structures dominates, say $^1|B^2\rangle$, and as δ_{AB} reaches the value $\delta_0 = 2\sqrt{K'_{AB}(K'_{AB} - K_{AB})}$, its energy is equal to that of S_0. This point represents a conical intersection of the S_0 and S_1 surfaces and is of considerable interest. It corresponds approximately to the cases of dissociated CH_3-NH_3^+ and twisted $CH_2 = NH_2^+$ discussed above. For even larger values of δ_{AB}, the S_0 state is described by the hole-pair structure $^1|B^2\rangle$, with a small admixture of $^1|A^2\rangle$, and the S_1 state is described by the dot-dot structure $^1|AB\rangle$. This situation corresponds to dissociated or twisted geometries of molecules with a sigma or pi dative bond, respectively.

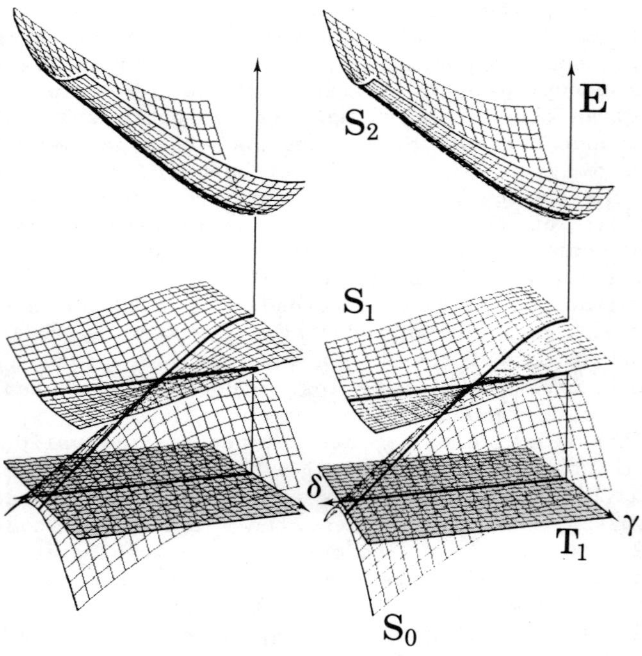

Figure 6. The energies of the S_0, S_1 and S_2 states of a biradical relative to its T_1 state, as a function of the perturbation parameters δ and γ (2-electron 2-orbital model). Reproduced with permission from ref. 1.

In general, γ_{AB} and δ_{AB} are both non-zero, and one deals with a non-symmetrical biradicaloid. Now, the S_0-S_1 crossing is avoided and such situations will clearly be preferred in the lowest singlet state, while heterosymmetric biradicaloid geometries will be favored in the S_1 state.

The energies obtained by solution of the full 3×3 Hamiltonian matrix for the singlets are shown in Figure 6 for one particular choice of K_{AB} and K'_{AB} values and provide a pictorial summary of the discussion of state energy differences.

6. An illustration: photochemical processes in oligosilanes.

Much of the current scientific interest in polysilanes $(SiRR')_n$, polymers with a pure silicon backbone, has centered around their very unusual electronic properties coupled with sensitivity to radiation. Even though these materials have a completely saturated sigma-bonded backbone, they all absorb strongly in the near UV region. The position and strength of this transition are strongly substituent dependent and vary with molecular weight and with conformation of the polymer backbone. Irradiation leads to rapid molecular weight degradation with the extrusion of reactive fragments. This process results in efficient bleaching of the intense initial optical absorption, and the materials have been shown to act as photoresists. They have other interesting properties: they are charge conductors, polymerization photoinitiators, ceramic precursors and are also of interest in non-linear optics. Much additional detail on the properties of polysilanes is available in a recent review.[2]

Qualitative considerations. The detailed nature of the electronic states of polysilanes and in particular, the detailed mechanism of their photochemical degradation, are not known. The intense absorption band is polarized along the chain direction and this has been used to assign it as a $\sigma \rightarrow \sigma^*$ transition, in agreement with semiempirical and *ab initio* calculations. According to these calculations and experiments, singlet excitation is localized on approximately planar all-trans segments of the long polymer chain, separated by gauche kinks, but can jump very rapidly from one such segment to another. Since the excitation energy decreases with increasing length of an all-trans chain segment in an asymptotic fashion up to perhaps a dozen Si atoms or so, and does not change much beyond, the sense of excitation migration is from shorter to longer segments, and fluorescence occurs primarily from the longer segments regardless of the initial excitation wavelength.

The primary photochemical processes that have been postulated to account for experimental results are

(i) Homolytic chain scission

$$\text{-SiRR'-SiRR'-SiRR'-SiRR'-} \xrightarrow{h\nu} \text{-SiRR'-SiRR'}\bullet + \bullet\text{SiRR'-SiRR'-}$$

(ii) Chain cleavage by 1,1-elimination

$$\text{-SiRR'-SiRR'-SiRR'-SiRR'-} \xrightarrow{h\nu} \begin{cases} \text{-SiRR'-}\ddot{\text{Si}}\text{R} + \text{R}'_2\text{RSi-SiRR'-} \\ \text{-SiRR'-}\ddot{\text{Si}}\text{R'} + \text{R}_2\text{R'Si-SiRR'-} \end{cases}$$

(iii) Chain abridgement by 1,1-elimination

$$\text{-SiRR'-SiRR'-SiRR'-SiRR'-} \xrightarrow{h\nu} \ddot{\text{Si}}\text{RR'} + \text{-SiRR'-SiRR'-SiRR'-}$$

Processes (i) and (ii) cause a rapid decrease in molecular weight, process (iii) does not. It has been observed that the degradation of molecular weight is slowed down by triplet quenchers in some cases and it has been proposed that process (i) proceeds via the triplet state of the polymer. At very low temperatures, the triplet state is stable and phosphoresces weakly. Fluorescent emission from the singlet is much stronger and at low temperatures its yields approach unity, while the photochemical degradation is slowed down considerably. Process (ii) appears to be of lesser significance than (i) and (iii), and yields only minor products. Process (iii) is wavelength dependent and has been observed only with wavelengths shorter than 300 nm. Although it could in principle be due to a reaction from a higher excited state, it has been proposed that the simplest way to account for this result is to postulate that only the excitation of the shorter all-trans chain segments yields this reaction, in competition with energy transfer to the longer chain, while the lowest excited states of the longer segments, which are at lower energy, do not. Processes (i) and (ii) appear to proceed at all wavelengths, either because they occur in segments of all lengths or because they occur from the longer segments, to which excitation is transferred efficiently.

These results raise some interesting spectroscopic and photochemical questions of a theoretical nature. (i) First, why is the singlet absorption of polysilanes shifted to such long wavelengths relative to polyethylene? In spite of some published views to the contrary, this has little to do with bond strengths. Although the Si-Si bond in $(CH_3)_3Si\text{-}Si(CH_3)_3$ is actually stronger than the central C-C bond in $(CH_3)_3C\text{-}C(CH_3)_3$, the lowest singlet-singlet transition in the former requires at least 30 kcal/mol less energy than that in the latter, and possibly much less.

A simple qualitative answer is provided by the analysis of the states of a two-electron two-orbital system discussed in sections 4 and 5 (Figures 1, 7). In the dissociation limit for a non-polar bond, the S_0 and T_1 states are described by dot-dot structures. At shorter separations, the S_0 state is stabilized, T_1 destabilized, and it is reasonable to expect that the $S_0 \rightarrow T_1$ excitation energy should be lower for the weaker bond. However, the situation is different for the $S_0 \rightarrow S_1$ excitation energy. In the dissociation limit, this excited state attempts to correlate with a hole-pair structure, i.e., an ion pair, and although this correlation may be thwarted by a crossing with a Rydberg state, it still roughly defines the location of the S_1 curve.

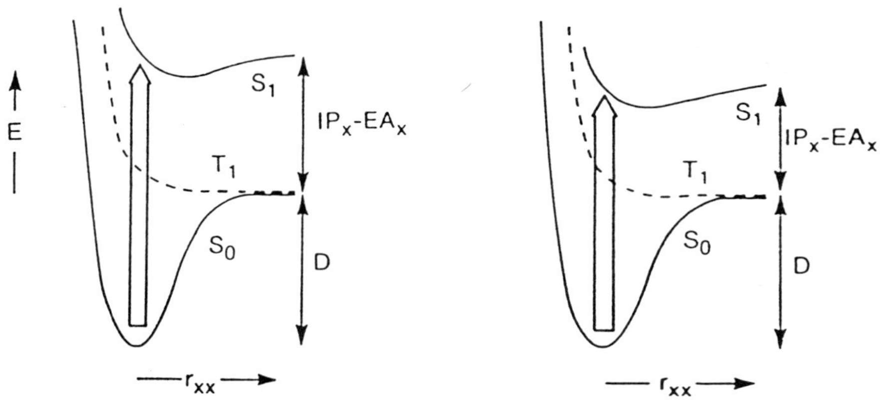

Figure 7. Dissociation curves for S_0, S_1 and T_1 states of two non-polar σ bonds X-X of equal strength D, differing in their "softness" IP_X - EA_X (schematic). Reproduced with permission from ref. 2.

The energy of the ion-pair dissociation limit is higher by IP_X - EA_X than that of the radical pair limit X• + •X, where IP is the ionization potential and EA the electron affinity of the radical X. Since the gradual descent of the S_1 surface in energy as the bond length is decreased, to yield a weakly binding minimum in the S_1 state, is similar for various choices of X, it appears that the difference of the $S_0 \rightarrow S_1$ excitation energy and the bond strength will be dictated primarily by the quantity IP_X - EA_X. This, of course, is much lower for X = Si than for X = C, and it can be said that the comparison of the $S_0 \rightarrow S_1$ excitation energies simply reveals the "zwitterionic" nature of the $^1\sigma\sigma^*$ state and the electropositive nature of silicon relative to carbon.

(ii) A second interesting question is related to the behavior of the saturated silicon chain as its length increases. This causes a reduction in the $S_0 \rightarrow S_1$ excitation energy and in the ionization potential, as expected for a conjugated system. Such a behavior is indeed anticipated from Sandorfy's classical picture of sigma conjugation[4], in which the electronic structure of a saturated chain is described by a chain of overlapping sp^3 hybrid orbitals on the carbon (or silicon) atoms, with strongly negative vicinal and more weakly negative geminal resonance integrals, isoconjugate with the pi system of a polyene with strongly alternating bond lengths (Figure 8). The question is, why is such sigma conjugation strongly affected by the conformation of the chain when sigma bonds are supposed to have cylindrical symmetry? After all, in Sandorfy's model conformational effects vanish!

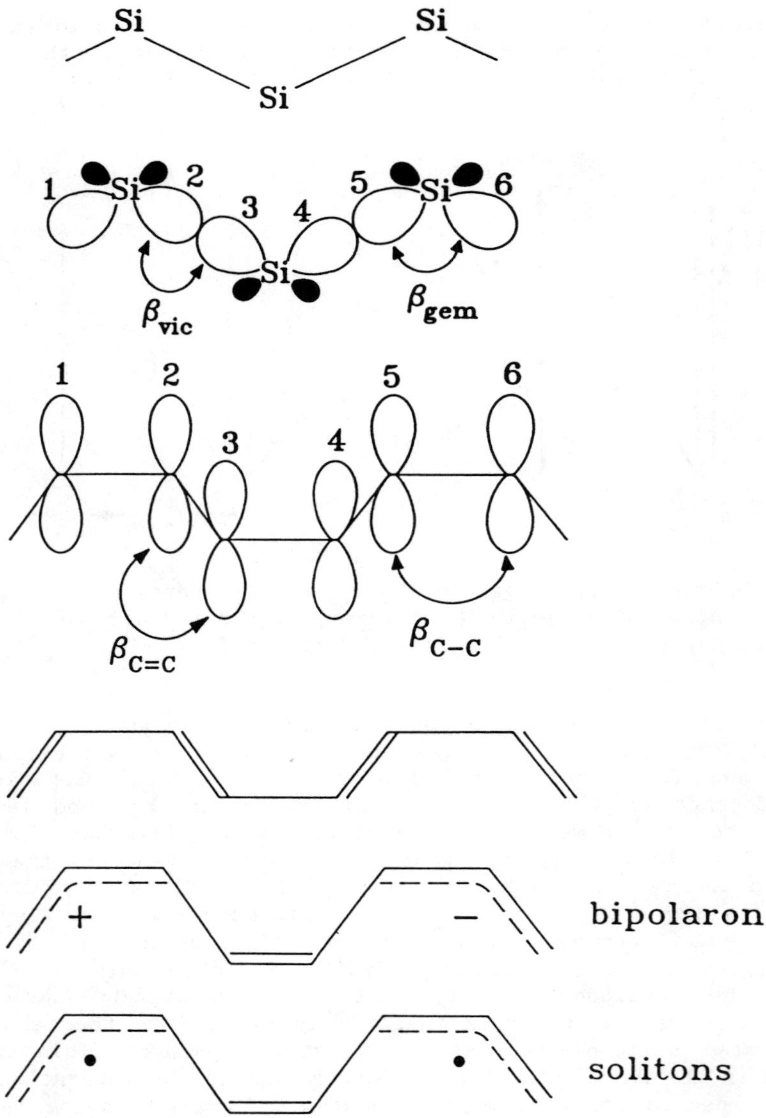

Figure 8. σ-Conjugation in a saturated silicon chain compared with π-conjugation in a polyene.

An answer to this question has been given in terms of the perturbation of the energies of the molecular orbitals resulting from the conjugation between the sp^3 hybrids (Figure 9). While the geminal and vicinal 1,2 and the 1,3 interactions are not affected by rotation about the

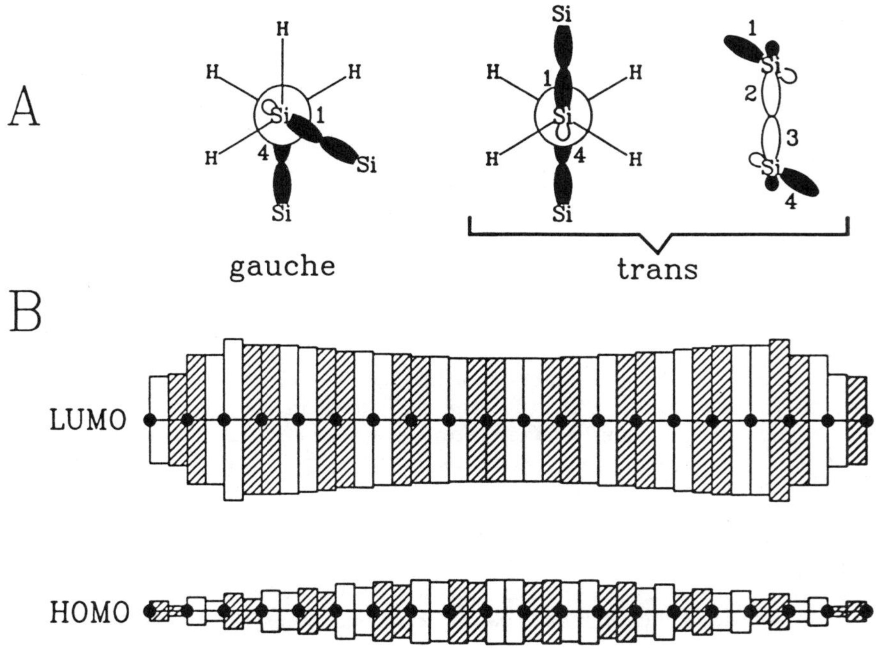

Figure 9. A: 1-4 interaction of sp^3 orbitals used for Si-Si bonding in polysilane. B: coefficients on the Si sp^3 orbitals in the HOMO and the LUMO of all-trans $Si_{20}H_{42}$ as calculated in the INDO/S approximation (length of bars, sign indicated by shading). The positions of Si atoms are indicated by black dots.

Si-Si bonds, the 1,4 interactions are. In an all-trans polysilane chain they are of anti-periplanar nature, the 1,4 overlap is negative and the corresponding resonance integral positive. As rotation around the common axis of hybrid orbitals 2 and 3 occurs, the 1,4 overlap gradually approaches zero and then grows until it is distinctly positive for the cis-planar arrangement. At this limit, the 1,4 resonance integral is negative. An inspection of the form of the highest occupied molecular orbital (HOMO) of the sigma conjugated chain shows that it has considerable p character and that its nodal properties are such that the introduction of 1,4 interactions will raise its energy in the all-trans conformation and have little effect in the all-gauche conformation. The situation is somewhat reminiscent of that encountered for the lone pairs in trans versus cis azo compounds. The energy of the lowest unoccupied molecular orbital (LUMO) is affected less by the 1,4 interaction, since it has considerable s character, immune to changes in the dihedral angle, but it is still increased somewhat in the all-trans relation to the all-gauche geometry. The net result is a decrease in the HOMO-LUMO gap at the all-trans geometry. Efficient localization of electronic transitions in the all-trans segments of long polysilane chains is indeed clearly predicted by the results of semiempirical calculations.

Ab initio calculations. We shall now illustrate the use of *ab initio* methods for obtaining additional information for these relatively complex systems. We wish to illustrate three points. First, the kind of questions that can be answered by these techniques: choice of small model systems, investigation of the nature of spectroscopic excited states, qualitative examination of photochemical reaction paths. Second, the choice of the cheapest computational method that is appropriate for a task at hand: closed-shell SCF + CI, GVB + CI, CAS-SCF + CI, different basis sets. Third, the validation of the semiempirical procedure, which alone can be used for long chains, and an identification of its limitations. The calculations were performed on the University of Texas Cray X-MP/24 computer using the GAMESS program.

The models for long-chain polysilanes $(SiRR')_n$ used in this work were the simple oligosilanes Si_2H_6 to Si_5H_{12}, and the dimethyl derivative $H_3SiSi(CH_3)_2SiH_3$.

1. Vertical excitation energies. Earlier semiempirical calculations suggest strongly that the lowest excited singlet state in polysilanes is of $^1\sigma\sigma^*$ nature and is connected to the ground state by a large transition moment. Polarization measurements leave little if any doubt that this indeed is the correct assignment of the intense observed transition, but the vexing question remains as to the location of the lowest singlet state of $\sigma\pi^*$ character (i.e., antisymmetric with respect to the plane of the Si atoms in the all-trans chain). The semiempirical calculations predict the transition to this state to have very low intensity and to lie a little above the $^1\sigma\sigma^*$ state. The experimental observations do not exclude the possibility that weak absorption bands might be hidden under the intense long-axis polarized $\sigma \rightarrow \sigma^*$ transition, but it is clear from the measured very short lifetime and high quantum yield of fluorescence that it originates in the strongly allowed $^1\sigma\sigma^*$ state and that this must represent the S_1 state in the longer chromophoric segments in the polymer. It is also clear that the S_1 state is of valence and not of Rydberg character in these longer chains. Relatively little is known about the spectra of the short members of the oligosilane series; the first intense broad bands in their vapor spectra peak at >>53 000 cm^{-1} (Si_2H_6), >53 000 cm^{-1} (Si_3H_8), 48 800 cm^{-1} (n-Si_4H_{10}) and 46 500 cm^{-1} (n-Si_5H_{12}) and are shifted to lower frequencies by permethylation.

Before exploring photochemical paths it was deemed important to ascertain whether the *ab initio* method of calculation, and in particular the basis set chosen, are capable of reproducing correctly the main observed features of the spectra and whether the results agree qualitatively with those of the semiempirical calculations.

Since these calculations were performed at the ordinary geometries near the ground-state equilibrium [Si-Si = 2.344 Å, Si-H = 1.48 Å, Si-C = 1.87 Å, C-H = 1.08 Å, α(SiSiSi) = 111.2°, α(HSiSi) = 110°, α(HCH) =

109°], a closed-shell SCF calculation was performed, followed by single and double CI calculations in an active space restricted to 6 - 10 highest occupied MO's and all the virtual MO's. Typical numbers of configurations ranged from 20,000 to 50,000. An expected shortcoming of this simple approach will be excitation energies that are too high, since the ground state will be described better than the excited states.

Huzinaga's double-zeta basis set (DZ) was used. The effect of the addition of Si d orbitals with an exponent of 0.395 was examined (DZ + d), as was the effect of the addition of diffuse s and p functions with exponents of 0.017 and 0.014, respectively (DZ + R).

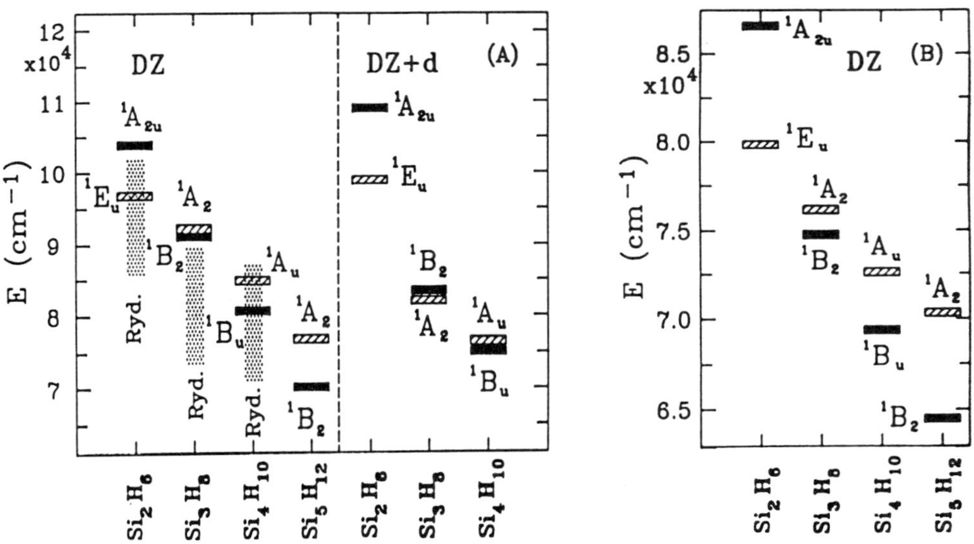

Figure 10. *Ab initio* vertical excitation energies for Si_nH_{2n+2} (n = 2-5). A: Results of SCF + CI calculations using DZ basis, DZ + d basis and DZ + sp (Rydberg) basis sets. B: Results of MR-CI calculations (DZ basis set).

The results are summarized in Figure 10A and it is seen that the excitation energies indeed are all too high. This was improved somewhat by adding p functions on the H atoms (in Si_3H_8) and more substantially by performing multi-reference CI calculations (MR-CI, Figure 10B). However, the qualitative nature of the wavefunctions did not change.

Figure 10 shows that in disilane the first $^1\sigma\pi^*$ (^1E_u) state lies well below the lowest $^1\sigma\sigma^*$ ($^1A_{2u}$) state. As the length of the silicon chain increases, the energies of both states drop. That of the $^1\sigma\sigma^*$ state decreases much faster than that of the $^1\sigma\pi^*$ state, and their order eventually changes, so that in long enough chains, $^1\sigma\sigma^*$ is calculated to correspond to S_1. The exact crossover point depends on the basis set used (the first oligosilane with a S_1 state of the $^1\sigma\sigma^*$ type is Si_3H_8 in a DZ calculation and Si_4H_{10} in DZ + d), but is probably not very reliable anyway. At this level of calculation, it is of only secondary interest. There seems to be no doubt that the method yields S_1 states of $^1\sigma\sigma^*$ and not $^1\sigma\pi^*$ character for oligosilane chain chromophores of lengths interesting for the understanding of polysilanes.

The order of the valence and the Rydberg states is another matter of interest. Again, the quality of the calculation, particularly for the longer chains, is clearly insufficient for quantitative statements but the trend is obvious. The energies of the Rydberg states drop approximately as the ionization potentials when the chain length increases, while the $^1\sigma\sigma^*$ state energies decrease much faster. Extrapolation to chains of six and more Si atoms leaves no doubt that the valence $^1\sigma\sigma^*$ state will be the lowest singlet even in the gas phase.

The conclusions drawn from these results for the parent oligosilanes thus extrapolate well to observations on peralkylated polysilanes and also agree with results of the semiempirical calculations, thus providing support for the use of the latter for the longer chains. An interesting prediction obtained from the *ab initio* results is that the lowest $^1\sigma\pi^*$ state should lie not much above the $^1\sigma\sigma^*$ state in the chain segments of lengths commonly believed to be the active chromophores in the polysilanes, 5 - 30 Si atoms.

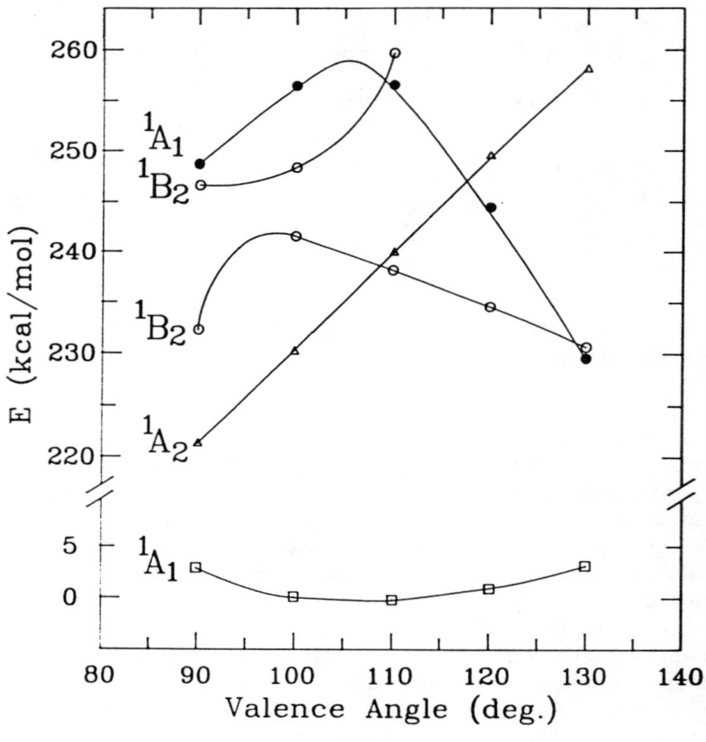

Figure 11. Calculated changes in the ground and excited singlet state energies in Si_3H_8 as a function of the Si-Si-Si valence angle.

The effect of methyl substitution can be evaluated qualitatively by comparison of the results for trisilane and 2,2-dimethyltrisilane. For the latter, the $^1\sigma\sigma^*$ excitation is red-shifted by about 12,000 cm^{-1} relative to the former and is calculated to be 4,000 cm^{-1} below the lowest $^1\sigma\pi^*$ excitation, which is now nearly degenerate with the second $^1\sigma\sigma^*$ excitation. This selective stabilization of the $^1\sigma\sigma^*$ states relative to $^1\sigma\pi^*$ upon alkylation is readily rationalized using simple perturbation theory. It provides an additional safety margin to the assignment of S_1 in alkylated polysilanes $(SiRR')_n$ as a $^1\sigma\sigma^*$ state.

The effect of variation in the valence angle α(SiSiSi) within the range 90 - 140° was explored in the case of Si_3H_8. As shown in Figure 11, this causes relatively little change in the S_0 energy but has a large effect on the excited states. An increase of the angles stabilizes the $^1\sigma\sigma^*$ state and destabilizes the $^1\sigma\pi^*$ state. At valence angles near 110° the two states of Si_3H_8 are essentially degenerate, at smaller angles the $^1\sigma\pi^*$ is the lowest, and at larger angles the $^1\sigma\sigma^*$ is the lowest. Since in the polysilane polymers the valence angle α(SiSiSi) is believed to be roughly 116°, this provides another argument in favor of the $^1\sigma\sigma^*$ below $^1\sigma\pi^*$ ordering of their excited states.

The effect of the variation in the dihedral angle α(SiSiSiSi) investigated for Si_4H_{10}, at the SCF + CISD level with an active space of 3 highest occupied and 13 lowest unoccupied MO's, is extremely small in the S_0 state. Even under conditions of rigid rotation the barrier is only about 3 kcal/mol. No attempt has been made to find the true saddle point but there is no doubt that the rotation is nearly free.

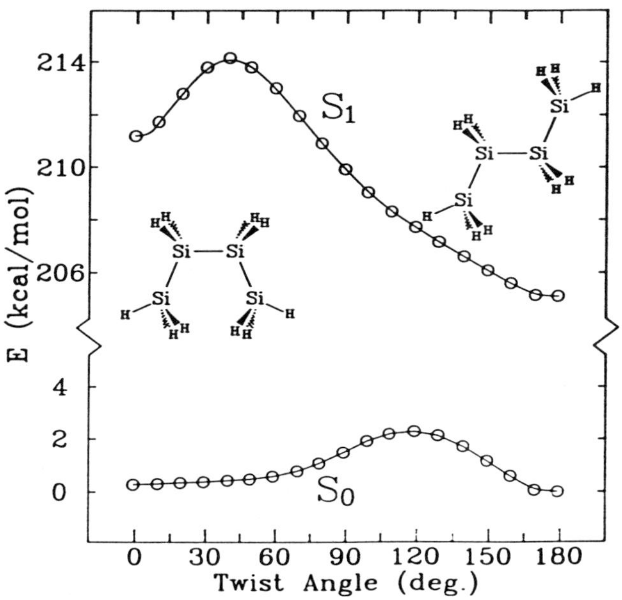

Figure 12. Calculated effect of variation of the Si-Si-Si-Si dihedral angle on the S_0 and S_1 energies of Si_4H_{10}.

On the other hand, in the S_1 state the trans geometry is favored significantly over gauche (Figure 12), in agreement with the results of semiempirical calculations. At the all-trans geometry, the lowest state is of $^1\sigma\sigma^*$ character. Upon twisting, this state rises in energy and near a dihedral angle of 40° undergoes an avoided crossing with a $\sigma\pi^*$ state, whose energy drops and which is responsible for a local minimum in the S_1 state at all-cis geometry. The results thus support the prior ideas about the nature of the localization of electronic excitation in polysilanes.

2. The homolytic scission path in Si_3H_8. This represents a two-orbital two-electron active space problem. The singlet space was explored at the level of one-pair GVB natural orbital calculation followed by CI with single and double excitations from the three highest occupied to the nine lowest unoccupied MO's. The triplet state calculations were done at the UHF level. The results shown in Figure 13 were obtained for a cut through the nuclear configuration space along which the structure of dissociating fragments, $Si_2H_5\cdot$ and $SiH_3\cdot$, was kept constant while their separation increased. This was considered adequate for a qualitative discussion of the likely photocleavage, although it clearly cannot provide numerically correct dissociation energies. The results are simple and correspond exactly to expectation based on the simple model of biradicaloid structure (Figure 1). The S_0 state is bound, the T_1 state is strongly dissociative, and the S_1 state, which may contain a shallow minimum, tends to a higher dissociation limit.

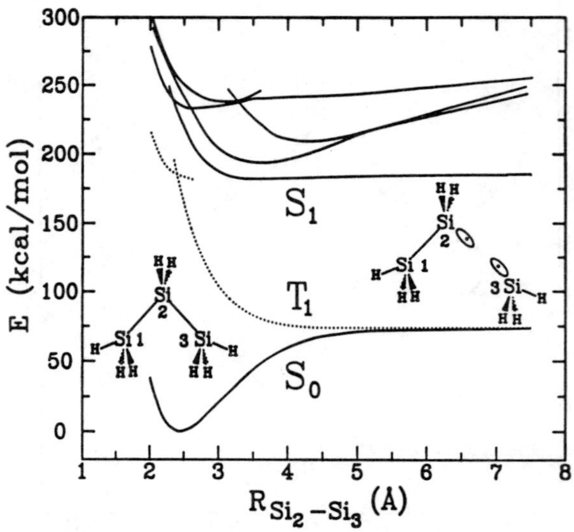

Figure 13. *Ab initio* potential energy curves for the homolytic cleavage of a Si-Si bond in Si_3H_8.

Since the limit will not change much upon going to longer chains, while the vertical excitation energy drops substantially, the longer silicon chains will have a bound S_1 state, in good agreement with their high fluorescence quantum yields. One concludes that the homolytic dissociation path is unlikely to be followed in the S_1 state, but is very probable in the T_1 state.

3. The chain abridgement path in Si_3H_8. This path was modelled by a simultaneous removal of the central SiH_2 group along a direction perpendicular to the Si(1)-Si(3) line, and contained in the Si(1)Si(2)Si(3) plane, and approach of the two SiH_3 groups with three-fold symmetries, so as to gradually form SiH_2 and Si_2H_6. The active MO space is now spanned by the two orbitals of the two SiH_3 groups and the two orbitals of the SiH_2 group that are responsible for the two Si-Si bonds in Si_3H_8 and for the Si-Si bond and the two non-bonding orbitals in Si_2H_6 + SiH_2. The calculations were therefore performed at the CAS-SCF level, using the two highest occupied MO's and the two lowest unoccupied levels as the active space. This was followed by CISD calculations using the same reference space.

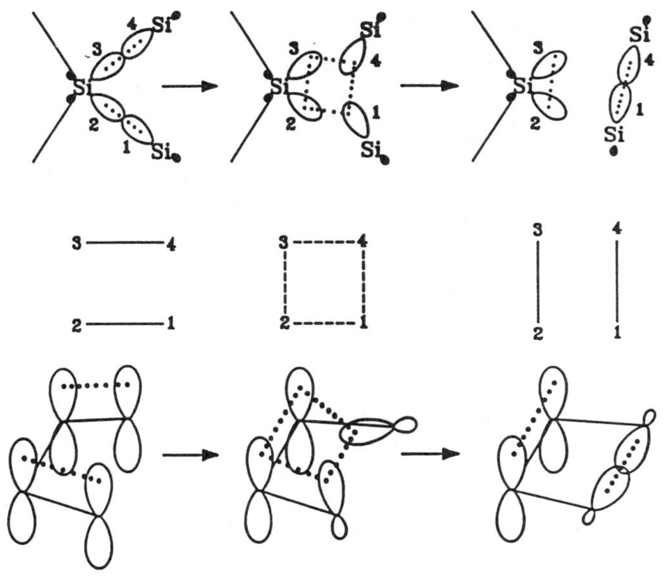

Figure 14. Comparison of the AO interactions in the photochemical chain abridgement in a polysilane (top) and in the disrotatory electrocyclic ring closure of *cis*-butadiene (bottom).

Figure 14 shows the orbitals involved and demonstrates that the problem is isoelectronic with the disrotatory electrocyclic closure of 1,3-butadiene to cyclobutene. The course of the singlet states along the reaction path is indeed just that anticipated from the latter case (Figure 15). The S_0 state of Si_3H_8 attempts to correlate with a doubly excited state of the products, consisting of ground-state Si_2H_6 and doubly excited SiH_2 with both electrons in the 3p orbital on Si. The S_0 state of the product would originate from a doubly HOMO to LUMO excited state of the starting materials, and the would-be crossing is avoided, resulting in a barrier in the S_0 and minimum in the S_1 surface. The thermal ground-state reaction would clearly choose a less symmetrical path, but in the photochemical process, the presence of the "pericyclic minimum" in S_1 along the path chosen provides a driving force.

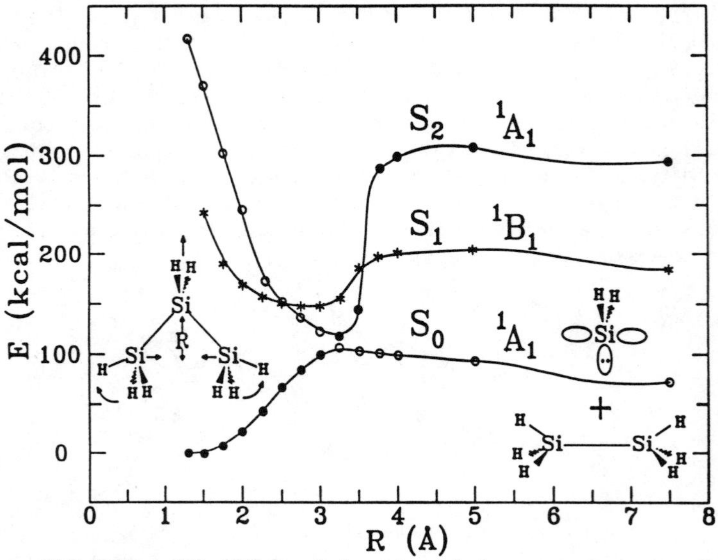

Figure 15. CAS-SCF + CI (DZ basis) potential energy curves for the reductive elimination of $\ddot{S}iH_2$ from Si_3H_8. The reaction coordinate is shown in the inset.

The optically strongly allowed vertical S_1 state, HOMO → LUMO in character, correlates with a singly excited state of the products, consisting of ground-state Si_2H_6 and SiH_2 in its lowest excited singlet state, 3s → 3p in character. Although it is of much less energy in the region of the pericyclic minimum, it does not represent the S_1 but only the S_2 state, and the analogy to 1,3-butadiene ring closure[5] is perfect. The resulting diagram is quite typical for a pericyclic process of the uncharged-perimeter kind (4N electrons, 4N centers). We expect this path to represent a feasible singlet photochemical process, in which a vertically singly excited molecule first moves to the pericyclic minimum in S_1, then returns to S_0 and proceeds back to the starting point or on to products.

The region of the pericyclic minimum is not likely to change much upon proceeding to longer Si chains. However, the vertical S_0 → S_1 excitation energy will drop and for long enough chains a barrier may well separate the singly excited vertical S_1 state from the pericyclic minimum, located in a doubly excited biradicaloid S_1 state. This would account for the wavelength dependence of their photochemical process. For short chains, motion to the pericyclic minimum in S_1 is downhill and occurs very fast. In longer ones, it requires thermal activation and other processes will compete more successfully.

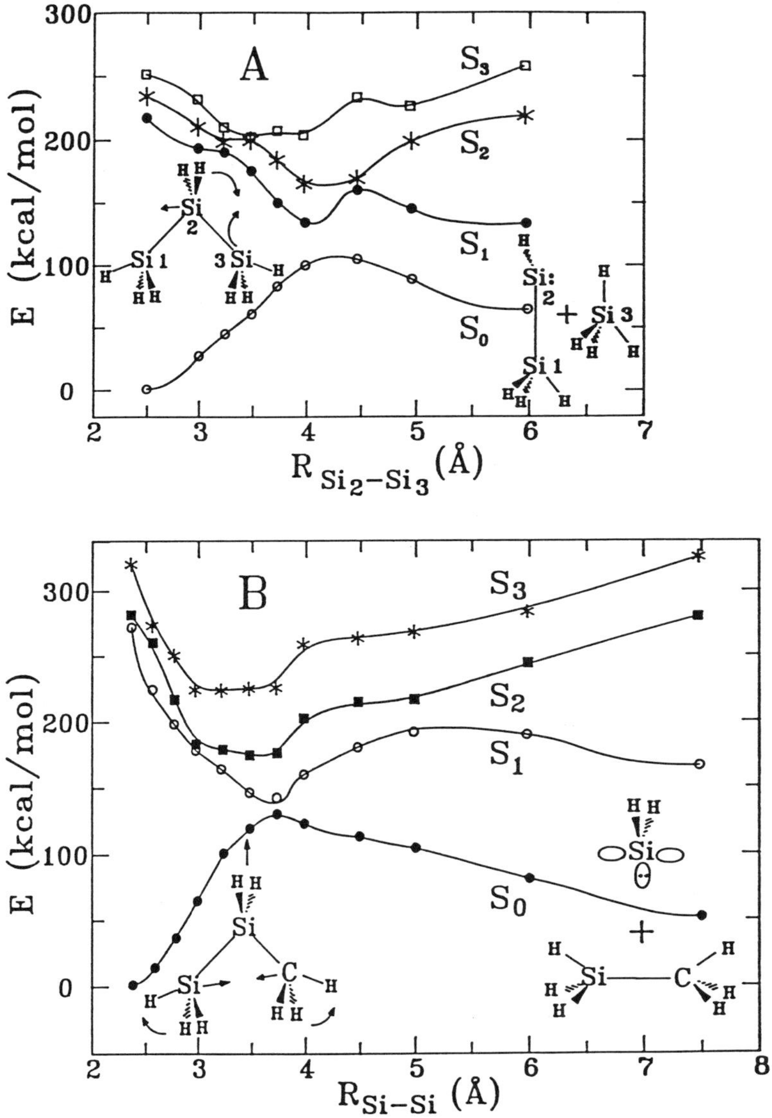

Figure 16. CAS-SCF + CI (DZ basis) potential energy curves for chain scission by reductive elimination. A: $H_3Si\text{-}SiH_2\text{-}SiH_3 \rightarrow SiH_4 + H_3Si\text{-}\ddot{S}iH$, B: $H_3Si\text{-}SiH_2\text{-}CH_3 \rightarrow H_3Si\text{-}CH_3 + \ddot{S}iH_2$. The reaction coordinate is shown in the inset.

4. Chain scission by 1,1-elimination. The processes $Si_3H_8 \rightarrow SiH_4 + H_3Si\text{-}SiH$ (Figure 16A) and $H_3Si\text{-}SiH_2\text{-}CH_3 \rightarrow H_3Si\text{-}CH_3 + SiH_2$ (Figure 16B) were explored in the same way as the closely related process $Si_3H_8 \rightarrow Si_2H_6 + SiH_2$ just described. It is seen that the results are very similar although the course of the potential energy curves is now more complicated mainly due to lower symmetry. The similarity is hardly a surprise, considering that all of these processes follow isoelectronic paths. Once again, these should be typical singlet photochemical processes. The intrinsically lower symmetry may lower the barrier between the vertical and the pericyclic minima in S_1, minimizing the effects of chain length.

5. Summary of photochemical paths in polysilanes. From the results of the calculations, it appears qualitatively that upon vertical excitation into the $\sigma\sigma^*$ S_1 state, the molecule will fluoresce in competition with intersystem crossing into a dissociative T_1 state (recombination or disproportionation may follow once the radical pair develops singlet character) and will travel to one of the pericyclic minima that correspond to the various 1,1-elimination processes, the latter being favored for the shorter chains.

This illustrates the kind of conclusions and interpretations currently possible quite easily for photochemical processes with a relatively small active space (in the present case, 20-dimensional for singlets and 15-dimensional for triplets). An attempt to proceed in a more quantitative fashion, even if only for the simple model compounds and not for the long silicon-chain compounds themselves, would require orders of magnitude more effort.

Acknowledgement. This work was supported by grants from the National Science Foundation (CHE 8796257) and the Air Force Office of Scientific Research (AFOSR 87-0001).

References

1. J. Michl and V. Bonačić-Koutecký, *Electronic Aspects of Organic Photochemistry*, Wiley, New York, 1990.
2. R. D. Miller and J. Michl, *Chem. Rev.* **89**, 1359 (1989).
3. V. Bonačić-Koutecký, J. Koutecký, and J. Michl, *Angew. Chem., Int. Ed. Engl.* **26**, 170 (1987); J. Michl and V. Bonačić-Koutecký, *Tetrahedron* **44**, 7559 (1988).
4. C. G. Pitt. (1977) In A. L. Rheingold (ed.), *Homoatomic Rings, Chains and Macromolecules of Main-Group Elements*, Elsevier, New York, Chapter 8 and references therein.
5. D. Grimbert, G. A. Segal, and A. Devaquet, *J. Am. Chem. Soc.* **97**, 6629 (1975).

NEW OPPORTUNITIES FOR THE ORGANIC CHEMIST IN THE COMPUTER AGE

ALAN R. KATRITZKY and STEPHEN J. CATO
Department of Chemistry
University of Florida
Gainesville, FL 32611
USA

ABSTRACT. A number of uses of computers in organic chemistry are studied by example. Principal Component Analysis studies of aromaticity in heterocycles and the use of statistical methods to study GC retention factors are covered. Semiempirical methods are used to study the tautomeric equilibria of heteroaromatic compounds in solution. The review ends with a brief overview of molecule and reaction databases.

1. Introduction

This survey of some of the new opportunities for the organic chemist in the computer age commences by a new look at a very old problem, that of aromaticity. It goes on to consider the whole question of structure activity relationships, and particularly, the prediction of physical properties first by using rather simple statistical methods, and then by using much more sophisticated approaches.

We then go on to consider how light can be shed on the classical problem of carbonium ion structure by using a newly available, easy-to-use, semiempirical calculation. The application of such calculations has mostly been restricted to the gas phase, but recent developments bring with them the possibility of rather simple extension to media of high dielectric constant, such as aqueous solution. This will be illustrated by some work on the tautomeric equilibria of heteroaromatic compounds.

We conclude by considering the help that organic chemists can now obtain using data bases, by reference both to the commercial availability of starting materials and to reaction databases.

2. Aromaticity

Aromaticity as a qualitative constant is very well-established in teaching and research in organic and particularly in heterocyclic chemistry. The concepts of aromatic, antiaromatic and nonaromatic are well-understood and relatively easy to explain. However, it is quite clear that aromaticity is not only a qualitative, but also a quantitative concept. Thus, as shown in Scheme 1, some compounds are clearly more aromatic than others, and it is quite obvious that a quantitative aromaticity scale would be extremely useful. However, it is here the difficulties really start, and it has been impossible to isolate any single well-defined measureable or calculable parameter that characterizes quantitatively the degree of aromaticity of a molecule or ion that is even moderately generally acceptable.

C. Ögretir and I. G. Csizmadia (eds.),
Computational Advances in Organic Chemistry: Molecular Structure and Reactivity, 355–404.

1 Aromaticity: Some are More Equal Than Others

furan less aromatic than benzene, shows some diene character (Diels-Alder).

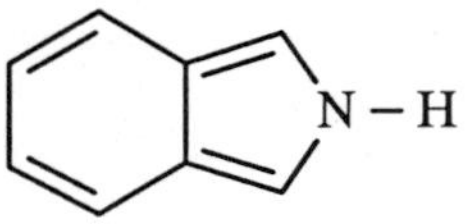

isoindole

<

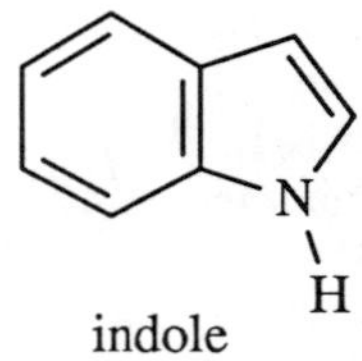

indole

anthracene shows diene character, phenanthrene does not

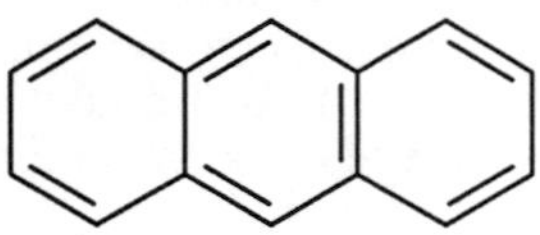

anthracene

<

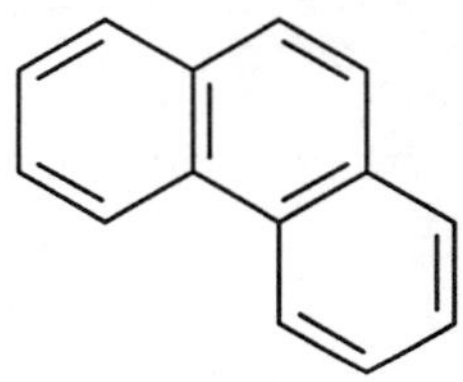

phenanthrene

hydrogenation of naphthalene to tetralin relatively easy

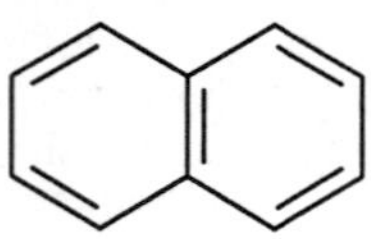

naphthalene

<

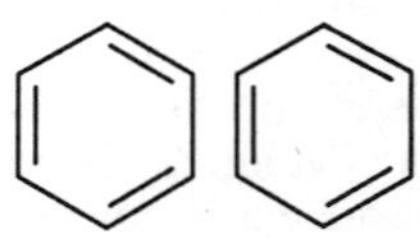

two benzenes

Hence a quantitative aromaticity scale would be useful

The difficulties have arisen from a variety of grounds, the most important of which are:

(i) Experimental measures of any one parameter are frequently not available for a full range of compounds;

(ii) Errors in experimental measurements arise from the need to use data obtained by different investigators at different locations, or from the fact that they often represent small differences between large measured values.

(iii) Most theoretical MO methods require the assumption of parameters for heteroatoms and there is no general agreement on the values to be taken for these parameters.

(iv) A particular uncertainty, affecting theoretical and experimental methods alike, is the need for comparison of actual aromatic compounds with nonaromatic models; the precise definition of these models is not easy.

Because of such difficulties, the concept of aromaticity has remained in an unsatisfactory state from a quantitative viewpoint. Recently, because of this, we investigated the situation by means of Principal Component Analysis.

Principal Component Analysis has been explained many times (for references see [1]); its basic equation is given in Scheme 2. The first Principal Component is defined as the best summary of preliminary relationship exhibited in the data. The second Component may be viewed as the second-best linear combination of variables that accounts for the most residual variants when the effect of the first component is removed from the data. Subsequent components are similarly defined until all the variants in the data are exhausted.

2 Principal Component Analysis

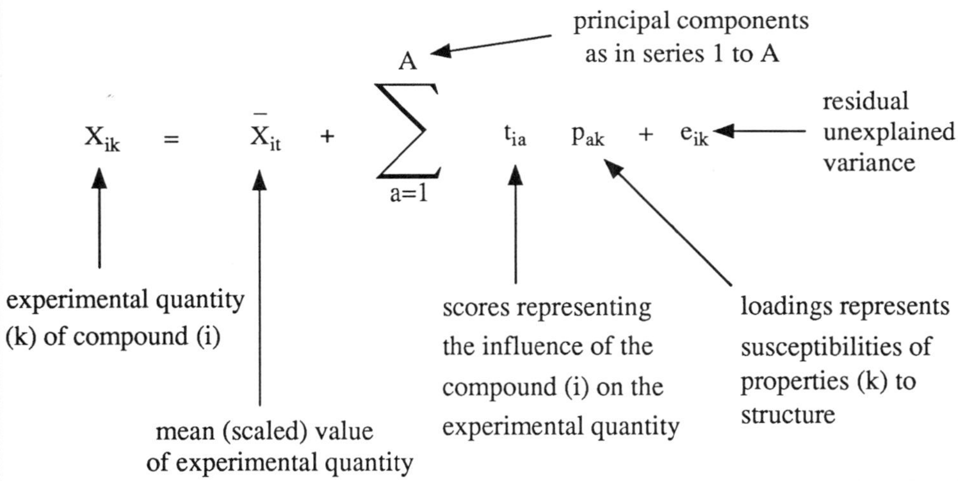

We chose a model set of sixteen objects, given in Scheme 3, and treated it using the SIMCA method. As physical properties, we adopted various geometrical, energetic and magnetic criteria.

3 A Model Set of Sixteen Objects

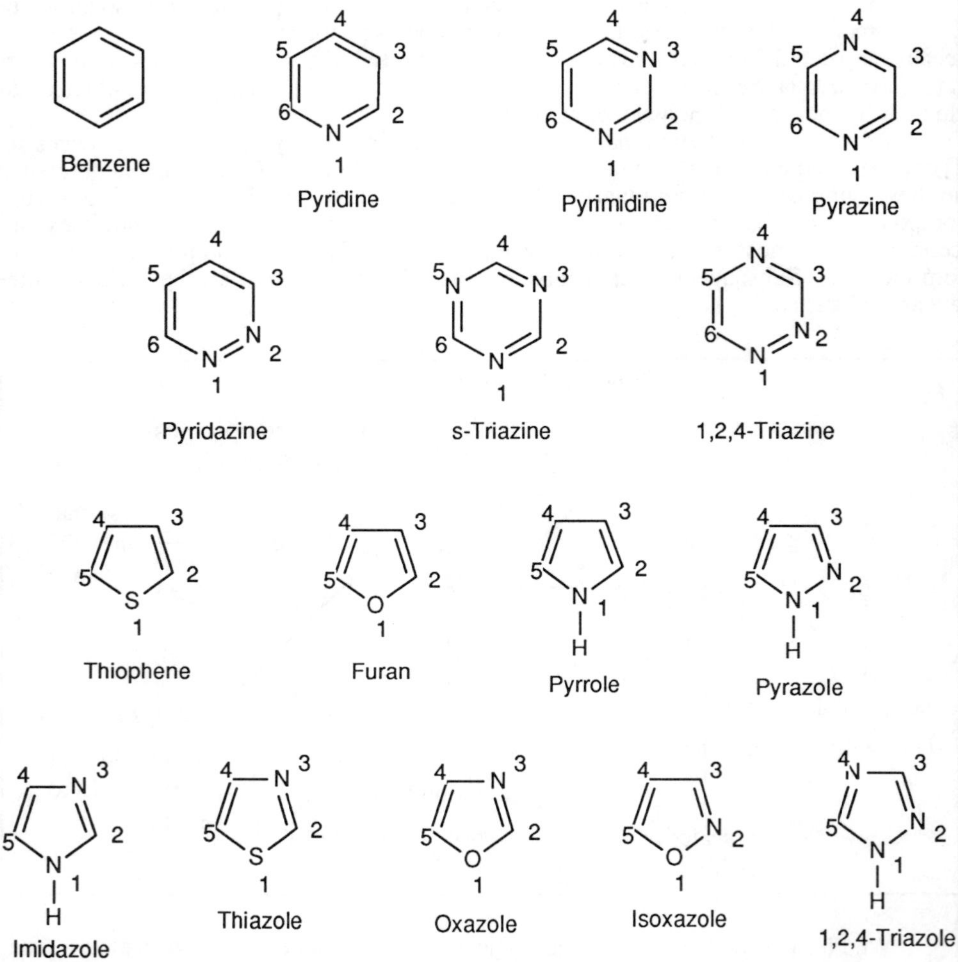

The geometrical criteria are shown in Scheme 4. They include two Bird aromaticity indexes: (i) that derived using experimental bond lengths, and (ii) using AM1 calculated bond lengths. Further geometric criteria include the Jug RC method which defines the weakest link in the ring, and Pozharskii's index which measures the differences in bond order.

The different thermodynamic stability of aromatic systems is the basis for the development of energy criteria. The quantitative estimates we have chosen (Scheme 5) are two Dewar (DRE and DRE′ calculated using AM1), the Hess-Schaad resonance energies, and the experimental and calculated heats of formation.

4 **Geometrical Criteria**

I_5 and I_6 -- the aromaticity indexes of 5- and 6-membered rings

$$I = 100\,\frac{(1-V)}{V_k} \qquad \text{where} \qquad V = \frac{100}{N_o}\sqrt{\frac{\Sigma(N-N)^2}{n}}$$

1. From experimental bond lengths (I_{sc})
2. From calculated bond lengths (AM1) (I'_x)

RC = the magnitude of a ring current (RC) as determined by the weakest link in the ring (lowest values of bond order) (Jug's index)

ΔN = the differences in bond order:

$$\Delta\bar{N} = \frac{\Delta\Sigma N}{n} \qquad \text{(Pozharskii's index)}$$

5 **Energetic Criteria:**

DRE = the Dewar resonance energy (polyene bond energies not available for N-N, N=N, N-O, or N-S)

DRE′ = the Dewar resonance energy calculated using AM1 heats of formation

HSRE = the Hess-Schaad resonance energy (Hückel π-bond energies not available for N=N, N-O, or N-S)

$\Delta H_{f\,exp}$ = the experimental heat of formation (where available)

$\Delta H_{f\,am1}$ = the calculated heat of formation (AM1)

Compounds have been considered as aromatic when induced diamagnetic π-electron ring currents are maintained within the molecule. The magnetic criteria that we have considered are (Scheme 6) the molar diamagnetic susceptibility, the dimagnetic susceptibility to exaltation, and the average ^{15}N chemical shift.

6 Magnetic Criteria

χ_M = the molar magnetic susceptibility (available experimental values)

$$\chi_M = N\mu_o \ (\xi + \mu^2_m \ 3kt)$$

Where ξ = the molecular magnetizability

μ_M = the magnetic dipole moment

μ_o = the vacuum permeability

N = the number per unit volume

Λ = the diamagnetic susceptibility exaltation (available experimental values)

$$\Lambda = \chi_M - \chi_M'$$

^{15}N n.m.r. = the average N-15 chemical shift

The data set of 12 characteristics and 16 compounds used is shown in Scheme 7. To determine the number of statistically significant eigenvectors, the method of cross-validation is used. The number of principal components is selected as three. The results on the PC analysis are given in Scheme 10. It is seen that two Principal Components account for 74% and three for 87% of the variance.

To assess the predictive power, the data matrix containing i objects and k independent variables per object was treated as follows: each data point, x_{ik}, was omitted successively one at a time and a PC analysis was carried out in each case on the remaining set of data. A new value for the data point omitted was then recalculated using the new values of the scores and the new values of the loadings. These recalculated values, when compared with the input values, demonstrate the predictive power of the model. They are plotted against the input values and R^2 is determined for the 1:1 line to determine a measure of fit. The resulting plots are shown in Schemes 9-14.

7 Data Set for Monocyclic Compounds

Compound	1 I_x	2 I_x'	3 RC	4 ΔN	5 DRE	6 DRE′	7 HSRE	8 ΔH_f	9 ΔH_{fcal}	10 χ_M	11 Λ	12 ^{15}N
1 Benzene	100.0	100.0	1.774	0.00	3.77	3.84	0.065	19.8	22.0	54.8	17.9	-
2 Pyridine	85.7	84.1	1.571	0.10	3.85	4.12	0.058	34.6	32.0	49.2	18.3	-63.5
3 Pyrimidine	84.3	82.4	1.498	0.13	3.37	3.60	0.050	47.0	46.8	43.1	18.2	-84.5
4 Pyrazine	88.8	90.9	1.561	0.06	2.85	4.00	0.049	46.9	44.4	37.6	12.7	-46.1
5 Pyridazine	78.9	73.4	1.468	0.18	-	-	-	66.4	55.4	40.5	8.7	20.2
6 *s*-Triazine	100.0	100.0	1.483	0.00	-	3.33	0.040	54.0	60.1	37.9	19.0	-98.5
7 1,2,4-Triazine	86.1	71.9	1.398	0.19	-	-	-	52.4	69.5	-	-	-6.7
8 Thiophene	66.0	67.7	1.533	0.25	1.08	0.96	0.032	27.6	40.1	57.4	17.8	-
9 Furan	43.0	35.2	1.105	0.47	0.72	0.63	0.007	-8.3	3.0	43.1	13.9	-
10 Pyrrole	59.0	66.5	1.344	0.24	1.40	2.15	0.039	25.9	40.1	47.6	14.5	-224.6
11 Pyrazole	73.0	74.2	1.316	0.18	-	-	0.055	43.3	66.5	42.6	15.5	-126.5
12 Imidazole	64.0	68.0	1.306	0.21	2.57	3.34	0.042	30.6	51.0	44.3	17.2	-169.0
13 Thiazole	64.0	59.4	1.283	0.32	1.87	1.32	0.034	37.0	49.9	50.6	17.0	-57.4
14 Oxazole	38.0	37.8	1.075	0.40	-	2.07	0.007	-3.7	12.6	39.2	16.0	-127.0
15 Isoxazole	47.0	48.1	1.042	0.36	-	-	-	18.8	44.4	37.5	14.3	2.7
16 1,2,4-Triazole	81.0	74.7	1.273	0.17	-	-	0.050	46.3	74.0	-	-	-131.0

8 Results of PC Analysis on Sixteen Objects

Principal Component(s)	Percentage of Variance Accounted For
t_1	48.4
$t_1 + t_2$	65.2 (48.4 + 16.8)
$t_1 + t_2 + t_3$	73.7 (48.4 + 16.8 + 8.5)

9-14: Plots of Recalculated vs. Original Data for Monocycles

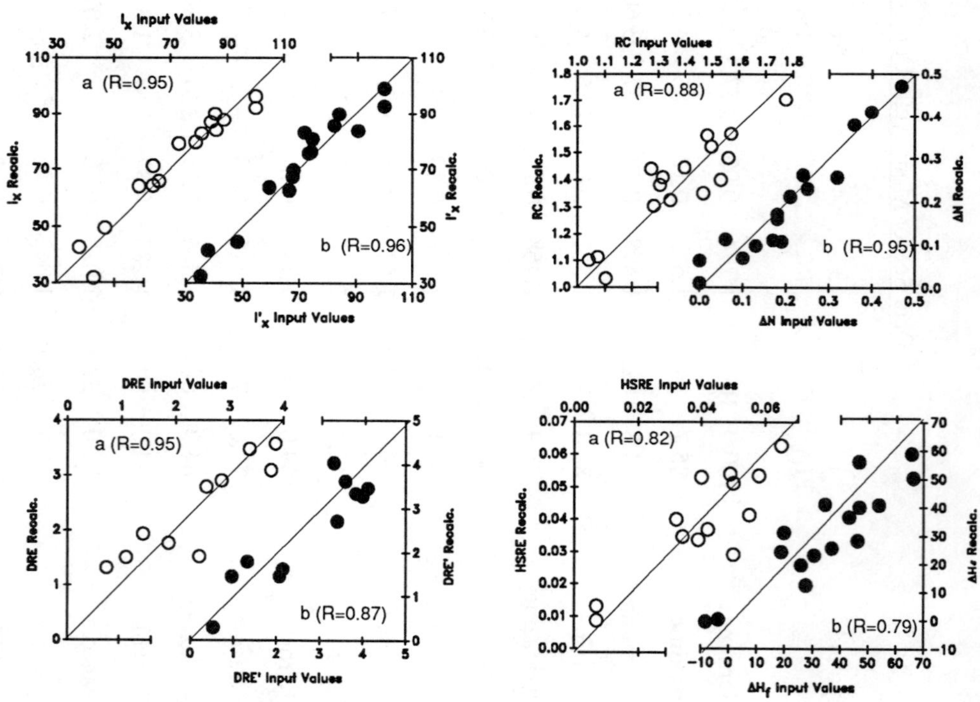

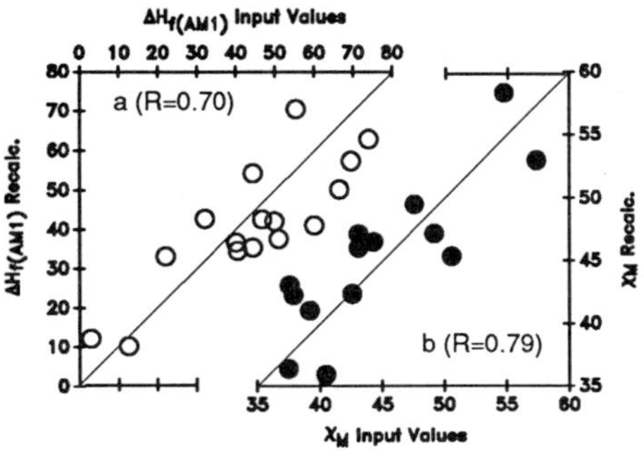

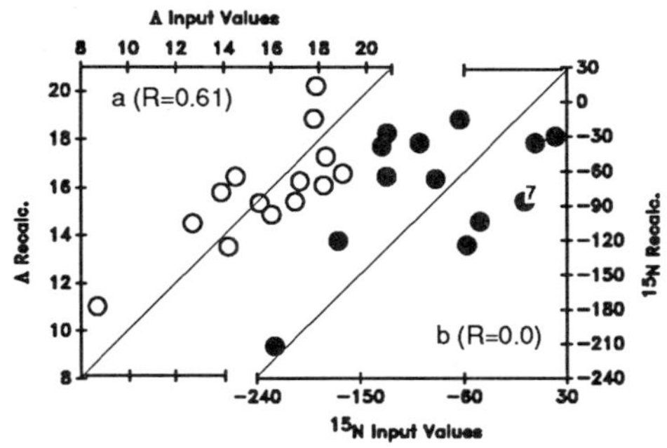

The PC scores for all 16 compounds are plotted graphically in Scheme 15, with the Principal Component scores t_1 and t_2 as the ordinate and abscissa, respectively, and the score t_3 for the third PC shown within the circle.

15

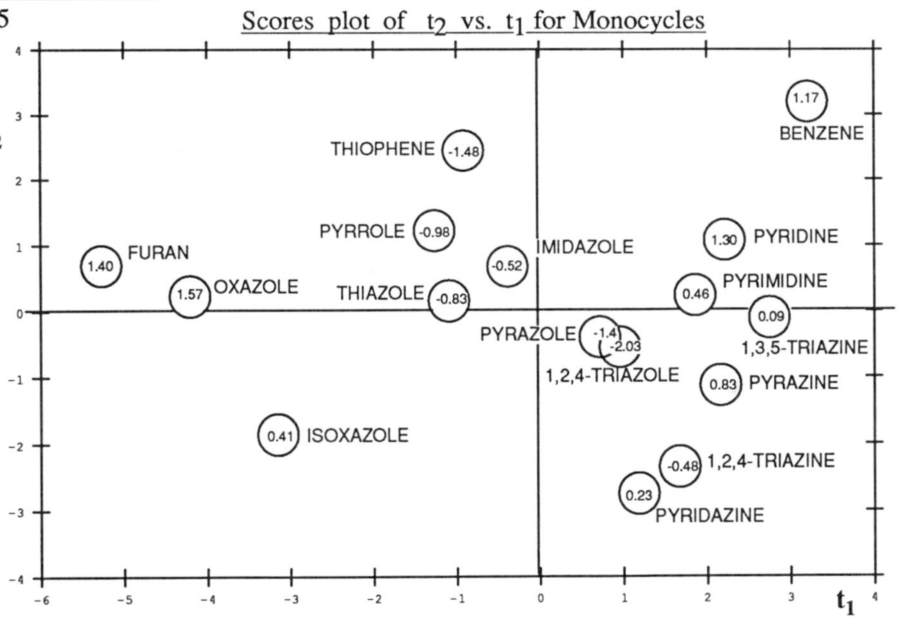

Scheme 16 shows that in the three-dimensional space defined by their principal components scores t_1, t_2, and t_3, the compounds 1-16 can be divided into four or five groups. Benzene and pyridine have all three t_1, t_2, and t_3 values positive. Di- and triazines are placed in another quartile with only t_2 negative. Azines and azoles with two nitrogen atoms in adjacent positions in the ring form a separate group with both t_2 and t_3 negative. The final group contains five-membered rings with nitrogen and/or sulfur heteroatoms (moderately negative t_1 negative and t_3 positive).

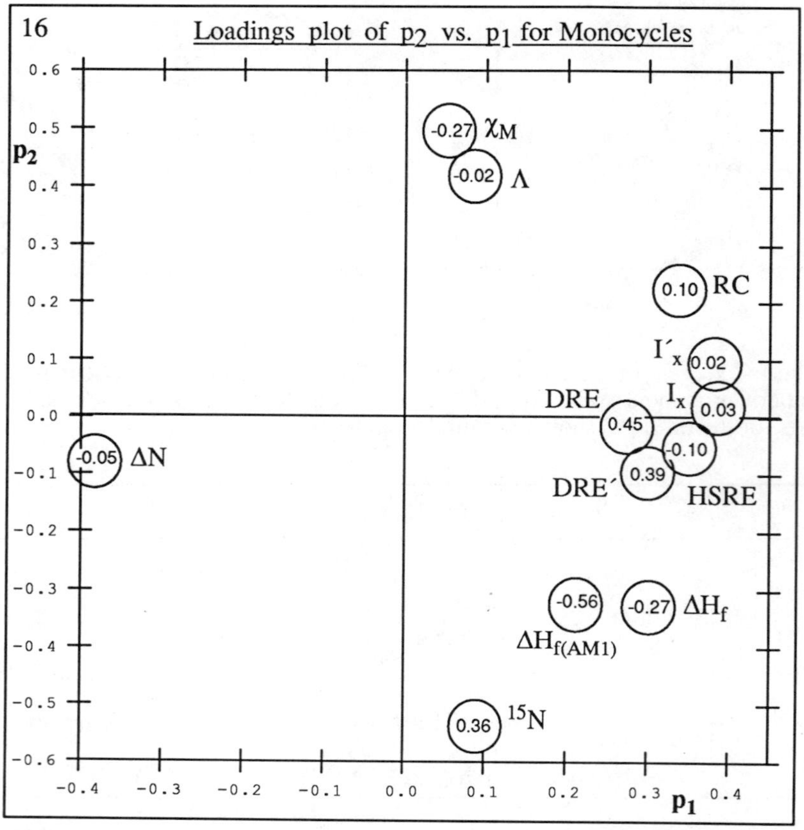

The PC loadings for the 12 characteristics are given in Scheme 18. In the three-dimensional space defined by their PC loadings, the characteristics can be divided into three main groups: those (Group A) dominated by the p_1 GC loadings (I_x, I'_x, ΔN, DRE, DRE′, HSRE) and with small p_2 and p_3 loadings; the second group (B) is comprised of the magnetic parameters χ_M, Λ, and ^{15}N, and is marked by a very small p_1 component but large positive p_2 and variable p_3 components; the remaining characteristics form the third group. Two of these groups, A and B, are almost orthogonal to each other, and can be related to the concepts of "classical" and "magnetic" aromaticity.

With 48.4% of the total variance in the data set, eigenvector number one is comprised predominantly of I_x, I'_x and ΔN. DRE, DRE′, HSRE, RC, and ΔH_f are also moderately relevant. For $\Delta H_{f(AM1)}$ low utility and for χ_M, Λ, and ^{15}N zero utility in the model (zero modeling power) were observed. For the second PC four variables (I_{sc}, I'_{sc}, RC and ΔN) are highly relevant and those remaining have moderate modeling powers. For third PC variables have similar, high utility in the model, with the magnetic variables having somewhat lower utility.

In conclusion, we believe that in this work, we have been able to considerable extent to resolve the apparent impasse between classical and magnetic aromaticity and to provide a firm basis for the consideration of aromaticity as a quantitative concept. In fact, there are at least two types of aromaticity. The best available measure of classical aromaticity is provided by the Bird-I-5-6 parameter, and this parameter correlates well for ΔN and for DRE. The second type of aromaticity is magnetic aromaticity, which is best measured by the molar magnetic susceptibility χ_M.

These two types, classical and magnetic, of aromaticity, are orthogonal. Other aromatic characteristics are inferred by both "classical" and by "magnetic" aromaticity to varying extents. Hence, we believe that there are at least two quantitative aromaticity scales.

More recently [2], we have extended this work to bicyclic compounds. The bicycles studied are shown in Scheme 17, and the corresponding data set in Scheme 18.

17

Bicycles Studied

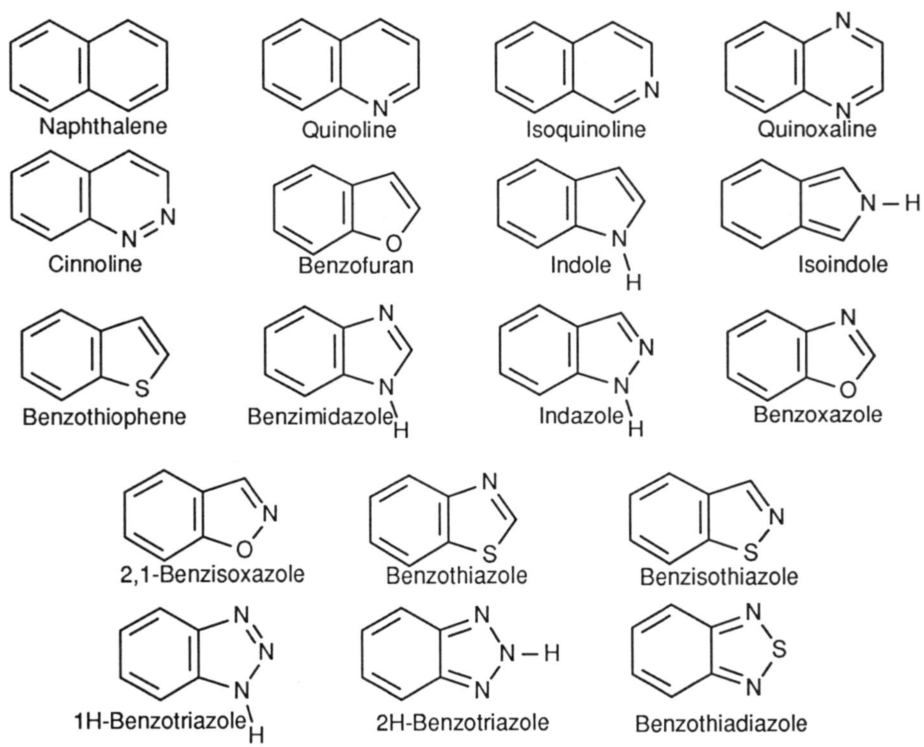

18 Data Set for Bicyclic Compounds

Compound	1 I_x	2 I_x'	3 RC	4 ΔN	5 DRE	6 DRE′	7 HSRE	8 ΔH_f	9 ΔH_{fcal}	10 χ_M	11 Λ	12 ^{15}N
1 Naphthalene	77.3	76.9	1.648	0.18	3.36	3.35	0.055	36.1	40.8	91.9	36.2	-
2 Quinoline	73.0	73.9	1.383	0.18	3.30	3.38	0.052	50.0	52.1	86.0	36.3	-67.5
3 Isoquinoline	72.1	78.2	1.447	0.14	3.41	3.65	0.051	50.0	50.2	83.9	34.2	-69.3
4 Quinoxaline	71.9	70.5	1.368	0.21	2.81	3.20	0.048	63.0	65.6	-	-	-50.1
5 Cinnoline	70.8	61.3	1.268	0.27	-	-	-	81.0	74.3	-	-	42.9
6 Benzofuran	45.4	56.6	1.059	0.33	2.03	0.76	0.036	6.5	20.8	79.8	31.8	-
7 Indole	70.0	66.5	1.311	0.23	2.38	1.47	0.047	44.6	55.2	85.0	33.1	-253.6
8 Isoindole	72.0	68.9	1.378	0.22	1.16	0.76	0.029	56.0	61.7	-	-	-218.0
9 Benzothiophene	57.0	76.1	1.515	0.18	2.48	1.18	0.044	40.0	53.9	93.3	34.9	-
10 Benzimidazole	71.0	65.3	1.255	0.28	3.09	1.39	0.050	43.4	67.1	71.9	26.0	-186.4
11 Indazole	68.9	70.8	1.292	0.23	-	-	0.056	60.7	82.3	72.7	26.8	-130.7
12 Benzoxazole	38.0	44.8	1.100	0.35	-	0.90	0.031	-	31.3	74.4	32.4	-131.5
13 2,1-Benzisoxazole	54.0	54.3	1.084	0.30	-	-	-	-	69.4	71.6	29.6	-7.5
14 Benzothiazole	57.0	65.7	1.223	0.23	-	1.35	0.047	-	64.1	88.5	36.1	-64.7
15 Benzisothiazole	68.0	69.6	1.280	0.20	-	-	-	-	73.1	85.1	32.7	-76.0
16 1H-Benzotriazole	67.0	58.9	1.178	0.28	-	-	-	83.0	104.2	-	-	-67.9
17 2H-Benzotriazole	73.0	68.6	1.417	0.26	-	-	0.067	-	117.0	-	-	-80.7
18 Benzothiadiazole	56.2	46.4	1.072	0.36	-	-	-	-	93.1	-	-	51.9

The results of PC analysis are given in Scheme 19. Here, two significant PC are found, which together explain 63% of the variance. The PC loadings and PC scores are plotted in Schemes 20 and 21, respectively.

19 Results of PC Analysis for Bicyclic Compounds	
Principal Component(s)	Percentage of Variance Accounted For
t_1	37.5
$t_1 + t_2$	63 (37.5 + 25.5)

20 Loadings Plot of p_1 vs. $-p_2$ for Bicyclic Compounds

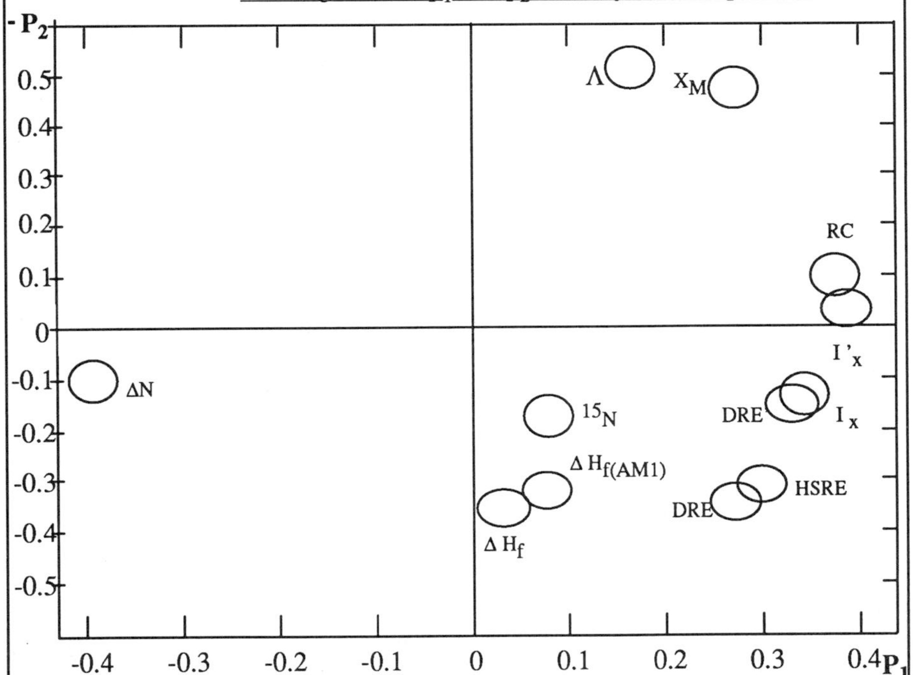

21 Scores plot of t_1 vs. $-t_2$ for Bicyclic Compounds

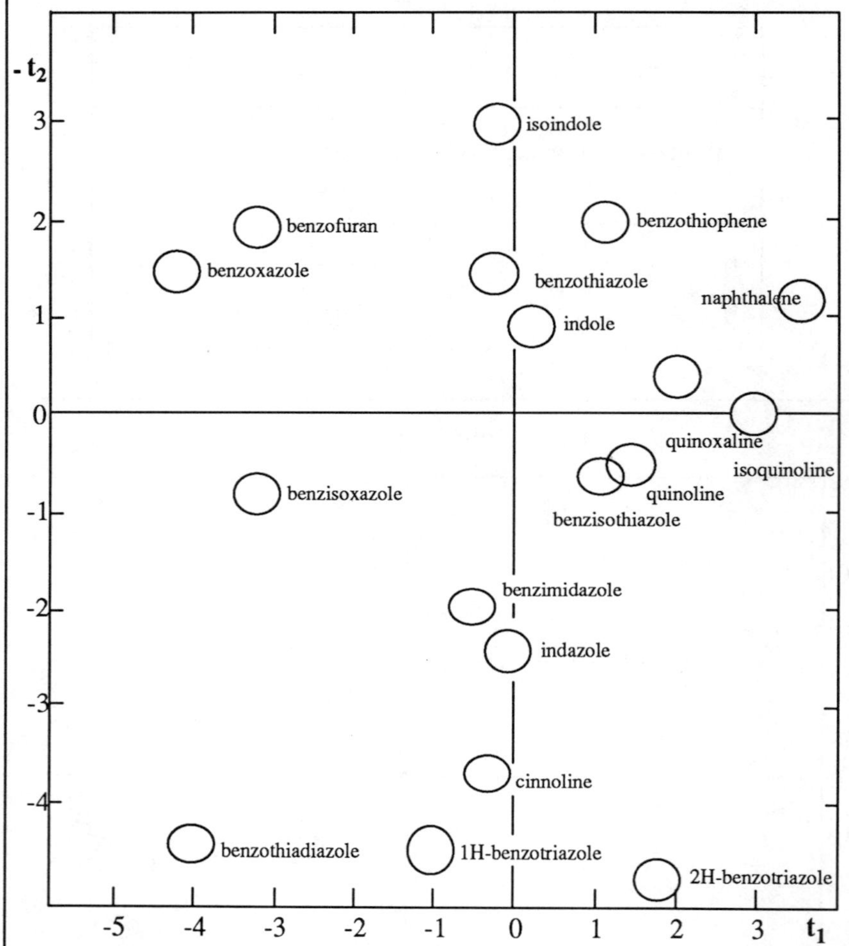

Although the monocyclic and bicyclic compounds form two completely independent data sets, we find most significant correlations between both the loadings and the scores of the two sets. Scheme 22 compares the loadings for mono and bicyclic derivatives: a close correspondence is found. Scheme 23 shows plots of the individual p_1 and p_2 for bicyclic vs. monocyclic compounds, and gives the correlation coefficients.

22 Loadings of Bicycles Compared to Monocycles -- Plot

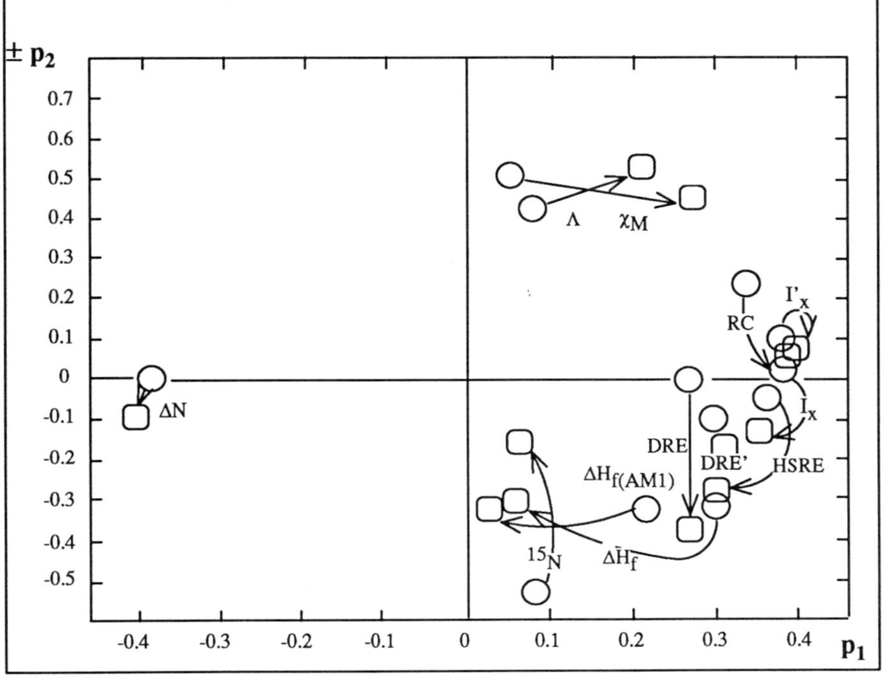

23 Comparison of Loadings for Monocycles and Bicycles

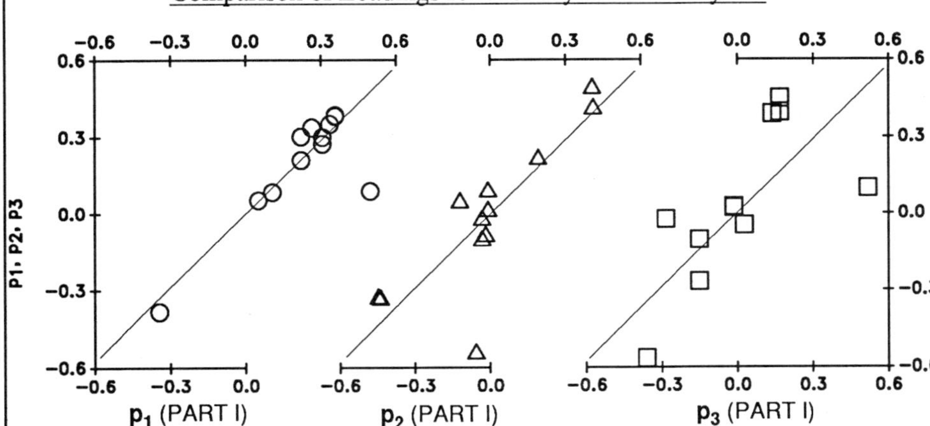

CORR. COEFF.: 0.98 CORR. COEFF.: 0.93 CORR. COEFF.: 0.74

In Scheme 24, the scores for the benzoheterocycles are compared with those for the corresponding monoheterocycles. Very good agreement is again found. Scheme 25 shows the plots for individual t_1 and t_2 for bicyclic vs. monocyclic compounds.

24

Scores of Bicycles Compared to Monocycles -- Plot

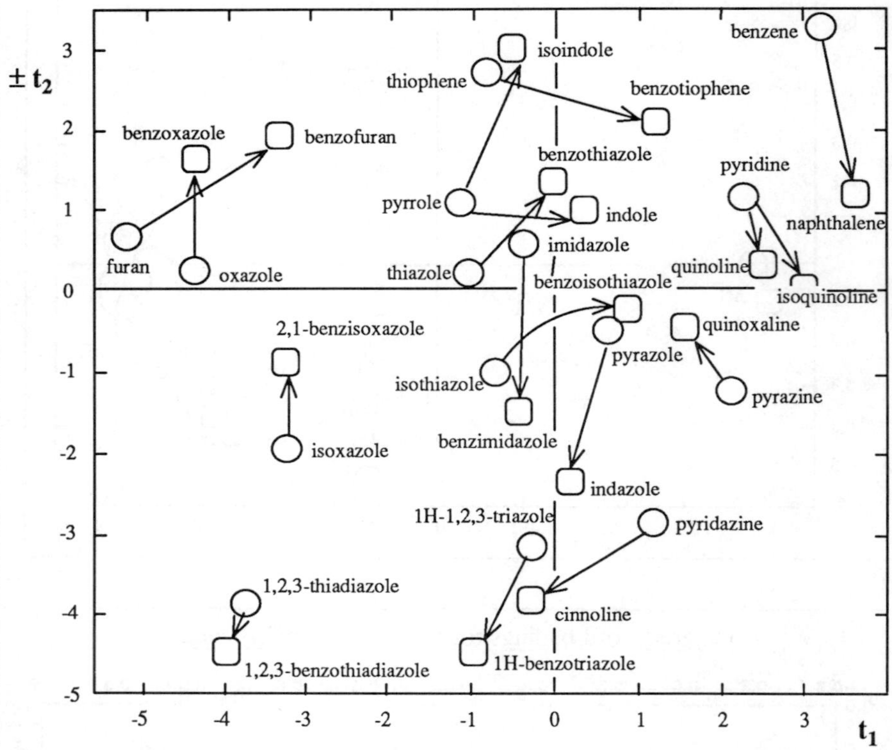

25 Comparison of Scores for Monocycles and Bicycles

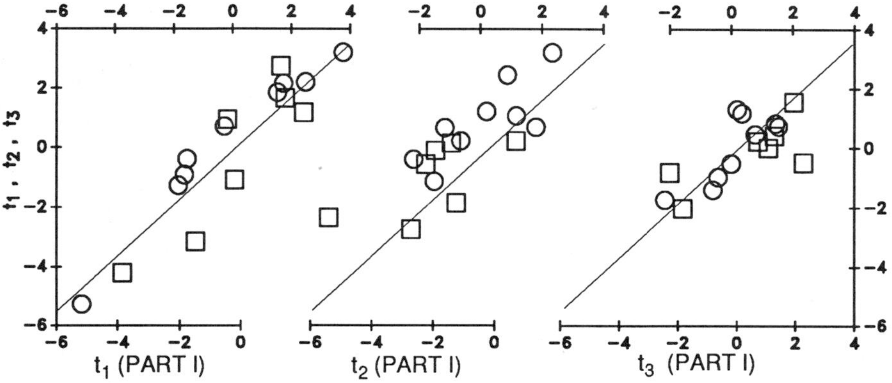

CORR. COEFF.: 0.93 (0.97)   CORR. COEFF.: 0.73 (0.79)   CORR. COEFF.: 0.68 (0.71)

Again, the predictive power was determined by dropping each individual data point and calculating it from the resulting model. The plots of predicted vs. observed data are shown in Schemes 26-31.

26-31 Plots of Recalculated vs Original Data for Bicycles

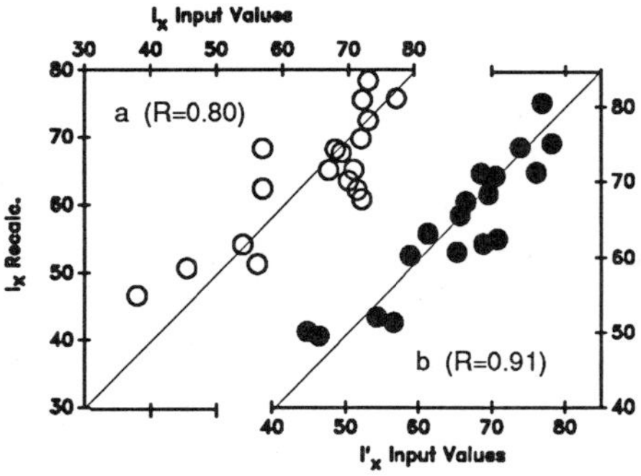

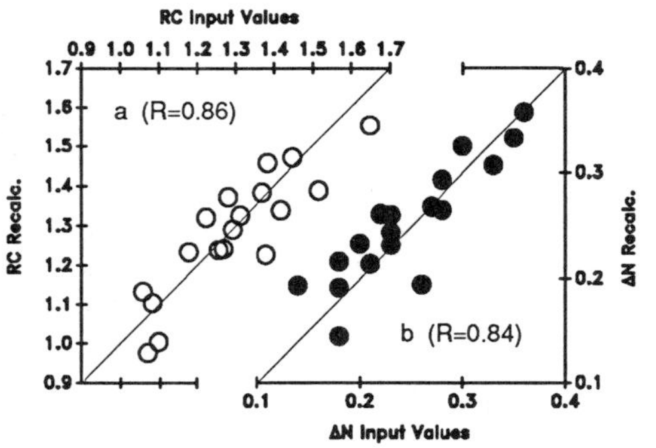

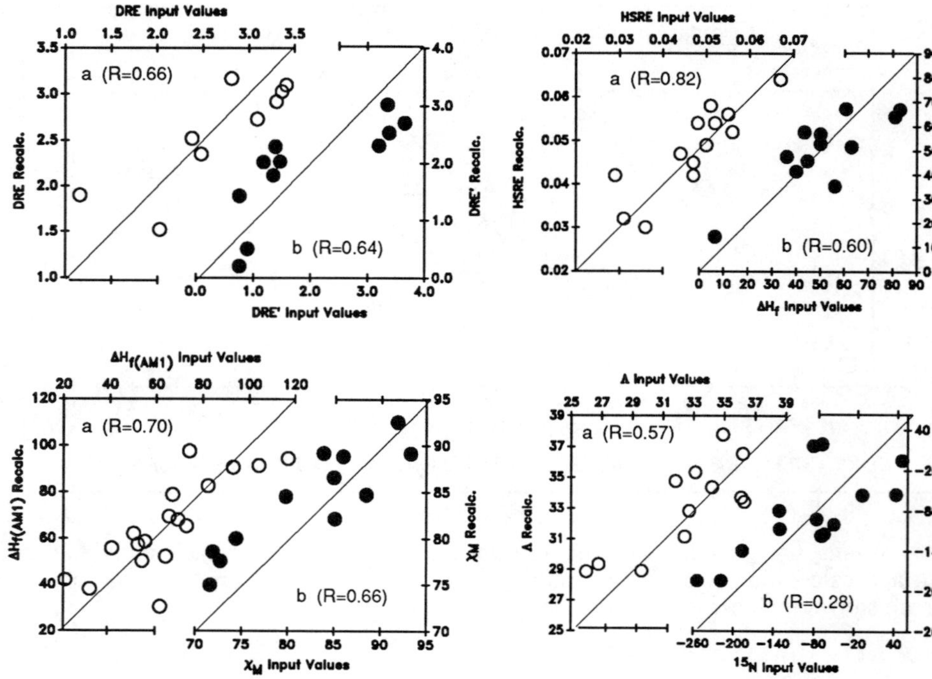

3. Structure-Activity Relationships

In a major ongoing project in our laboratory, we were faced with the analysis of complex mixtures containing up to 30 different compounds derived from a single precursor at high temperatures. We found that such mixtures could be satisfactorily treated by the GC/MS method. The peaks were resolved, and from the fragmentation patterns, knowing the precursor, and having some idea of the mechanism, we were able to assign the chemical structures to all our products. However, before we could carry out a quantitative analysis, it was nesessary to know the response factors for all the compounds, i.e. the relationship between the integrated area under the GC curve for the individual peaks, and the compound structure. Many of the compounds formed as products were unavailable, and in some cases, even unknown. Thus, we turn to the possibility of investigating relationships between response factors for GC analysis and chemical structure.

An illustration of the problem is given by the analysis of the 10 compounds shown in Scheme 32. The corresponding GC trace of the mixture is shown in Scheme 33, and it can be seen that the response factors vary considerably.

32 Compounds in a Complex Mixture for GC/MS Analysis

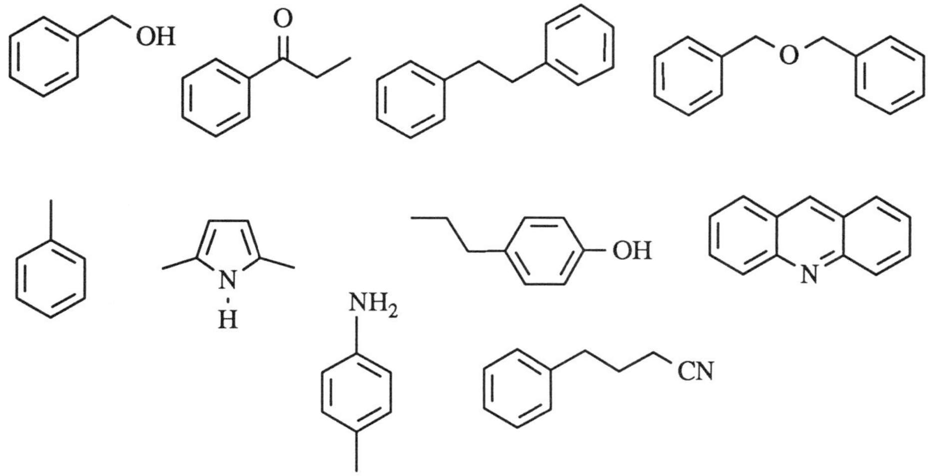

The definition of a response factor is given in Scheme 34, and we decided to use Principal Component Analysis to find out if there were any reliable dependence of the RF on the structural descriptors in terms of atomic groupings [3].

34 Definition of Response Factors

$$RF_{Dietz} = \frac{(\text{wt of standard}) \times (\text{area of compd})}{(\text{wt of compd}) \times (\text{area of standard})}$$

use PCA (principal component analysis) to find dependence of RF on structural parameters

We took a set of 120 different compounds and measured their response factors under standard conditions. We used these response factors as the dependent variable in the Principal Component Analysis and used variables 2-18 of Scheme 35 as the explanatory variables, all of which were simply read off from the chemical structure.

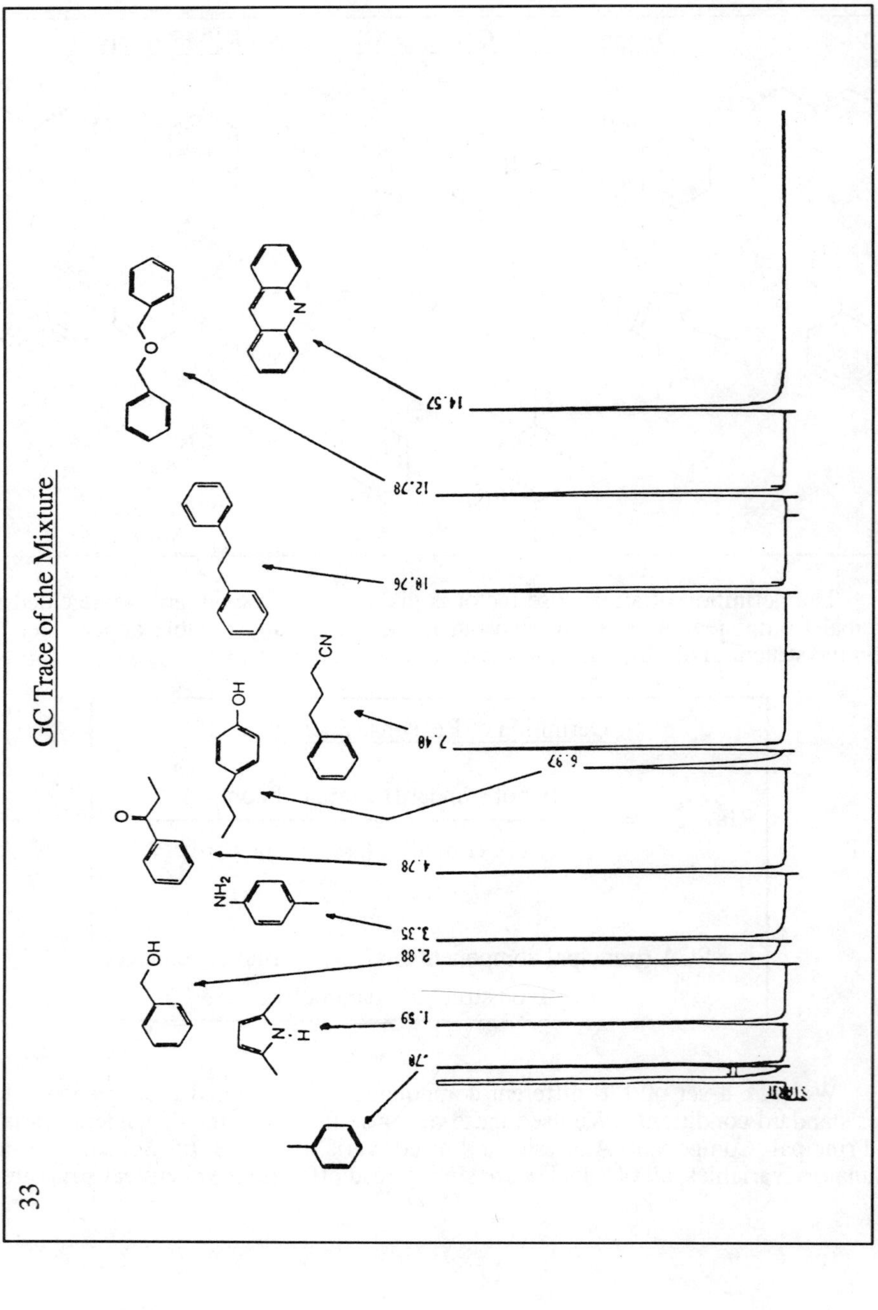
33
GC Trace of the Mixture
START
.78
1.59
2.88
3.35
4.78
6.97
7.48
10.76
12.78
14.57
OH
NH2
CN
N
O

35 Weights and Averages of the Variables Included in the PLS Analysis

Variables 2-18 are the "explanatory" variables (X Block); variable 1 is the dependent variable

No.	Symbol	Variable	Weights[a]	Averages[b]
1	RF_{Dietz}	Response Factor	5.6106	3.9179
2	MW	Molecular Weight	0.0244	3.3323
3	C	No. of C Atoms	0.3167	2.1715
4	H	No. of H Atoms	0.1510	1.5040
5	O	No. of O Atoms	1.4428	0.8945
6	N	No. of N Atoms	1.2766	0.8298
7	S	No. of S Atoms	4.1897	0.2514
8	C=C	No. of C=C Double Bonds	0.6605	2.0014
9	C=0	No. of C=O Double Bonds	2.8075	0.3369
10	C=N	No. of C=N Double Bonds	1.6467	0.7081
11	C≡N	No. of C≡N Triple Bonds	3.6676	0.2934
12	Rings	No. of Rings	1.8165	2.1800
13	$-CO_2H$	No. of Carboxyl Groups	4.1897	0.2514
14	-OH or -SH	No. of Hydroxyl Groups	2.4875	0.4975
15	-CHO	No. of Aldehyde Groups	4.1897	0.2514
16	$C-CO_2-C$	No. of Ester Groups	4.0775	0.2031
17	$-NH_2$	No. of Amino Groups	2.8675	0.4015
18	C-O-C or C-S-C	No. of Ether or Sulfide Groups	3.0619	0.3674

[a] Factor required to autoscale the variables to unitary variance.
[b] Arithmetic means of the variables.

The results of the Principal Component Analysis are shown in Scheme 36. This leads to the equation given in Scheme 37, which now enables the prediction of response factors for novel compounds.

The success of the method is shown graphically in Scheme 38.

36 Relevance of Individual Descriptors in the PLS Model[a] as Described by the Loadings (P_k) and by the Modeling Powers (ψ_c) after 1,2 and 3 PLSC[c].

Var.[e]	P_{1k}	ψ_{1k}	P_{2k}	ψ_{2k}	P_{3k}	ψ_{3k}
2	0.25	0.07	-0.51	0.48	0.06	0.48
3	0.37	0.18	-0.42	0.49	0.14	0.51
4	0.44	0.27	-0.26	0.38	-0.14	0.40
5	-0.38	0.19	-0.26	0.29	-0.18	0.32
6	-0.33	0.14	0.18	0.18	-0.30	0.25
7	0.15	0.02	-0.25	0.10	-0.14	0.10
8	0.04	0.00	-0.23	0.05	0.45	0.20
9	-0.29	0.10	-0.11	0.12	0.28	0.17
10	-0.34	0.15	0.08	0.15	-0.02	0.15
11	0.00	0.00	0.05	0.00	-0.04	0.00
12	-0.08	0.02	-0.19	0.04	0.48	0.20
13	-0.24	0.07	-0.28	0.16	-0.07	0.16
14	0.04	0.00	0.05	0.00	-0.04	0.00
15	-0.15	0.02	-0.02	0.02	-0.16	0.03
16	-0.12	0.01	-0.04	0.01	0.16	0.02
17	-0.14	0.02	0.11	0.03	-0.37	0.12
18	0.12	0.01	-0.28	0.10	-0.05	0.10

[a] The percentage of Y variance explained by the X block after three components is 84%

[b] Proportionality coefficients $b_1 = 0.500$, $b_2 = 0.260$, $b_3 = 0.248$

[c] For the definition of variables, see Scheme 35

37 Prediction of Response Factors

RF values (y_{ia}) can now be predicted from the PLS model parameter relationship

$$RF = 0.991 - 0.000908\, x_2 + 0.00234\, x_3 + 0.00276\, x_4 - 0.112\, x_5 - 0.0711\, x_6 - 0.160\, x_7 + 0.00337\, x_8 - 0.0434\, x_9 - 0.0479\, x_{10} - 0.0777\, x_{11} + 0.00481\, x_{12} - 0.323\, x_{13} - 0.0536\, x_{14} - 0.206\, x_{15} - 0.0459\, x_{16} - 0.166\, x_{17} - 0.0675\, x_{18}$$

By inserting the appropriate x value for each descriptor, response factors can now be calculated for any compound using this range of functional groups.

38 Plot of RF-Dietz obs. Against RF-Dietz calcd.

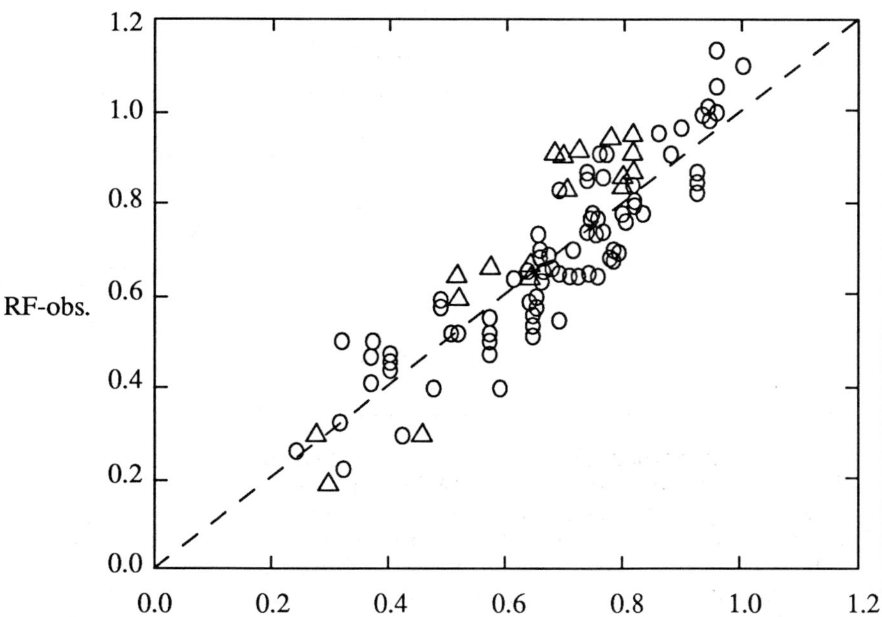

The studies discussed so far were carried out using the SIMCA (Soft Independent Modeling of Class Analogy) of Wold [4]. This PC-based program is capable of many multivariate data analysis including principal component analysis and pattern recognition techniques. However, all the descriptors used must be provided by the user.

In an attempt to find somewhat more sophisticated molecular descriptors that would enable a more precise prediction of the RF value, we are now turning [5] to the ADAPT program.

39

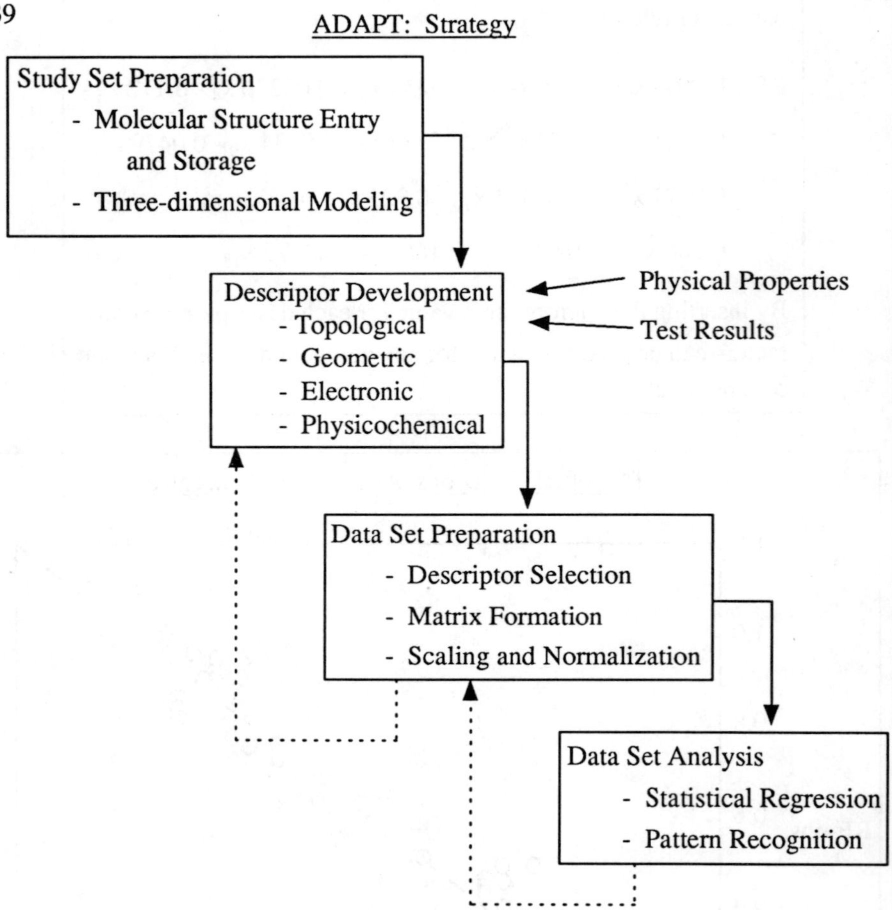

ADAPT (Automated Data Analysis *via* Pattern-recognition Techniques [6]), written by Peter Jurs and distributed by Molecular Design Ltd, combines the kind of statistical analysis possibilities given by SIMCA mentioned above, with the ability to calculate many possible descriptors for the molecules concerned. ADAPT is a very powerful program that can help the user make intelligent analyses of the properties of interest. The operational strategy falls into four basic sections (Scheme 39). Firstly, data for the molecules, including structure and properties, is entered and 3D modeling is performed. Once the data set is prepared, descriptors are developed. These describe various properties of the molecule, many of which the program can calculate (Scheme 40). Physical results, outside calculations, etc., can also be added at this stage. At the third stage, each descriptor is scaled to normalize the data. The user has many options -- maybe the square root or the inverse of the function is needed. Finally statistical or pattern recognition methods are used to analyze the data. All these steps feed back to their predecessors to help refine the model. We are currently using the ADAPT method to study the RF data set in more detail.

40 ADAPT: Descriptors Generated

WHOLE MOLECULE BASED

Topological:

- Path Lengths
- Molecular Connectivity
- Molecular Fragments
- Symmetry Descriptor

Electrophysico:

- Sigma Charge
- Simple Hückel
- Molecular Reactivity

Geometric:

- Principal Axes
- Moments of Inertia
- Molecular Volume
- Surface Area

SUBSTRUCTURE BASED

Topological:

- Substructure All path
- Substructure Counts

Electrophysico:

- Electronic descriptors
- Molar Refractivity

OTHER

- Nearest Neighbor

4. Bond Energies and Reaction Mechanisms

During the past decade, extensive investigations of nucleophilic substitution at saturated carbon atoms in solution, using charged pyridinium substrates and neutral pyridines as leaving groups, have considerably enhanced our understanding of these reactions. Evidence has been presented that such reactions can occur by five distinct mechanisms [7,8].

There are major advantages in the use of a neutral leaving group such as pyridine over an anionic leaving group. Unimolecular reactions of a neutral substrate dissociate to two ionic species and this involves charge creation in the transition state, and they therefore only proceed at a measureable rate in media of high dielectric constant. Advantageously, substrates with neutral leaving groups undergo unimolecular reactions in media of low dielectric constant, because no change is created in the transition state.

We have now extended this work to the gas phase, utilizing Fourier transform ion cyclotron resonance mass-spectroscopy. We find that pyridinium cations do indeed undergo fragmentation to neutral pyridine and the carbocation in the gas phase, as shown in Scheme 41 [9]. Moreover, these pyridinium cations are stable in the gas phase unless they are energized by collisions, and it is possible to measure the collisionally activated dissociation energies by observation of the appearance potentials.

41 Fragmentation to Pyridine and Carbocation

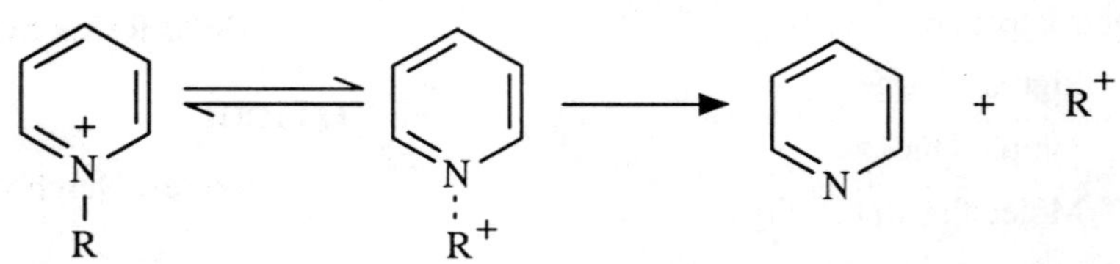

In addition to the fragmentation pathway to neutral pyridine and the carbocation, a second pathway yields parent pyridinium cation and an olefin [10]. This pathway is shown in Scheme 42, together with a suggested mechanism. It will also be noted that in Scheme 41, an intermediate ion-molecule pair is depicted. The significance of this is discussed later.

42 Fragmentation to Pyridinium Cation and Alkene

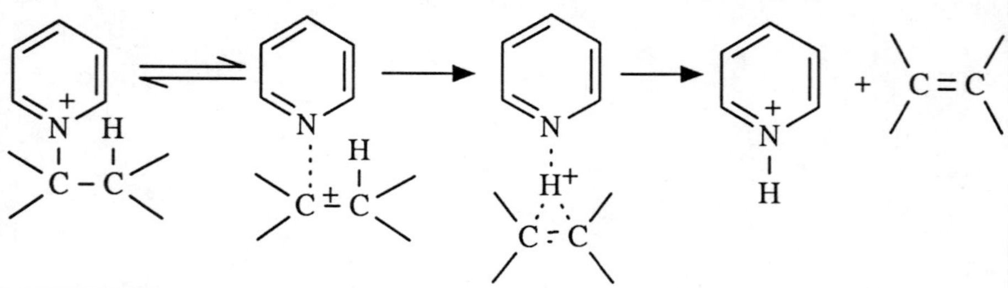

43

Basis of Method

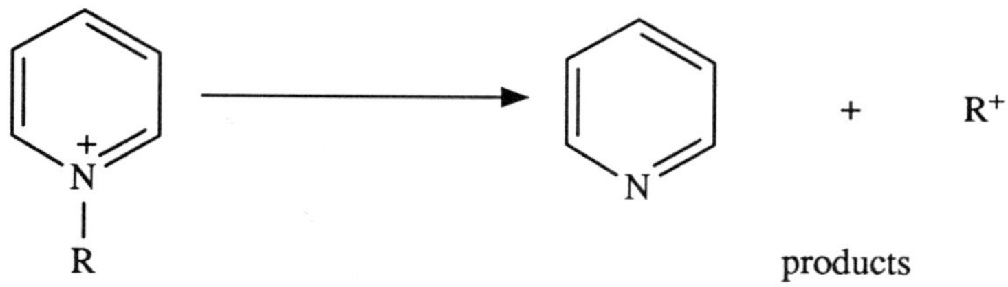

reactant

products

1. (a) Calculate by AM1 ΔH_f for reactant

   (b) Calculate by AM1 total ΔH_f for products

   (c) Measure activation energy for reaction

2. Compare (a) + (b) with (c)

Note: (a) + (b) should be greater than (c).

The basis of the method used is shown in Scheme 43. In the ion cyclotron resonance, it is possible to measure the activation energy needed for the dissociation.

The kinetic energy required for dissociation was measured as demonstrated in Scheme 44 by measuring the percentage of collisionally activated dissociation as a function of the relative kinetic energy. The lines in Scheme 44 were fitted using the least squares method, and calculations of the x intercept gave estimates of the appearance energy for fragmentation.

We then calculated the heat of formation values using the AM1 method of Dewar [11] for the starting pyridinium cation for the carbocation produced and for pyridine, and compared the energy difference between the products and the starting material with the appearance potential, i.e. the energy which has to be put into the system to get the reaction to work.

For those pyridiniums where the carbocation cannot, or is unlikely to, undergo any rearrangement to a more stable system, we found that the appearance potentials were invariably higher than the differences so calculated. Some examples are shown in Scheme 45.

44 Kinetic Energy available via CAD vs $[R^+]/([R^+] + [PyR^+])$

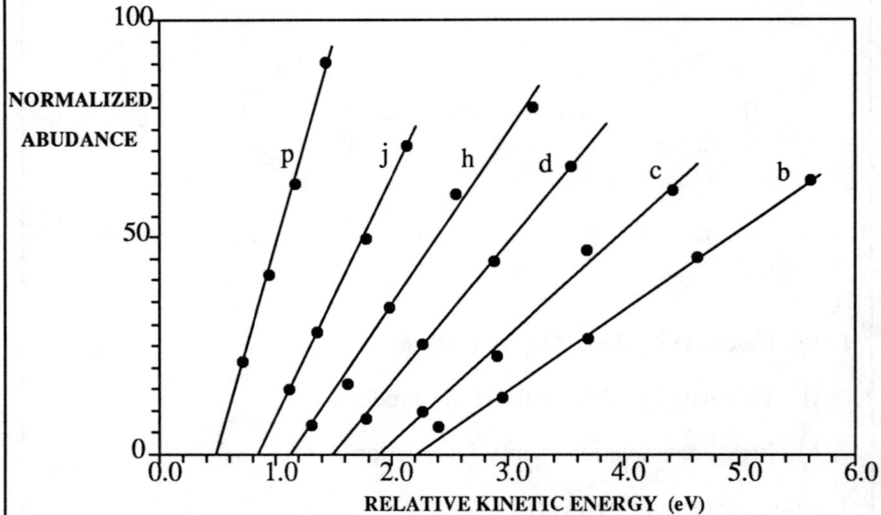

Key: b: R = CH_2=$CHCH_2$ c: R = $CH_3OCH_2CH_2$ d: R = *p*-$NO_2C_6H_4CH_2$
h: R = 1-adamantyl j: R = *p*-$CH_3C_6H_4CH_2$ p: R = *p*-$CH_3OC_6H_4CH_2$

45 Calculation of Appearance Potentials Using AM1 ΔH_f Values

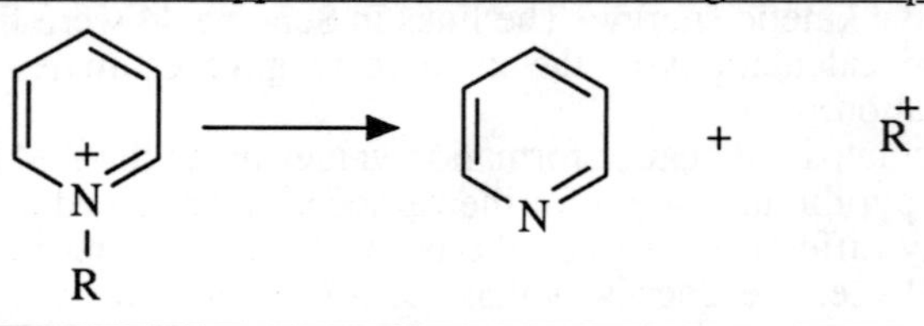

R	PyR^+	R^+	Py	difference	AP
Methyl	186.1	252.3	32.1	98.1	> 96
Allyl	204.0	226.3	32.1	54.4	67
Cyclopentyl	168.2	185.9	32.1	49.8	42
1-Adamantyl	155.2	168.6	32.1	45.5	43
Trityl	289.9	258.0	32.1	0.2	< 9

By contrast, for other N-substituted pyridinium cations, the appearance potentials were found to be considerably less than those that would be required to produce an unrearranged carbocation. However, in all these cases, we found that rearrangement pathways were available to the carbocation that would reduce its heat of formation such that the appearance potential would provide enough energy. The situation is illustrated for various N-benzylpyridinium cations in Scheme 46.

46 Relation Between Appearence Potentials and $\Delta\Delta H_f$ Values for N-Benzylpyridinium

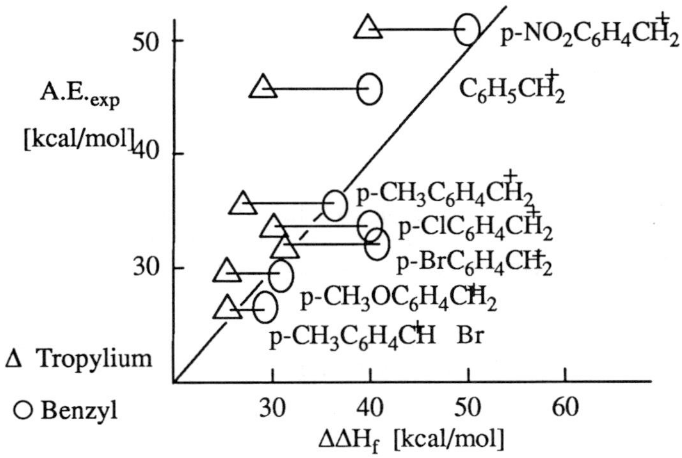

In the benzyl cation itself, and in most of its *p*-substituted derivatives, the energy of the appearance potential is sufficient to produce the unrearranged carbocation. However, for the *p*-bromo and *p*-chloro analogs, we believe that rearrangement is taking place in the ion-molecule intermediate stage (cf. Scheme 41) of the incipient benzyl cation to the corresponding tropylium cation, which has a considerably lower heat of formation. These rearrangements are shown in Scheme 47.

47 Rearrangement of Benzyl Cations

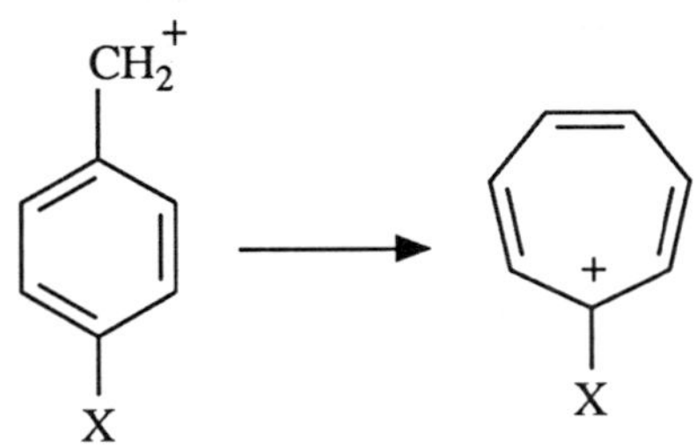

The situation is more complex for β-phenylethyl cations. Here, the initial cation can rearrange, either into an α-phenylethyl cation or into a spirobicyclic cation, or into a methyltropylium cation (see Scheme 48).

48 Rearrangement of β-Phenylethyl Cations

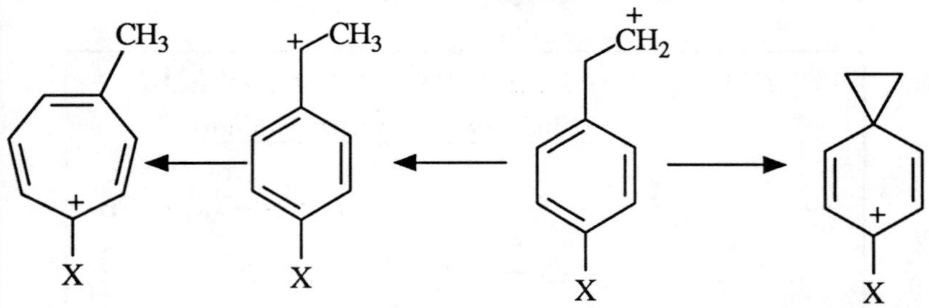

The appearance potential for four of these compounds, the parent and three *p*-substituted derivatives, are shown in Scheme 49. The β-phenylethyl derivatives probably rearrange in all cases on the way towards the tropylium or the α-phenylethyl derivative, and such rearrangement is particularly extensive in the case of the *p*-methyl and *p*-methoxy derivatives.

49 Relation Between Appearance Potential and $\Delta\Delta H_f$ Values for N-(β-Phenylethyl)pyridinium

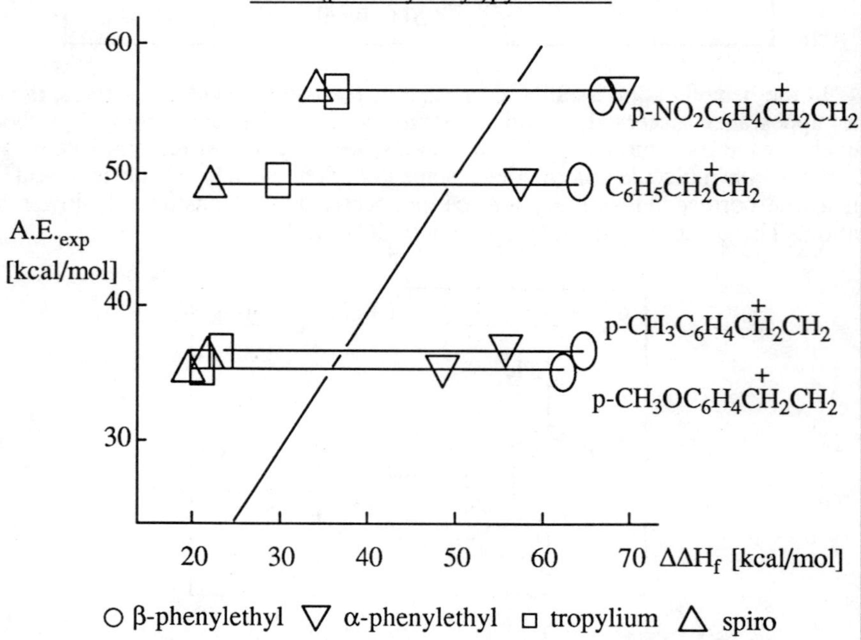

It is quite fascinating to study further carbocation rearrangements by this method in the gas phase. For example, the β-methoxyethyl cation clearly rearranges to the α-methoxyethyl cation. The phenylthiomethyl cation rearranges right through to the protonated thiotropone, as is shown in Scheme 50.

50 Further Ion-Molecule Pair Rearrangements (i)

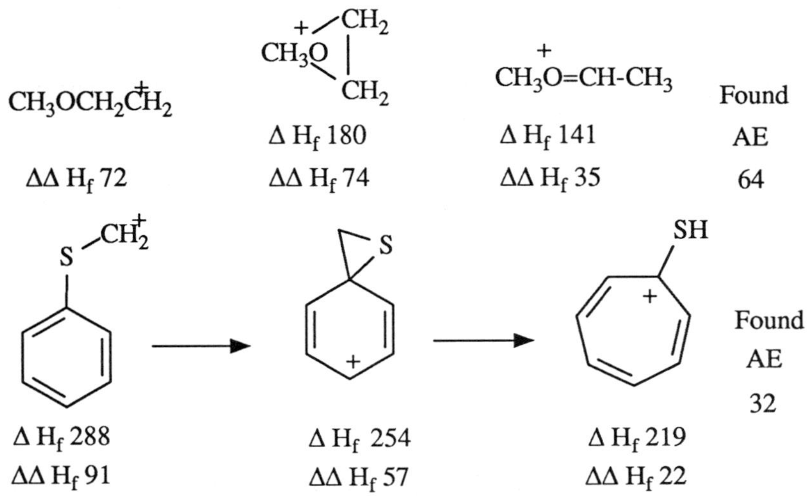

A rather different situation is found in the collisionally activated dissociation of N-substituted pyridinium cations to olefin and the parent pyridinium cation. In these cases, considerable activation barriers were found from the appearance potentials, and comparison of these with the heats of formation of starting material and product. Some of these appearance potentials are shown in Scheme 51.

51 Formation of Olefin and Pyridinium Cations

Considerable activation barriers found, which decrease as number of C-atoms in R increases.

R = Et	(67 kcal/mol)
R = Pr^i, Pr^n	(55-52 kcal/mol)
R = Bu^n, Bu^i, Bu^s or Bu^t	(52-44 kcal/mol)
R = 3-pentyl or cyclopentyl	(43-42 kcal/mol)

Little dependence on primary, sec or tert nature of R shows that R^+ character does not dominate in TS (contrast the fragmentation to Py + R^+).

It is very striking that there is much less dependence on the primary, secondary, or tertiary nature of the N-substituent in the starting pyridinium cation, and much more dependence on the number of carbon atoms in that N-substituent. Thus, for example, all four N-butylpyridinium cations show activation barriers which are within 5 kcals of each other. This quite clearly shows that carbocation character does not dominate in the transition state. Looking at Scheme 52 (which duplicates Scheme 42), it is for this reason that we depict the reaction coordinate as involving intermediates in which the pyridine nitrogen atom becomes bonded to the hydrogen before the bonding to the carbon is lost.

52

Fragmentation to Pyridinium Cation and Alkene

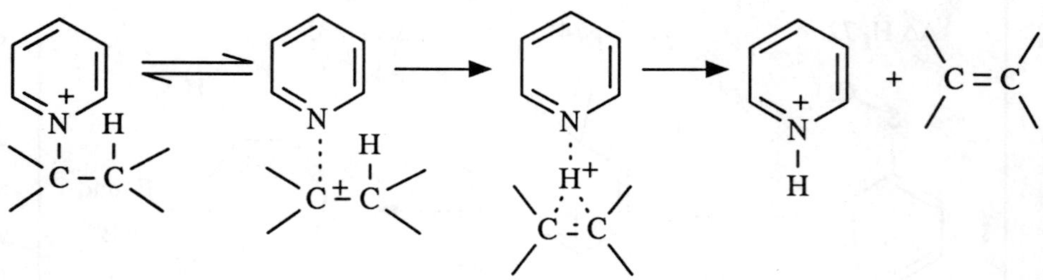

The picture for the isomeric N-butylpyridinium cations is now rather complete, and the energy profile is shown in Scheme 53. The energies for the starting cations and the products (on the left hand side, the carbocation and neutral pyridine; on the right hand side, olefin and pyridinium) are all calculated from AM1 heats of formation. The transition states are added into the picture using the appearance potentials that have been discussed.

Another illustration of the unimportance of carbocation character in the transition state is demonstrated by the fact that the formation of ethyl acrylate shows relatively little difference for α-pyridinioproprionate and the corresponding β derivative (see Scheme 54). In solution, the β derivative would undergo facile elimination *via* reverse Michael mechanism, whereas the α derivative would be rather stable towards elimination.

53 Energy profile for the dissociation of N-butylpyridinium cations Py^+R into Py and R^+ and into Py^+H and olefin.

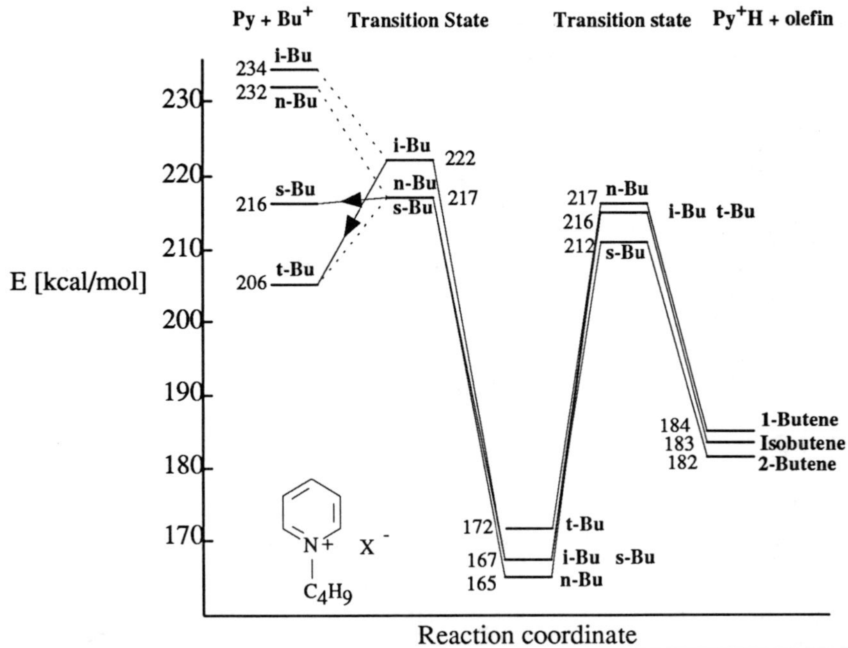

54 Formation of Ethyl Acrylate

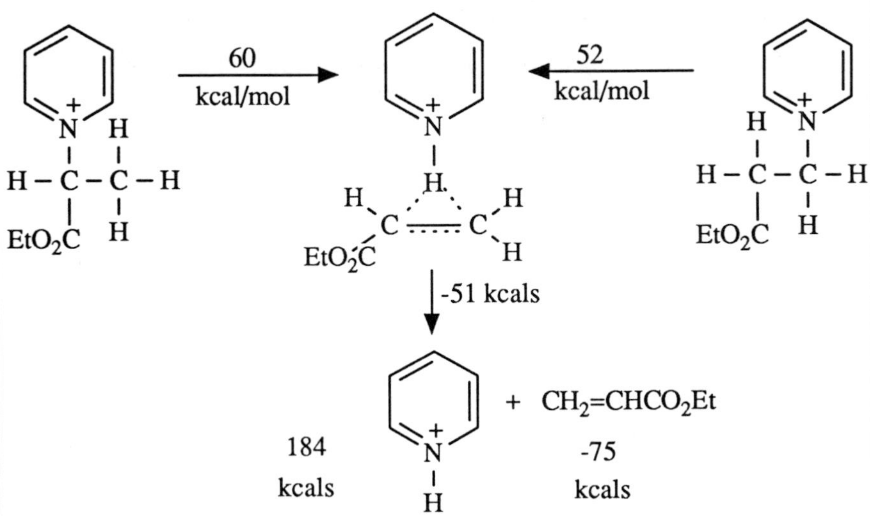

Similar work has been carried out on a series of N-benzyloxypyridiniums [12]. Two pathways are found (Scheme 55). In Path A, the appearance potentials clearly show rearrangement of the benzyl cation to tropylium. However, much larger AEs are found for Path B, and no rearrangement of $PhCH_2O^+$ cation is dictated by the calculations.

55 Dissociation Pathways for N-Benzyloxypyridinium Cations

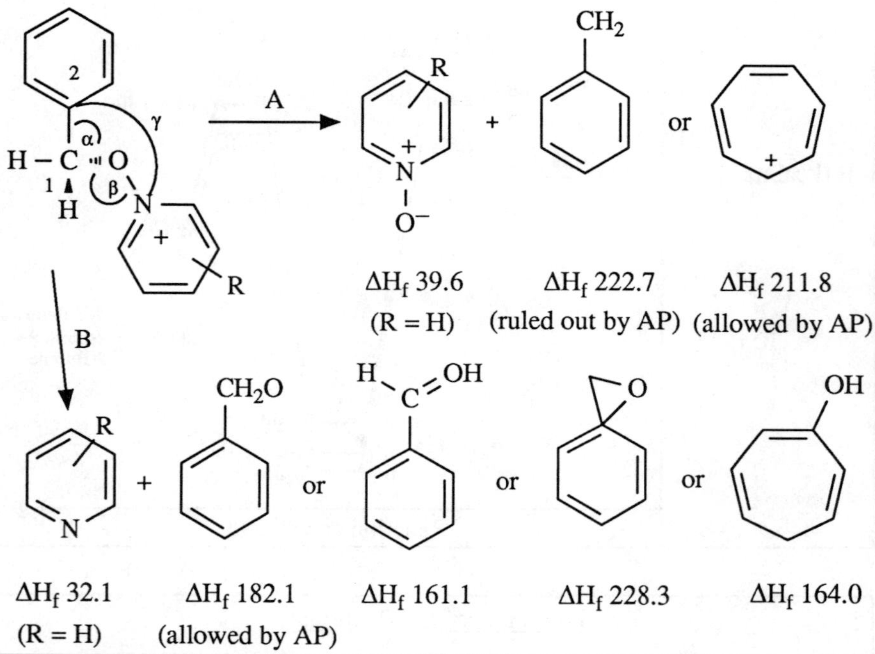

Additional novel dissociation pathways for pyridinium cations have also been studied [13]: some of these are described in Scheme 56.

56 Additional Pathways for N-Substituted Pyridinium Fragmentation

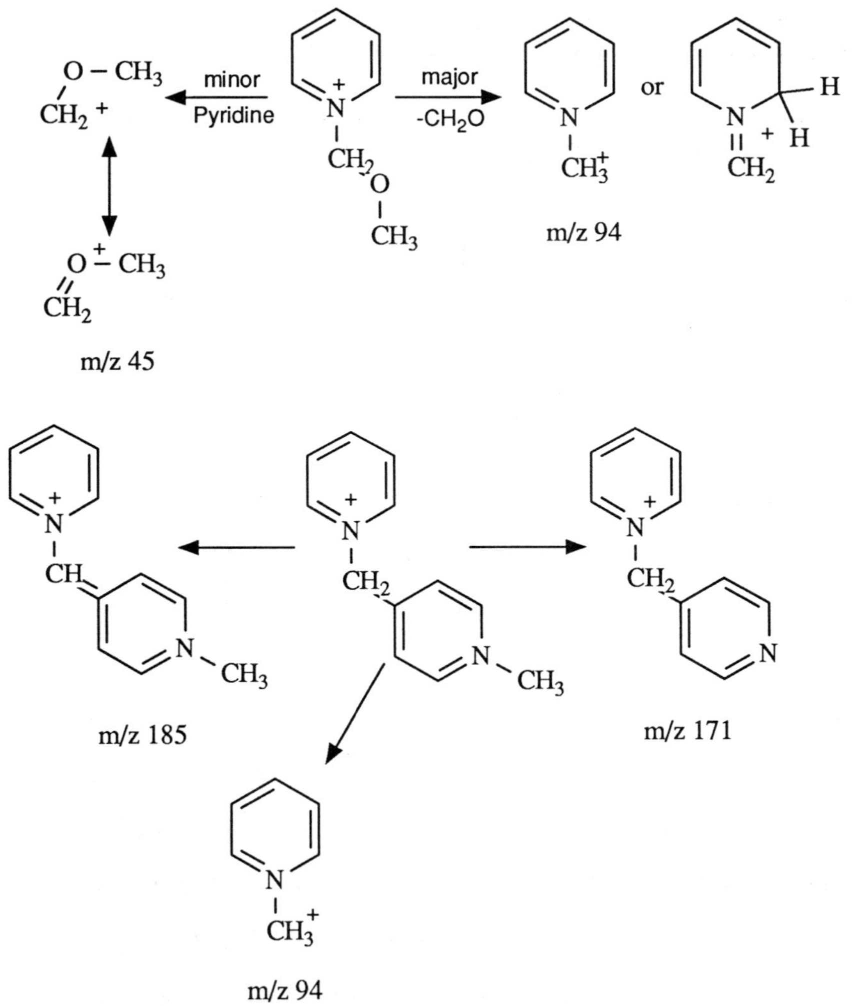

5. Solution Equilibria

The great utility of the AM1 semiempirical method for molecular orbital calculations is very strikingly illustrated by its application to problems of prototropic tautomerism. Tautomerism calculations have provided a severe test of molecular orbital methods. As shown in Scheme 57 for the 2-hydroxypyridine/2-pyridone equilibrium, the hydroxy form is favored in the vapor phase, by approximately 0.4 kilocalories. All the previous semiempirical methods, and many *ab initio*, methods give (a wide variety of) results widely diverging from this experimentally determined value. At the *ab initio* level, it is only at the MP 2/6-31 G* that a satisfactory agreement is found. However, the agreement is just as satisfactory using the AM1 semiempirical method.

Still more impressive is the fact that the AM1 method can be corrected for media of high dielectric constant by applying a simple application of reaction field theory. Thus, we obtain for a medium of dielectric constant 78 corresponding to water, a value in very good agreement with the aqueous solution experimental value (see Scheme 57) [14].

57 Calculation of the Tautomerism of 2-Hydroxypyridine to 2-Pyridone

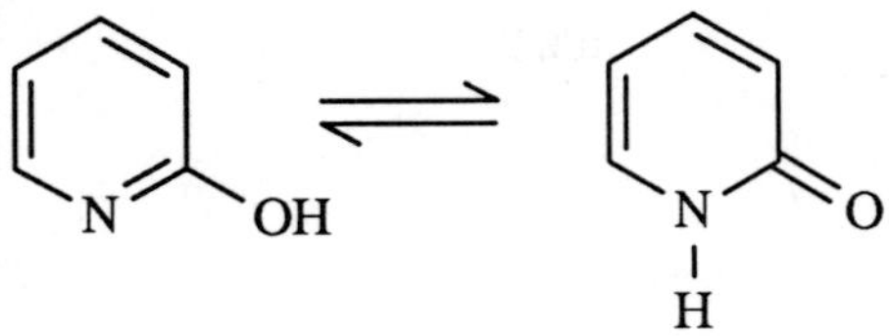

Hydroxy Form	Oxo Form	Experimental or Theoretical Method
favored by 0.4 kcals	--------	vapor phase exp. value
--------	favored by 15.4 kcal	STO - 3G ab initio
--------	favored by 14.2 kcal	MINDO/2
--------	favored by 3.7 kcal	MINDO/3
favored by 1.7 kcal	--------	3-21 G ab initio
favored by 11.3 kcals	--------	CNDO/2
favored by 0.6 kcal	--------	MP2/6-31 G* ab initio
favored by 0.55 kcal/mol	--------	AM1 (gas phase)
--------	favored by 3.8 kcal/mol	AM1 (corrected to $\mathcal{E} = 78$)
--------	favored by 4.2 kcal/mol	Aq. solution exp. value

In Scheme 58, the other rather fragmentary experimental gas phase data for similar tautomeric equilibria is compared with the AM1 values. It is seen that the agreement is very satisfactory.

58 Other Compounds with Gas Phase Data

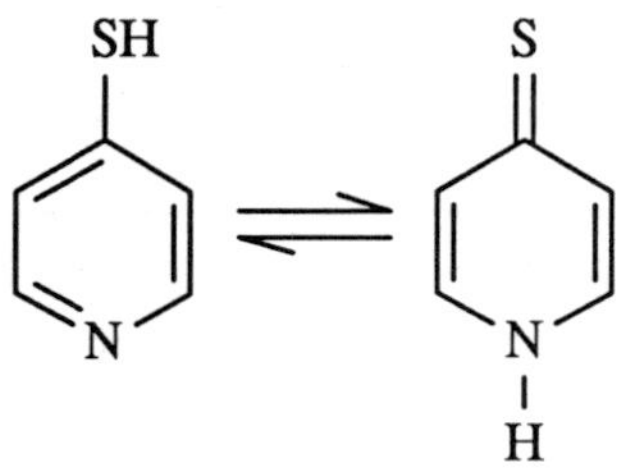

In vapor phase, log K_T estimated as -1.0 in favor of SH form. AM1 value is 1.7

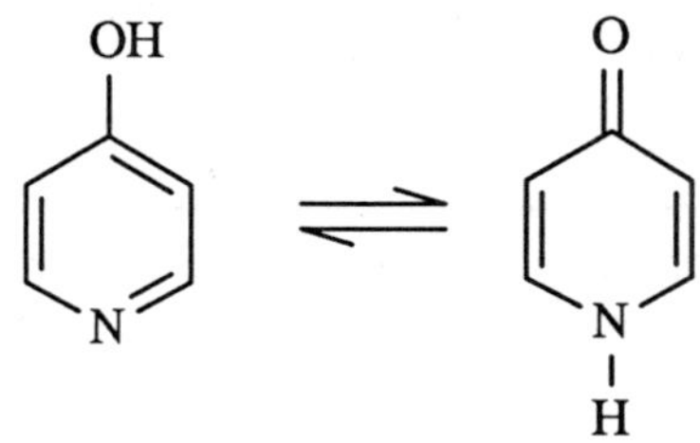

In vapor phase ΔΔH estimated at ca. 7 kcals/mol in favor of OH. AM1 value is 8.1 kcals

In Scheme 59, the basis is shown for the dielectric correction. This neglects specific interactions and merely assumes that the molecule is imbedded in a medium of constant dielectric constants in a circular cavity corresponding to the approximate volume of the molecule.

59 Basis of Dielectric Correction

$$E = \langle\psi | H_o | \psi\rangle - \frac{M^2}{2}\left\{\frac{2(\mathcal{E}-1)}{r_c^3(2\mathcal{E}+1)}\right\}$$

where $M = \langle\psi | \underline{\mu} | q\rangle$ and $\mathcal{E}$ = bulk dielectric constant

$$r_c = \text{cavity radius} = \sqrt[3]{\frac{3V_c}{4\pi N_a}}$$

solved using $[f_o - \mu g M]\,\varphi_i = \mathcal{E}_i\varphi_i$

K_T values calculated by the AM1 values with this rather simple correction for dielectric constant are compared with experimental values in Scheme 60. It will be seen that with one exception (for 3-hydroxypyridine), the predicted values are in very good agreement with those found. In Scheme 61, this is demonstrated in another way: the experimental values are plotted against AM1 values with and without the dielectric constant correction. Without the dielectric constant correction, many of the values lie very far from the line. With it, as already mentioned, there is only one real discrepancy.

60 Comparison of Experimental and Calculated log K_T Values for Aqueous Solution (Positive Values Indicate the Predominance of Pyridinoid Form)

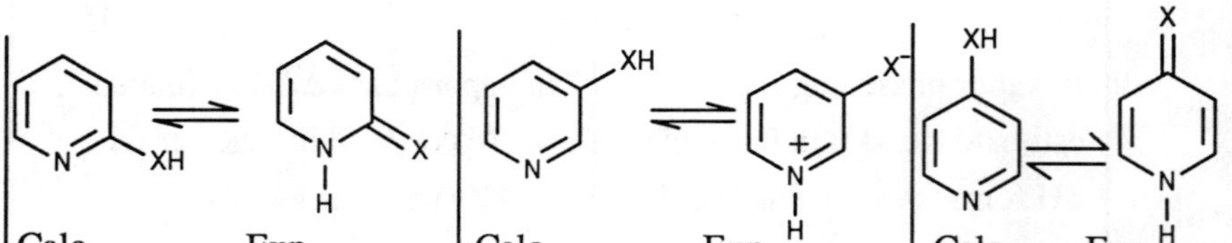

X	Calc.	Exp.	Calc.	Exp.	Calc.	Exp.
O	-2.8	-3.1	6.7	0.1	-3.0	-3.3
S	-4.7	-4.7	-2.4	-2.2	-7.3	-4.6
NH	7.8	6.0	20.0	(large)	7.2	8.7
CH_2	14.8	13.3	25.7	(large)	14.8	13.4

61 Experimental K_T Values for Aqueous Solution Plotted Against AM1 Values With and Without Dielectric Constant Correction

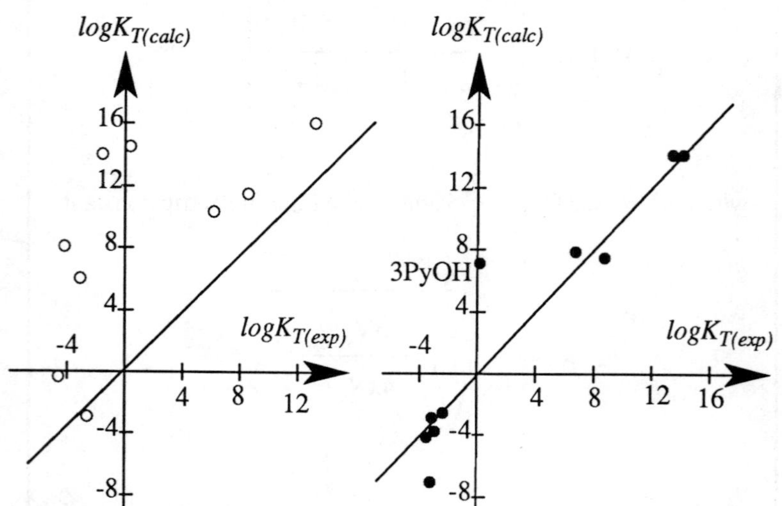

We have extended this work to an investigation of the tautomeric equilibria of heteroaromatic compounds with five-membered rings [15]. The results for hydroxyisoxazoles are shown in Scheme 62, and those for hydroxyoxazoles in Scheme 63. In all cases, the only experimental values available are those for solution. In some cases, these are merely qualitative.

The results are expressed by setting the calculated AM1 value for the heat of formation of the most stable tautomer as zero, both for the gas phase (no dielectric constant correction), and for aqueous solution (with the correction). 3-Hydroxyisoxazole is correctly predicted to exist in the hydroxy form in water, although in the gas phase the tautomers are predicted to be of almost equal stability (no experimental details).

For 4-hydroxyisoxazole, again, the hydroxy form is correctly predicted to be somewhat more stable in water than the oxo form, but the latter should be the more stable in the gas phase.

In the case of 5-hydroxyisoxazole, the 2H oxo form is correctly predicted to dominate in water, but AM1 indicates that the 4H oxo form should dominate in the gas phase by a large amount, and indeed, in the lower dielectric solvent constant, chloroform, that is already the case.

The 2-oxo form of 2-hydroxyoxazole should dominate under all conditions and is known to do so for solutions in water. For 4-hydroxyoxazole, the same is true of the 5H 4-oxo form.

In the case of 5-hydroxyoxazole, we expect the Zwitterionic form to dominate in water, but the oxo form in the gas phase. Here, the available experimental result is for chloroform solution, and it agrees with the gas phase prediction.

A much more complex equilibrium is that presented by 2-thiouracil. 2-Thiouracil exists in six tautomers which retain cyclic conjugation, and forms four monocations with cyclic conjugation, but just one stable dication (Scheme 64).

62

<u>Hydroxyisoxazoles: $\Delta\Delta H_f$ Values for Tautomers</u>

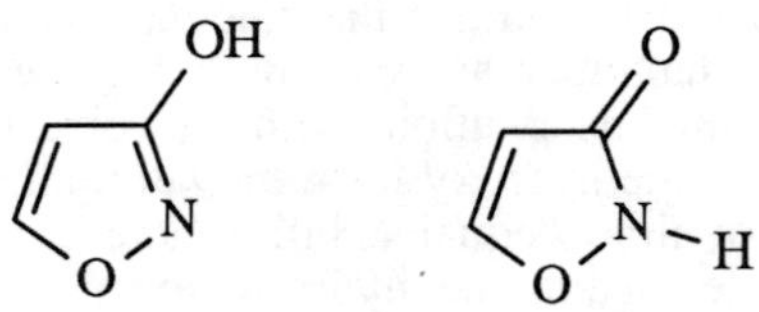

AMI (gas)	0	0.1
AMI (water)	0	2.6
Experimental (water)	dominant	-

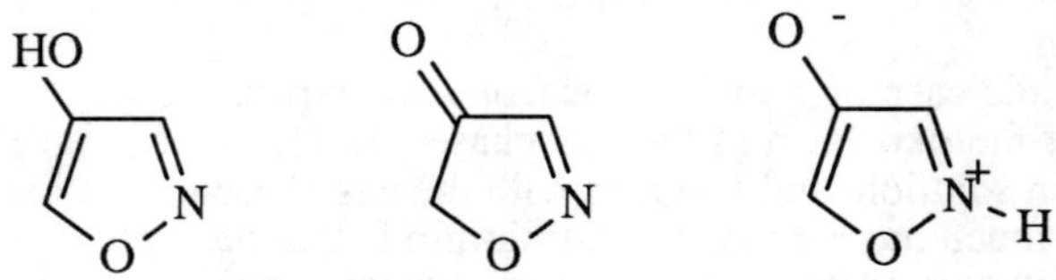

AM1 (gas)	10.4	0	46
AM1 (water)	0	0.3	21
Experimental (water)	dominant	-	-

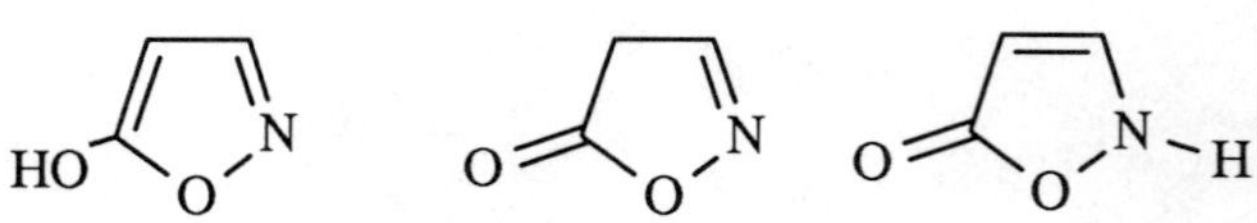

AM1 (gas)	14.8	0	15.4
AM1 (water)	17.4	1.1	0
Experimental (water)	-	-	dominant
Experimental ($CHCl_3$)	-	dominant	-

63 Hydroxyoxazoles: ΔH_f Values for Tautomers

	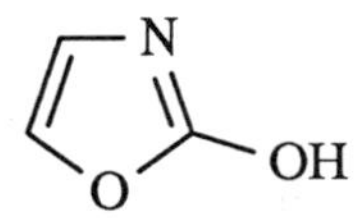	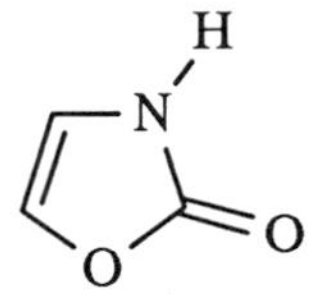
AM1 (gas)	13	0
AM1 (water)	23	0
Experimental (water)	-	dominant

	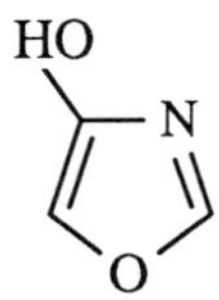	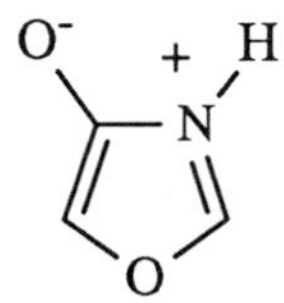	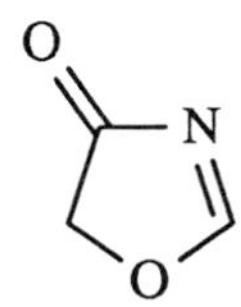
AM1 (gas)	14	23	0
AM1 (water)	18	8	0
Experimental (water)	-	-	dominant

	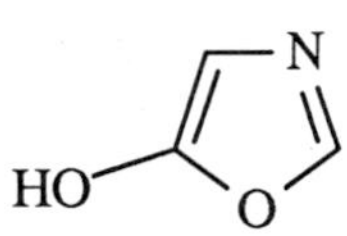	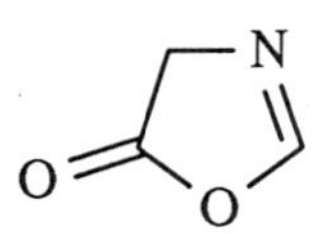	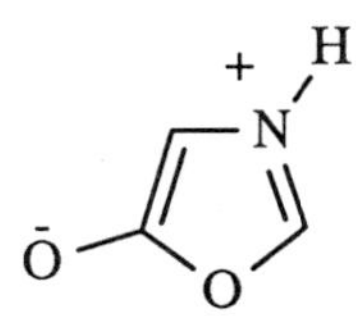
AM1 (gas)	13	0	19
AM1 (water)	28	17	0
Experimental ($CHCl_3$)	-	dominant	-

64 2-Thiouracil Species with Cyclic Conjugation

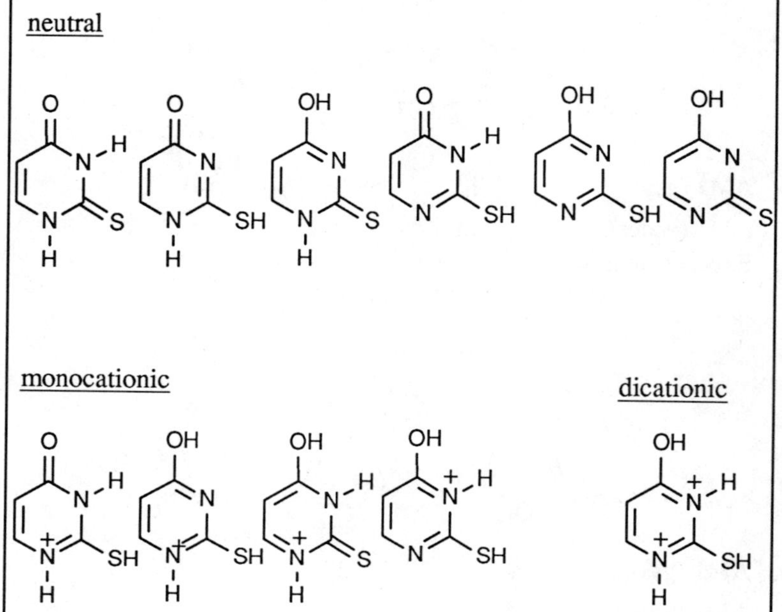

We approach the problem [16] by measuring the proton affinity of all of the monoalkyl models of thiouracil and of four of the possible six dialkyl models.

We then calculated by the AM1 method the ΔH_f values for all the neutral and monocationic species, and for their methyl derivatives [17]. A comparison of the measured and calculated values allowed the assignment of the structures to the various cations formed, and these led to the assignments of the measured energies differences. These were in fair agreement with the calculated values. In this way, an in-depth view of the tautomerism was enabled, leading to the energy diagram, as in Scheme 65.

65 Deduction of Proton Affinities of Individual Tautomers of 2-Thiouracil From PA Measurements

Tautomers

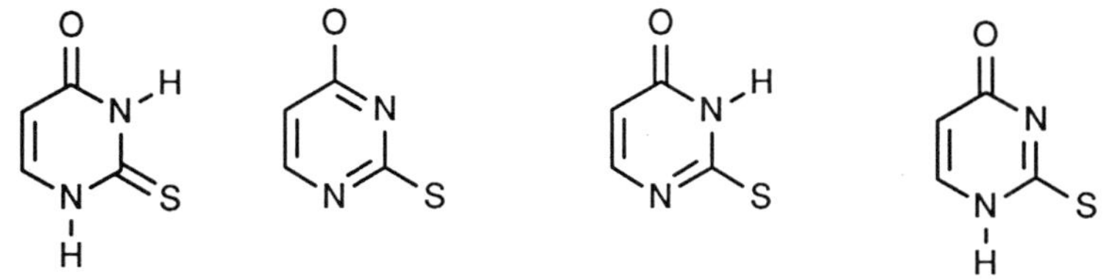

PA Deduced	209.1	217.6	220.3	227.8

Compound Measurement

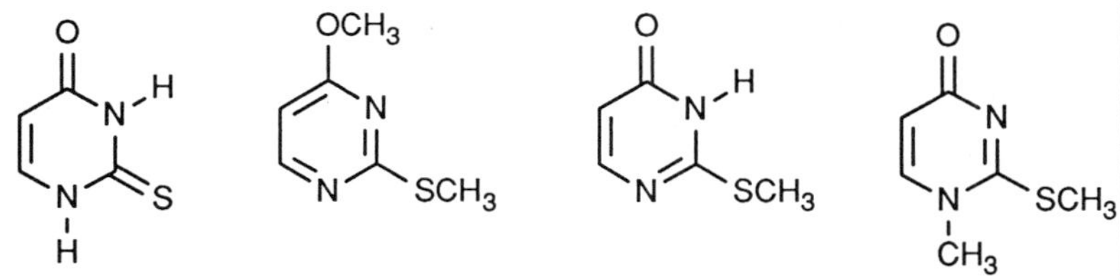

PA Measured	209.1	223.0	220.7	233.2
ΔNMe	------	------	------	-5.0
ΔSMe		-0.4	-0.4	-0.4
ΔMe	-------	-5.0	------	------
Relative PA	0	8.5	11.2	18.7
Relative ΔH_f	0	4.9	6.2	17.0

6. Searching for Starting Materials

The task of searching through catalogs to find suitable starting materials can now be considerably alleviated by using ChemQuest online, or the fine chemical directory in house software. As illustrated in Scheme 66, 61 chemical suppliers are now united in one data base. Many different search techniques are available (as shown in Scheme 67). The real power of the method is embodied in the ability to perform a substructure search to find all examples containing a given fragment. This allows very fast retrieval of all molecules containing a partcular backbone or ring system, or a search can be made for a particular functional group. Such a search would be extremely difficult and time-consuming to conduct manually using hardcopy catalogs.

66

Chemquest -- On line Orbit Infoline

Fine Chemical Directory -- MDL In-House Software

Finding and Ordering Chemicals -- the Easy Way

61 Chemical Suppliers in One Data Base (Update 89:1)

Aldrich	Alfa	Apin	B & J
Bader	Bayer	BDH	Bedoukian
Calbio	Carbolab	Chemalog	Columbia
Eachem	Echo Resins	Emkachem	Fairfield
Farchan	Fisher	Fisons	Fluorchem
Fluka	Frinton	H. W. Sands	Hoechst
Indofine	ISC	Janssen	JBL Sci
JPS	J. T. Baker	K & K	Kalichemie
Koch Light	Kodak	Lahr	Lancaster
May & Baker	Maybridge	Mitsubishi	P & B
Parish	PCR/SCM	Petrarch	Pierce
Prolabo	Pyraspec	RBI	Reidel
Salford	Schuchar	Shell	Sigma
SSF	Strem	TCI	Transworld
Vega	Vickers	Wiley	Wychem
Yarsley			

67 ChemQuest SEARCH TECHNIQUES

- COMPOUND DESCRIPTION
- CAS REGISTRY NUMBER
- MOLECULAR FORMULA
- EXACT STRUCTURE SEARCH
- SUBSTRUCTURE SEARCH

Taking as an example substructure search to find available benzotriazoles substituted in the benzene ring, the results are shown in Scheme 68, where it appears that in addition to the parent, there are seven C-substituted derivatives commercially available.

68 Structure Search: Available Benzotriazoles

Target Substructure:

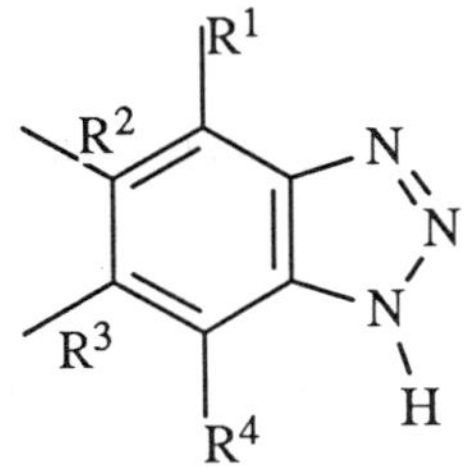

Structures Retrieved:

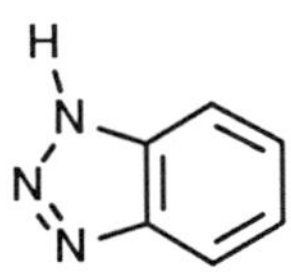

Benzotriazole

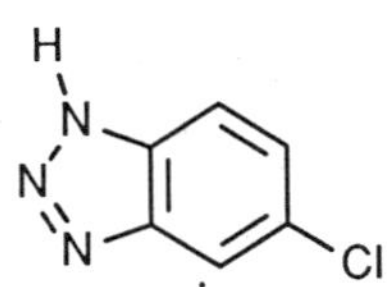

5-Chlorobenzotriazole

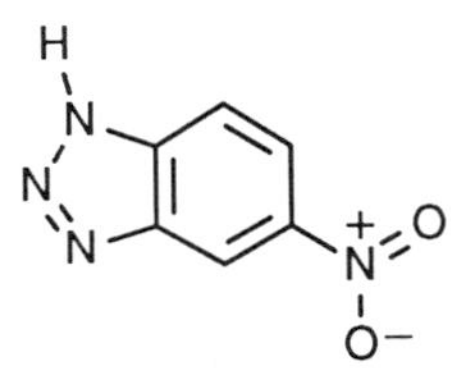

5-Nitrobenzotriazole

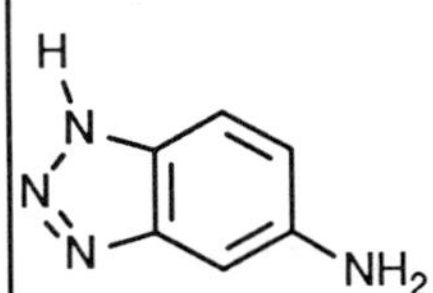

5-Aminobenzotriazole

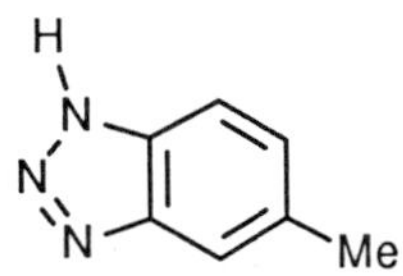

5-Methylbenzotriazole

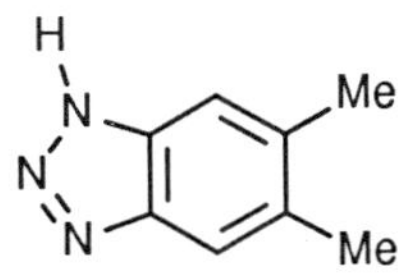

5,6-Dimethyl-1H-Benzotriazole

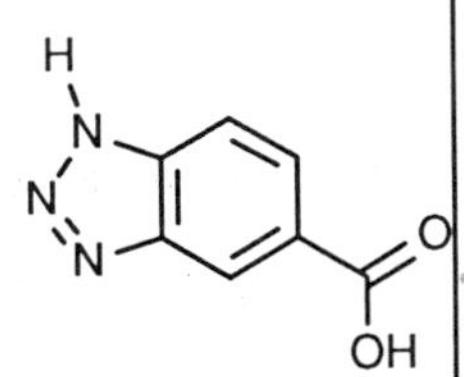

Benzotriazole-5-Carboxylic Acid 99%

Summing up the results of the search, available in N-unsubstitued C-substituted benzotriazoles are shown in Scheme 69. It is then possible to find out where these compounds can be purchased. The supplier and catalog number are indicated. Some price information is available in the online system. The results of this search are summarized for three derivatives in Scheme 70.

69

Available N-Unsubstituted C-Substituted Benzotriazoles

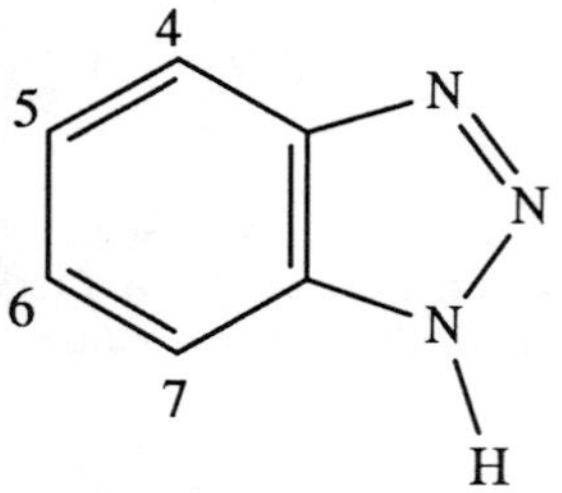

parent

4-Substituted : Me

5-Substituted : Cl, NO_2, NH_2, Me, CO_2H

5,6-Disubstituted : di-Me

70

Details of 5-Nitro, 5-Amino and 5-Carboxyl-Benzotriazoles

5-Aminobenzotriazole	K & K	K21269	3325-11-9
	Lancaster	LAN-6660	3325-11-9
5-Nitrobenzotriazole	Aldrich	17618-4	2338-12-7
	Chemalog	66-4020-00	2338-12-7
	Frinton	FR-0235	2338-12-7
	P & B	N07290	2338-12-7
	TCI	N0173	2338-12-7
Benzotriazole-5-carboxylic acid	Aldrich	30423-9	23814-12-2

7. Chemistry Databases

The FCD searching inhouse mentioned above is a specific example of the more general category of Chemical Databases. We are now making extensive use of these. FCD is available to search using MACCS (Molecular ACCess System) from Molecular Design Ltd. The program allows the storage and retrieval of molecular structures and associated information and data. The system is customizable, allowing the user to set up his own data fields, menus, and displays. The program can also act as a chemical front-end to a number of industry standard data management systems.

REACCS (REaction ACCess System) is to reactions what MACCS is to molecules. It can be used to search many commercially available databases -- as shown in Scheme 71 -- or the user may create his own databases.

71 REACCS Data Bases

Theilheimer: General synthetic methods 1946-1980 (M60748 R46818)

JSM (Journal of Synthetic Methods): General synthetic methods 1982 - (M40978 R25535)

CLF (Current Literature File): MDL's own database 1983 - (M39411 R25110)

Orgsyn (Organic Syntheses): Proven synthetic methods 1926 - (M5178 R4932)

FCDRX (Fine Chemicals Directory): Commercially available chemicals (M64976 R0)

PKFILERX (pK File): Citations of pK measurements (M10744 R0)

Chiras: Asymmetric synthesis 1975 - (R5500) ←— **NEW!**

REACCS can also be used to create your own proprietary reaction databases.

Scheme 72 illustrates the power of the method. A search is conducted for methods of reduction of aldehydes in the presence of ketones. A combination of features such as mapping of atoms in reactants to products, marking of reacting centers, and specification of hydrogen counts are used to locate the query successfully. A typical hit (there are not many) is shown in Scheme 73.

72 REACCS Searching Example: Selective Reduction of Aldehydes in the Presence of Ketones

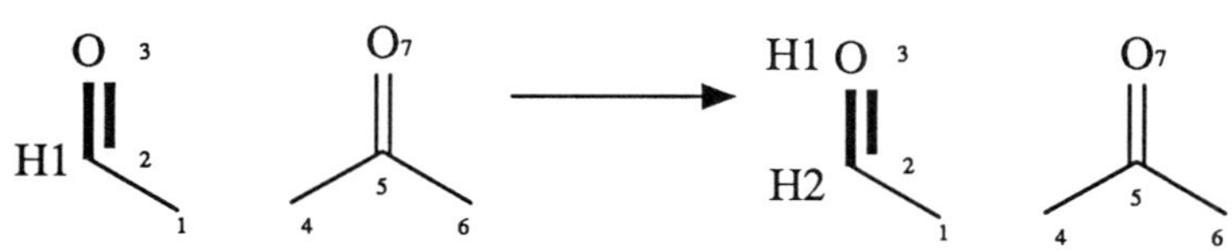

- Both fragments of starting material and products drawn as one molecule - looking for them in the same molecule
- Atom/atom mapping is used and reaction center bond is marked
- Hydrogen counts are specified

73

REACCS: Sample Hit

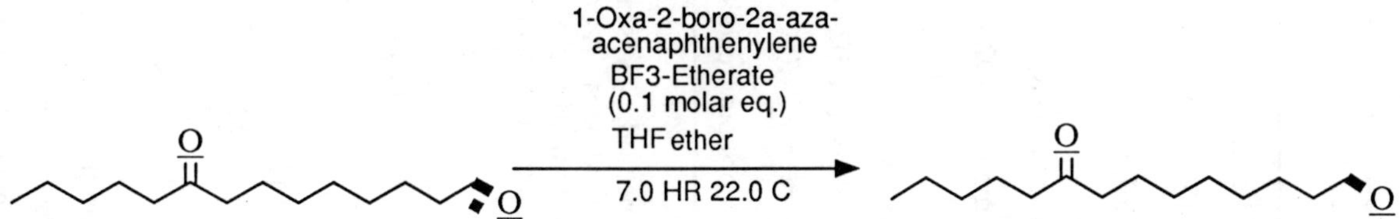

S. Kim, S. Yang, Tetrahedron Lett, 25(28), p. 2985, 1984.

regno: 11477	extreg - folder - path - step 91021698 18 A STEP	VARIATION 1 OF 1

74 Acknowledgements I

Collaborations with Other Research Groups

Professor Ernst Anders	University of Erlangen, F.R.G.
Professor Jim Deyrup	University of Florida
Professor John Eyler	University of Florida
Dr. Peter Johnson	ORAC Ltd., Leeds, UK
Professor Mati Karelson	Tartu University, Estonia, USSR
Professor Guiseppe Musumarra	University of Catania, Italy
Dr. Mike Siskin	Exxon, Clinton, New Jersey
Dr. Dennis Smith	MDL, San Leandro, California
Dr. John Stevens	3M Corporation, St. Paul, Minnesota
Professor Miroslaw Szafran	University of Poznan, Poland
Professor Mike Zerner	University of Florida

75
Acknowledgements II

Katritzky Research Group Members

Piotr Barczynski

Steve Cato

Andrzej Lapucha

Franz J. Luxem

Ramaiah Murugan

Stanislaw Rachwal

8. References

1 A. R. Katritzky, P. Barczynski, G. Musumarra, D. Pisano and M. Szafran, *J. Am. Chem. Soc.,* 1989, **111,** 7.

2 A. R. Katritzky, M. Szafran, P. Barczynski, and G. Musumarra, *Unpublished work.*

3 G. Musumarra, D. Pisano. A. R. Katritzky, A. R. Lapucha, F. J. Luxem, R. Murugan, M. Siskin, and G. Brons, *Tetrahedron Computer Methodology,* 1989, **2**, 17.

4 S. Wold and M. Sjöström, "SIMCA, a method for analyzing chemical data in terms of similarity and analogy," in *Chemometrics, Theory and Application* (B. Kowalski, Ed.). American Chemical Society Symposium Series 52, pp 243-282, 1977.

5 A. R. Katritzky, S. J. Cato, and A. R. Lapucha, *Unpublished Work.*

6 N. Ramiller, *American Laboratory,* June 1984.

7 A. R. Katritzky and G. Musumarra, *Chem. Soc. Rev.,* 1984, **13**, 47.

8 A. R. Katritzky and B. E. Brycki, *J. Phys. Org. Chem.,* 1988, **1**, 1.

9 A. R. Katritzky, C. H. Watson, Z. Dega-Szafran, and J. R. Eyler, *J. Am. Chem. Soc.,* 1990, in press.

10 A. R. Katritzky, C. H. Watson, Z. Dega-Szafran, and J. R. Eyler, *J. Am. Chem. Soc.,* 1990, in press.

11 M. J. S. Dewar, E. G. Zoebisch, E. F. Healy and J. J. P. Stewart, *J. Am. Chem. Soc.,* 1985, **107**, 3902.

12 A. R. Katritzky, Z. Dega-Szafran, C. H. Watson, and J. R. Eyler, *J. Chem. Soc. Perkin Trans. II.,* 1990, in press.

13 A. R. Katritzky, C. H. Watson, Z. Dega-Szafran, and J. R. Eyler, *Org. Mass. Spectrom.,* 1989, **24**, 1017.

14 M. M. Karelson, A. R. Katritzky, M. Szafran, and M. C. Zerner, *J. Org. Chem.,* 1989, **54**, 6030.

15 M. M. Karelson, A. R. Katritzky, M. Szafran, and M. C. Zerner, *J. Chem. Soc. Perkin Trans. II.,* 1990, 195.

16 A. R. Katritzky, G. Baykut, S. Rachwal, M. Szafran, K. C. Caster, and J. R. Eyler, *J. Chem. Soc., Perkin Trans. II.,* 1989, 1499.

17 A. R. Katritzky, M. Szafran, and J. Stevens, *J. Chem. Soc., Perkin Trans. II.,* 1989, 1507.

INDEX